gcse geography
edexcel B

Series editor:
Bob Digby

Cameron Dunn
Dave Holmes
Sue Warn
Dan Cowling
Catherine Hurst

OXFORD
UNIVERSITY PRESS

OXFORD
UNIVERSITY PRESS

Great Clarendon Street, Oxford OX2 6DP

Oxford University Press is a department of the University of Oxford.
It furthers the University's objective of excellence in research,
scholarship, and education by publishing worldwide in

Oxford   New York

Auckland Cape Town Dar es Salaam Hong Kong Karachi
Kuala Lumpur Madrid Melbourne Mexico City Nairobi
New Delhi Shanghai Taipei Toronto

With offices in
Argentina Austria Brazil Chile Czech Republic France Greece
Guatemala Hungary Italy Japan Poland Portugal Singapore
South Korea Switzerland Thailand Turkey Ukraine Vietnam

Oxford is a registered trade mark of Oxford University Press
in the UK and in certain other countries

Authors: Bob Digby, Cameron Dunn, Dave Holmes, Sue Warn, Dan
Cowling, Catherine Hurst

The moral rights of the authors have been asserted

Database right Oxford University Press (maker)

First published 2009. Second edition 2013

British Library Cataloguing in Publication Data

Data available

ISBN 978-0-19-839221-7

10 9 8 7 6 5 4

Printed in Malaysia by Vivar Printing Sdn Bhd.

Paper used in the production of this book is a natural, recyclable product made from
wood grown in sustainable forests. The manufacturing process conforms to the
environmental regulation s of the country of origin.

Acknowledgements
The publisher and authors would like to thank the following for permission to use
photographs and other copyright material:

Cover: Don Hammond/Design Pics/Corbis; Nicemonkey/Shutterstock

Illustrations by: Barking Dog Art; Born Dreaming; Steve Evans; Mark Walker

Photos: p6: Jim Mills/123rf; **p7t:** Tony Gentile/Reuters; **p7m:** Trinh Le Nguyen/
Shutterstock; **p7b:** David Roos/Shutterstock; **p8l:** JEAN-CLAUDE REVY, ISM/SCIENCE
PHOTO LIBRARY; **p8r:** Photolibrary; **p9:** IODP/JAMSTEC; **p11:** Winfried Stricker/
Fotolia; **p17:** Roger Ressmeyer/CORBIS; **p18:** Sayyid Azim/Associated Press; **p23:** Andy
Wong/Associated Press; **p24:** Bryan & Cherry Alexander Photography; **p26:** Courtesy
of SOHO/[instrument] consortium. SOHO is a project of international cooperation
between ESA and NASA; **p28:** Wolfgang Kaehler/Corbis; **p29:** The Print Collector/
Alamy; **p34:** Original photograph taken in 1928 of the Upsala Glacier. ©Archivo
Museo Salesiano/De Agostini Recent comparison image taken in 2004. ©Greenpeace/
Daniel Beltrá; **p35:** Warren Faidley/Corbis; **p39:** Robert Dear/Associated Press; **p40:**
Jacques Descloitres, MODIS Land Science Team/NASA; **p45t&l:** Photolibrary; **p45r:**
Pavel Filatov/Alamy; **p48l:** Bill Bachman/Alamy; **p48r:** Getty Images; **p50t:** Sergey
Uryadnikov/Shutterstock; **p50b:** Ralph Loesche/Shutterstock; **p57:** Kevin Schafer/
Corbis ; **p58t:** NASA Images / Alamy; **p58m:** Fallsview/Dreamstime; **p58b:** Paul
Hardy/Corbis; **p59t:** Stephanie Cabrera/zefa/Corbis; **p59b:** Craig Tuttle/CORBIS; **p62-
63:** NASA/Map adapted from the IPCC Fourth Assessment Report; **p67:** Margaret
Courtney-Clarke/Photolibrary; **p68:** Balkanpix.com/Rex Features; **p70:** Andy Z/
Shutterstock; **p72:** Pascal Deloche/Godong/Corbis; **p74l:** Ashley Cooper/Alamy; **p74r:**
GRAEME EWENS / SCIENCE PHOTO LIBRARY; **p76:** Cameron Dunn; **p77:** Flasshary;
p81t: John Farmar/Ecoscene; **p81b:** Simmons Aerofilms; **p83:** Getty Images; **p85:**
GRAEME EWENS / SCIENCE PHOTO LIBRARY; **p86:** John Boyes/World of Stock; **p87-
94:** Bob Digby; **p96t:** Webb Aviation; **p96b:** Neil Holmes Freelance Digital/Alamy;
p100: Kelly Dorset: www.flickr.com/delanthear; **p101:** Wendy North; **p102:** Dave
Thompson/Associated Press; **p103t:** Stringer UK/Reuters; **p103b:** Rimmington CE/
Photographers Direct; **p105:** Rotherham Investment and Development Office; **p106:**
geogphotos/Alamy; **p111:** Norbert Wu/Minden Pictures/Corbis; **p118:** SuperStock/
Alamy; **p119:** Ross Frid/Alamy; **p123:** REUTERS/Ho New; **p126:** Geophotos;
p127: Jacques Langevin/Sygma/Corbis; **p128:** Aldo Pavan/
Corbis; **p131t:** Martin Harvey/Corbis; **p131m:** Martin B Withers/FLPA; **p131b:**
Eric and David Hosking/Corbis; **p132:** Paul Mayall/Photographers Direct; **p133t:**
Steve Strike/Photographers Direct; **p133m:** Lisa Scope Young/Getty Images;
p133b: Warwick Kent/Getty Images; **p134t:** Richard McDowell/Alamy; **p134b:**
Susanna Bennett/Alamy; **p136:** Joern/Shutterstock; **p141:** Jeremy Hartley/Panos
Pictures; **p144t:** Tobias Schwarz/Reuters; **p144b:** Toby Melville/Reuters; **p145l:**
Associated Press; **p145r:** Aly Song/Reuters; **p146:** With kind permission from
the UN; **p147:** Elena Yakusheva/Shutterstock; **p154:** REUTERS/Yuriko Nakao;
p155: John Warburton-Lee Photography/Alamy; **p156:** Keith Dannemiller/Corbis;
p161: Alamy/Tim Graham; **p162:** Getty Images; **p164:** Danny Lehman/Corbis;
p166: Justin Sullivan/Staff/gettyimages; **p167l:** MARTIN WRIGHT / Still Pictures;
p167r: ACE STOCK LIMITED/Alamy; **p169:** Ariana Cubillos/ASSOCIATED PRESS;
p170: Hulton Archive/Stringer/Getty Images; **p172t&m:** worldmapper.org;
p172b: ben smith/shutterstock; **p173:** OUP Picture Bank; **p174:** Antoine Gyori/
AGP/Corbis; **p176:** Getty Images; **p179:** 'Albuisson, M., Lefevre M., and Wald L./
Mines ParisTech'; **p180:** Helene Rogers/Alamy; **p181:** Diaphor La Phototheque/
Photolibrary; **p182:** Alex Segre/Alamy; **p183:** BESTWEB/Shutterstock; **p184:** With
kind permission from Volkswagen; **p185:** UIG via Getty Images; **p186t:** A Dow/
ILO; **p186b:** REUTERS/Kham; **p187:** Getty Images; **p192:** Paul Chesley/Getty
Images; **p193:** Time & Life Pictures/Getty Images; **p194:** AFP/Getty Images; **p195:**
Matthew Wakem/Getty Images; **p196:** Image Scotland/Alamy; **p197:** Bloomberg
via Getty Images; **p199:** Joerg Boethling/Alamy; **p203:** Mark Boulton/Alamy;
p204: www.worldmapper.org; **p206:** ostill/Shutterstock; **p209:** Hein von Horsten/
Getty Images; **p210:** Russell Chapman; **p217:** Shamik Mehta; **p218:** Bob Digby;
p219: Alex Segre/Alamy; **p220:** Bob Digby; **p222r:** Hugh Penney/Photographers
Direct; **p225:** Ronald Sanderson/Tyne and Wear Archives and Museums; **p226:**
Robert Read/Alamy; **p227:** Mark Mercer/Alamy; **p228:** newcastlephotos.blogspot.
com; **p230:** Robert Brook/Alamy; **p231t:** Andrew Fox/Alamy; **p231b:** Conrad
Elias/Alamy; **p232:** Alamy; **p233:** Caro/Alamy; **p234:** Alex Segre/Alamy; **p237:**
Paul Thompson Images/Alamy; **p238:** Dan Atkin/Alamy; **p241:** Construction
Photography/Alamy; **p242:** Nathan King/Alamy; **p244:** Joe Doylem/Alamy; **p245:**
Joe March/Photographers Direct; **p246:** Roy Lane/Alamy; **p247:** ScotImage/
Alamy; **p248:** Doug McKinlay/Getty Images; **p250:** Planet Observer/Getty Images;
p251: Duncan Shaw/Alamy; **p252t:** Alamy/Fredrik Renander; **p252b:** Getty;
p254: Dinodia Photos/Alamy; **p255:** Tips Images/Tips Italia Srl a socio unico/
Alamy; **p256:** NASA; **p259t:** VIEW Pictures Ltd/Alamy; **p259b:** Bob Digby; **p260t:**
Guy Somerset/Alamy; **p260b:** Justin Kase zsixz/Alamy; **p261:** Bikeworldtravel/
Shutterstock.com; **p262:** AFP/Getty Images; **p263t:** Chris Mellor/Getty Images;
p263b: Getty/AFP; **p264t:** Getty Images; **p264b:** Heitor C. Jorge; **p265t:** John
W Banagan/Getty Images; **p265b:** dpa picture alliance/Alamy; **p266t:** Karen
Kasmauski/Getty Images; **p266b:** Borderlands/Alamy; **p267:** Marion Kaplan/
Alamy; **p268:** Aerial Archives/Alamy; **p269:** ALI HAIDER/epa/Corbis; **p270:** Bob
Digby; **p272t:** Bob Digby; **p272b:** World Pictures/Alamy; **p273:** Ashley Cooper/
Alamy; **p274t:** Ashley Cooper pics/Alamy; **p274b:** Mike Kipling Photography/
Alamy; **p275t:** imagebroker/Alamy; **p275b:** Photo Resource Hawaii/Alamy; **p276:**
Sue Cunningham Photographic/Alamy; **p277:** Eitan Simanor/Alamy; **p278:** Nigel
Pavitt/John Warburton-Lee Photography/Photolibrary; **p279:** Jorgen Schytte / Still
Pictures; **p280:** F1online digitale Bildagentur GmbH/Alamy; **p281:** Peter Davey/
FLPA; **p282:** Walt Anderson/Getty Images; **p283:** Ariadne Van Zandbergen/Africa
Image Library; **p284t:** Martin Wright/Still Pictures; **p284b:** Robert Nickelsberg/
Contributor/Time & Life Pictures/Getty Images; **p285:** Neil Cooper/Alamy; **p286t:**
Karen Kasmauski/Getty Images; **p286b:** UIG via Getty Images; **p287 from top:**
ALI HAIDER/epa/Corbis; Andy Z/Shutterstock; Kevin Schafer/Corbis; Getty Images;
p288: Kkaplin/Shutterstock; **p292:** NASA Earth Observatory/Robert Simmon,
using ALI data from the EO-1 Team and USGS Earthquake Hazard Program; **p293:**
2011 Getty Images; **p297:** Tupungato/Shutterstock; **p298 from top:** James
Jagger/Photographers Direct; David Hills; Nigel Francis Ltd/Robert Harding; David
Hughes/Robert Harding; **p301:** Bob Digby; **p303:** Reproduced by permission
of Ordnance Survey on behalf of HMSO. © Crown Copyright (2013). All rights
reserved. Ordnance Survey Licence number 100000249; **p304:** Bob Digby; **p310:**
Bob Digby; **p311:** Bob Digby; **p315:** Bob Digby

The publisher and authors would like to thank Russell Chapman for his
contribution to the first edition of this title, and for his contribution to chapter
15 in this edition.

Every effort has been made to contact the copyright holders of material
reproduced in this book. Any omissions will be rectified in subsequent printings
if notice is given to the publisher.

Contents

continued ...

Unit 1 Dynamic planet

✚ In this section you will learn about the theme of unit 1 – how our dynamic planet links together into one physical system which supports us, but can also bite back.

> **✚ Dynamic** means constantly changing.

Supporting life

As far as we know, ours is the only planet that supports life. The Earth is a survivor; it is 4.6 billion years old! The planet has slowly evolved into a complex life-support system that keeps humans, plants and animals alive. Four spheres – which really means layers – provide the vital services.

Sphere	What is it?	Why do we need it?
Atmosphere	The layer of gases which make up the air around us.	• Oxygen, needed for animals to breathe. • Carbon dioxide for plant growth. • The greenhouse effect to keep the planet warm. • Weather and climate.
Hydrosphere	The layer of water – seas, rivers, lakes, groundwater and ice – on the Earth's crust.	• Water for animal and plant life. • Water cycle which moves water around the planet.
Biosphere	The very thin layer of living things – plants, animals and humans – living on the crust.	• Plants and animals provide us with food. • Many living things can also be used to make medicines and as fuel.
Geosphere	The rocks of the Earth's crust, and deeper into the Earth towards the core.	• The Earth's core makes a magnetic field protecting us from space radiation. • We use rocks, minerals and fossil fuels as resources for fuel, building, and much more.

The four spheres are all linked together. The living layer, the biosphere, is the most important. It is also unique, because other planets do not have a biosphere. The biosphere interacts with the other spheres in a number of ways:

- Plants and animals turn rock and sediment (the geosphere) into soil.
- Plants are part of the water cycle (the hydrosphere), because of transpiration.
- Plants take in carbon dioxide, and give out oxygen, keeping the air (the atmosphere) breathable.

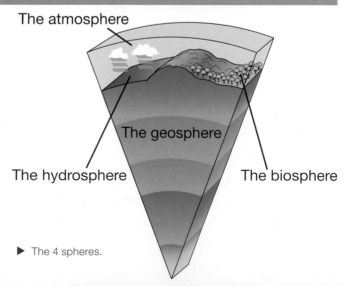

The atmosphere

The geosphere

The hydrosphere

The biosphere

▶ The 4 spheres.

Making life harder

The Earth does a very good job of supporting life, but sometimes it can harm humans. One part of the Earth is very restless – the geosphere. It can produce earthquakes, volcanoes and tsunami, which cause disasters when humans get in the way. The hydrosphere can also turn nasty too, when flooding threatens lives and homes. The atmosphere produces hurricanes, storms and tornadoes, which can cause havoc. The 'spheres' make our lives possible, but they can also threaten us.

On your planet

+ There may be between 2 and 80 million species in the world, but some scientists think half of them will be extinct in 100 years.

Planet in peril?

Humans are changing and damaging all four of our life-support spheres. This is surprising, because we depend on them so much. Damaging the spheres places the planet in danger, and might even threaten the future of the human race.

- Humans have polluted the atmosphere, increasing the amount of carbon dioxide in it by 40%.
- We have used up most of the fossil fuels buried in the geosphere, and will soon need to look for new energy sources.
- Deforestation, farming and pollution are causing extinctions in the biosphere faster than ever before.
- Human use of water means that many rivers and lakes are polluted. Water supplies in some places are drying up.

The rest of this unit explores the four spheres and explains what they are like. It looks at how they affect humans, and how humans affect them.

▲ Damaging the spheres.

your questions

1 Write down a definition of each of the four spheres.
2 How have you used the four spheres today? Write down one way you have used each sphere.
3 Draw a large version of this diagram. It will need to be a whole page.

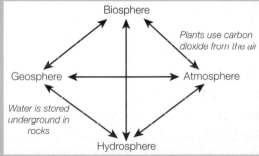

Biosphere

Plants use carbon dioxide from the air

Geosphere   Atmosphere

Water is stored underground in rocks

Hydrosphere

See if you can describe the links between the four spheres. Two have been done for you.
4 State four ways in which humans might be harmed by the spheres.
5 How could your life be made worse if humans keep damaging the spheres?

In this section you will learn about the interior of the Earth.

Journey to the centre

No one has ever seen the inside of the Earth. The deepest we have managed to get is 3.5 km in the Witwatersrand gold mine in South Africa. Miners would have to drill another 6365 km to reach the centre of the Earth.

What we know about the Earth's interior comes from direct and indirect evidence. We can get direct evidence from the Earth's surface. Indirect evidence, like earthquakes and material from space, also helps us to understand the Earth.

Examining the crust

The diagram on the right shows the structure of the Earth. The crust is the surface of the Earth. It is a rock layer forming the upper part of the **lithosphere**. The lithosphere is split into **tectonic plates**. These plates move very slowly, at 2-5 cm per year, on a layer called the **asthenosphere**.

> **+** The **lithosphere** is the uppermost layer of the Earth. It is cool and brittle. It includes the very top of the mantle and, above this, the crust.

There are two types of crust:

- **Continental crust** forms the land. This is made mostly of granite, which is a low density igneous rock. Continental crust is on average 30-50 km thick.
- Under the oceans is **oceanic crust**. This is much thinner, usually 6-8 km thick. It is also denser and made of an igneous rock called basalt.

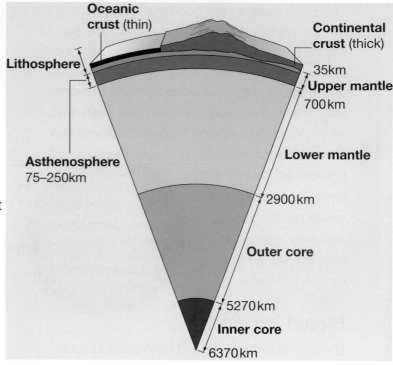

▲ **Geologists** (scientists who study the Earth and its structure) think the Earth has a layered structure, like an onion.

▲ Geologists collect samples of the crust and compare the two types: polished granite from continental crust (left) and basalt from oceanic crust (right).

The asthenosphere and mantle

The movement of the tectonic plates is evidence that there is a 'lubricating' layer underneath the lithosphere. This is the asthenosphere. You might think that this layer would be a liquid. But if it was a liquid, the heavy tectonic plates would sink into it. Geologists think the asthenosphere is partly molten rock and partly solid rock. It may be like very thick, dense, hot porridge!

The asthenosphere is in the top layer of the **mantle**. The mantle is the largest of the Earth's layers by volume, and is mostly solid rock. We know this because sometimes you can see the top of the mantle attached to an overturned piece of crust.

Earthquake waves tell us about the physical state of the Earth. They speed up, change direction or stop when they meet a new layer in the Earth. Some waves travel easily through the crust, mantle and inner core, but not through the outer core. This suggests that the outer core has a different physical state and may be liquid, not solid.

▲ The Japanese have built a 57 000-tonne international scientific drilling ship, the *Chikyu*, which will drill 7 km through the oceanic crust to reach the mantle – further than anyone has ever gone before.

Clues from space

At a depth of 2900 km, we can't sample the Earth's core. Geologists think that it is metal – mostly nickel and iron. Evidence for this comes from **meteorites**, which are fragments of rock and metal that fall to Earth from space. Most come from the asteroid belt between Mars and Jupiter. Meteorites come in several types:

- Stony meteorites, with a similar composition to basalt.
- Stony-iron meteorites, containing a lot of the mineral olivine.
- Iron meteorites, which are solid lumps of iron and nickel.

These meteorites may be fragments of the lithosphere, mantle and core of a shattered planet. Iron meteorites may show that the Earth's core is made up of iron and nickel.

Layer		Density (grams / cm³)	Physical state	Composition	Temp (°C)
Lithosphere	Continental crust	2.7	Solid	Granite	Air temp - 900
	Oceanic crust	3.3	Solid	Basalt	Air temp - 900
Mantle	Asthenosphere	3.4-4.4	Partially molten	Peridotite	900-1600
	Lower mantle	4.4-5.6	Solid		1600-4000
Core	Outer core	9.9-12.2	Liquid	Iron and nickel	4000-5000

your questions

1 Draw a cross-section of the Earth. You need a large circle divided into layers. Label the layers with details of density, temperature and physical state.
2 What are the main differences between the lithosphere and asthenosphere?
3 Exam-style question Describe the differences between the oceanic and continental crust. (4 marks)

✚ In this section you will learn how the Earth's core drives the process of plate tectonics.

Hot rocks

Inside the Earth it is hot. We know this because of:

- molten lava spewing from active volcanoes
- hot springs and geysers.

Heat from inside the Earth is called **geothermal** ('Earth-heat'). The heat is produced by the **radioactive decay** of elements such as uranium and thorium in the core and mantle. This raises the core's temperature to over 5000 °C.

✚ Some elements are naturally unstable and radioactive. Atoms of these elements release particles from their nuclei and give off heat. This is called **radioactive decay**.

The inner core is so deep and is under such huge pressure that it stays solid. The outer core is liquid because it is under lower pressure. As heat rises from the core, it creates **convection currents** in the liquid outer core and mantle (see below). These vast mantle convection currents are strong enough to move the tectonic plates on the Earth's surface. The convection currents move about as fast as your fingernails grow. Radioactivity in the core and mantle is the engine of plate tectonics.

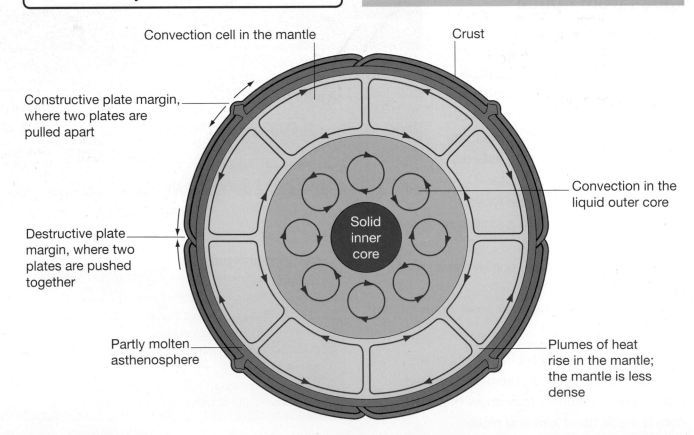

Convection cell in the mantle

Crust

Constructive plate margin, where two plates are pulled apart

Convection in the liquid outer core

Solid inner core

Destructive plate margin, where two plates are pushed together

Partly molten asthenosphere

Plumes of heat rise in the mantle; the mantle is less dense

Plumes

The parts of convection cells where heat moves towards the surface are called **plumes**. These are concentrated zones of heat. In a plume, the mantle is less dense. Plumes bring **magma** (molten rock) to the surface. If magma breaks through the crust, it erupts as **lava** in a volcano.

- Some plumes rise like long sheets of heat. These form **constructive plate boundaries** at the surface.
- Other plumes are like columns of heat. These form **hot spots**. Hot spots can be in the middle of a tectonic plate, like Hawaii and Yellowstone in the USA.

Magnetic field

The Earth is surrounded by a huge invisible magnetic field called the **magnetosphere**. This is a force field. It protects the Earth from harmful radiation from space and the sun (see right).

The Earth's magnetic field is made by the outer core. As liquid iron in the outer core flows, it works like an electrical dynamo. This produces the magnetic field.

On your planet
➕ Did you know that the Earth's magnetic field sometimes 'flips', so north becomes south and south becomes north?

▶ Sometimes you can see the magnetosphere. The northern lights (aurora borealis) form when radiation from space hits the magnetosphere and lights up the sky.

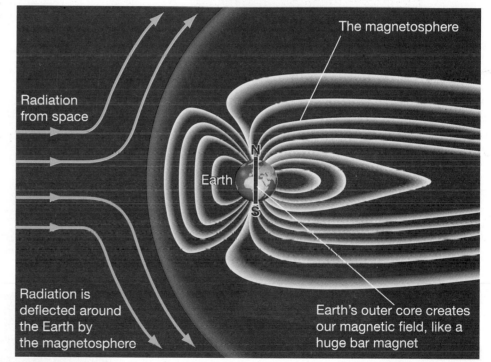

The magnetosphere

Radiation from space

Earth

N

S

Radiation is deflected around the Earth by the magnetosphere

Earth's outer core creates our magnetic field, like a huge bar magnet

your questions

1 Why is the centre of the Earth at over 5000 °C?
2 Look at the diagram of the Earth's convection currents. What happens to the crust at the top of one of the convection currents in the mantle?
3 Find out why Mars is a 'dead' planet with no plate tectonics. You might like to visit this website: http://mars.jpl.nasa.gov/classroom/students.html

4 Each set of words below has an odd one out. For each:
  a say which is the odd one out
  b explain your choice.
  - inner core, outer core, mantle, crust
  - convection, northern lights, plume, cell, current
  - lava, uranium, magma, geyser
5 **Exam-style question** Describe the different layers of earth's interior. (4 marks)

In this section you will find out how the Earth's tectonic plates have moved in the past, and about plate boundaries.

Pangea, the supercontinent

Scientists know that the continents were once all joined together. They formed a supercontinent called **Pangea**. The diagram on the right shows the position of the continents 250 million years ago. Identical rocks and fossils dating from this time have been found in West Africa and eastern South America. This tells us that Africa and South America were once joined. Pangea started to split apart about 200 million years ago. Since then, plate tectonics has moved the continents to the positions they are in today.

Moving plates

Today, the Earth's lithosphere is split into 15 large **tectonic plates** and over 20 small ones. These are like the patches that make up a football. The plates move very slowly on the asthenosphere. Where two plates meet, there is a **plate boundary**. There are three types of plate boundary, as shown on the map opposite:

- Constructive plate boundaries – formed when two plates move apart.
- Destructive plate boundaries – formed when two plates collide.
- Conservative plate boundaries – formed when two plates slide past each other.

Plate boundaries are where the 'action' is. Most earthquakes and volcanoes are found on plate boundaries.

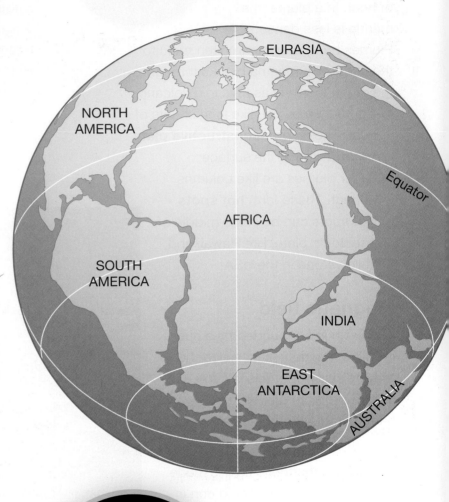

EURASIA

NORTH AMERICA

Equator

AFRICA

SOUTH AMERICA

INDIA

EAST ANTARCTICA

AUSTRALIA

On your planet

+ Every year, the distance between the UK and the USA grows by about 2 cm. This is because the mid-Atlantic constructive plate boundary creates new oceanic crust.

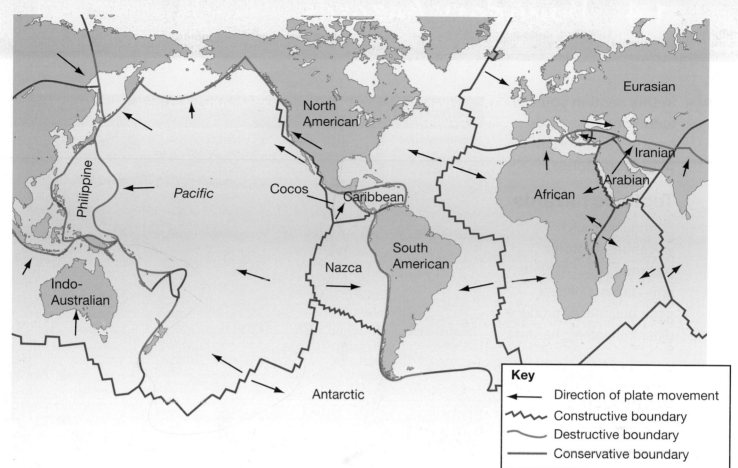

Key

⟵ Direction of plate movement

〰〰 Constructive boundary

〜 Destructive boundary

— Conservative boundary

Crust: old and new

Most continental crust is 3-4 billion years old. The oldest oceanic crust is only 180 million years old. Why the age difference?

New oceanic crust forms constantly at constructive plate boundaries:

- Convection currents bring magma up from the mantle.
- The magma is injected between the separating plates.
- As the magma cools, it forms new oceanic crust.
- The plates continue to move apart, allowing more magma to be injected.

Old oceanic crust is destroyed by **subduction** at destructive plate boundaries – it is 'recycled' by the Earth. Continental crust was formed billions of years ago, and has not formed since. It is less dense than oceanic crust, so can't be subducted and destroyed.

+ Subduction
describes oceanic crust sinking into the mantle at a destructive plate boundary. As the crust subducts, it melts back into the mantle.

On your planet

+ Look at the map of plates and plate boundaries above. The circle of plate boundaries around the Pacific Ocean is called the 'Pacific ring of fire', because it has many active volcanoes.

your questions

1 Compare the map of Pangea with a modern world map from an atlas. Describe how India has moved since the time of Pangea.
2 Look at the map of tectonic plates.
   a Which plate is the UK on?
   b Name a country which is split by two plates.
   c Name two plates that are moving apart.
   d Name two plates that are colliding.
3 **Exam-style question** Explain why earth's tectonic plates move. (6 marks)

➕ In this section you will learn which hazards happen at different plate boundaries.

Tectonic hazards

Earthquakes and volcanoes (**tectonic hazards**) occur at plate boundaries. Different plate boundaries produce different tectonic hazards.

Conservative boundaries

As plates slide past each other, friction between them causes earthquakes. These are rare but very destructive, because they are shallow (close to the surface). The San Andreas fault is shown below.

Plate boundary	Example	Earthquakes	Volcanoes
Conservative	San Andreas fault in California, USA. North American and Pacific plates sliding past each other.	• Destructive earthquakes up to magnitude 8.5. • Small earth tremors almost daily.	No volcanoes.
Constructive	Iceland, on the mid-Atlantic ridge. The Eurasian and North American oceanic plates pulling apart.	• Small earthquakes up to 5.0-6.0 on the Richter scale.	• Not very explosive or dangerous. • Occur in fissures (cracks in the crust). • Erupt basalt lava at 1200 °C.
Destructive	Andes mountains in Peru and Chile. Nazca oceanic plate is subducted under the South American continental plate.	• Very destructive, up to magnitude 9.5. • Tsunami can form.	• Very explosive, destructive volcanoes. • Steep sided, cone shaped. • Erupt andesite lava at 900-1000 °C.
Collision zone	Himalayas. Formed as the Indian and Eurasian continental plates push into each other.	• Destructive earthquakes, up to magnitude 9.0. • Landslides are triggered.	Volcanoes are very rare.

Collision zones

➕ **Collision zones** are a type of destructive boundary. They form mountain ranges like the Himalayas (see right). Two continental plates of low-density granite collide, pushing up mountains. Earthquakes happen on faults (huge cracks in the crust) in collision zones.

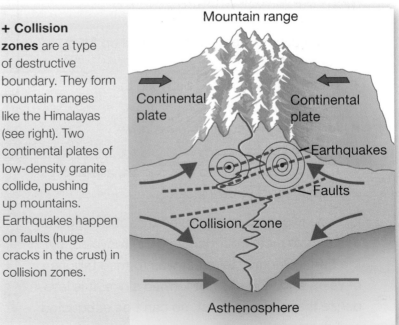

▲ A conservative plate boundary

▲ A collision zone boundary

Constructive boundaries

As plates move apart, magma rises up through the gap, as the diagram below shows. The magma is **basalt** and is very hot and runny. It forms **lava flows** and shallow sided volcanoes.

Earthquakes are caused by **friction** as the plates tear apart. These earthquakes are small. They don't cause much damage.

> **On your planet**
> ✛ Every year there are about 100 000 earthquakes strong enough to be felt. The largest earthquake recorded was a magnitude 9.5 in Chile in 1960.

> ✛ **Tectonic hazards** are natural events that affect people and property.

Destructive boundaries

As the plates push together (see the diagram below), oceanic plate is **subducted**. As it sinks, it melts and makes magma called **andesite** (after the Andes). Sea water is dragged down with the oceanic plate. This makes the magma less dense so it rises through the continental crust. The water erupts as steam making volcanoes very explosive.

Sinking oceanic plate can stick to the continental plate. Pressure builds up against the friction. When the plates finally snap apart, a lot of energy is released as an earthquake. These earthquakes can be devastating, especially if they are shallow.

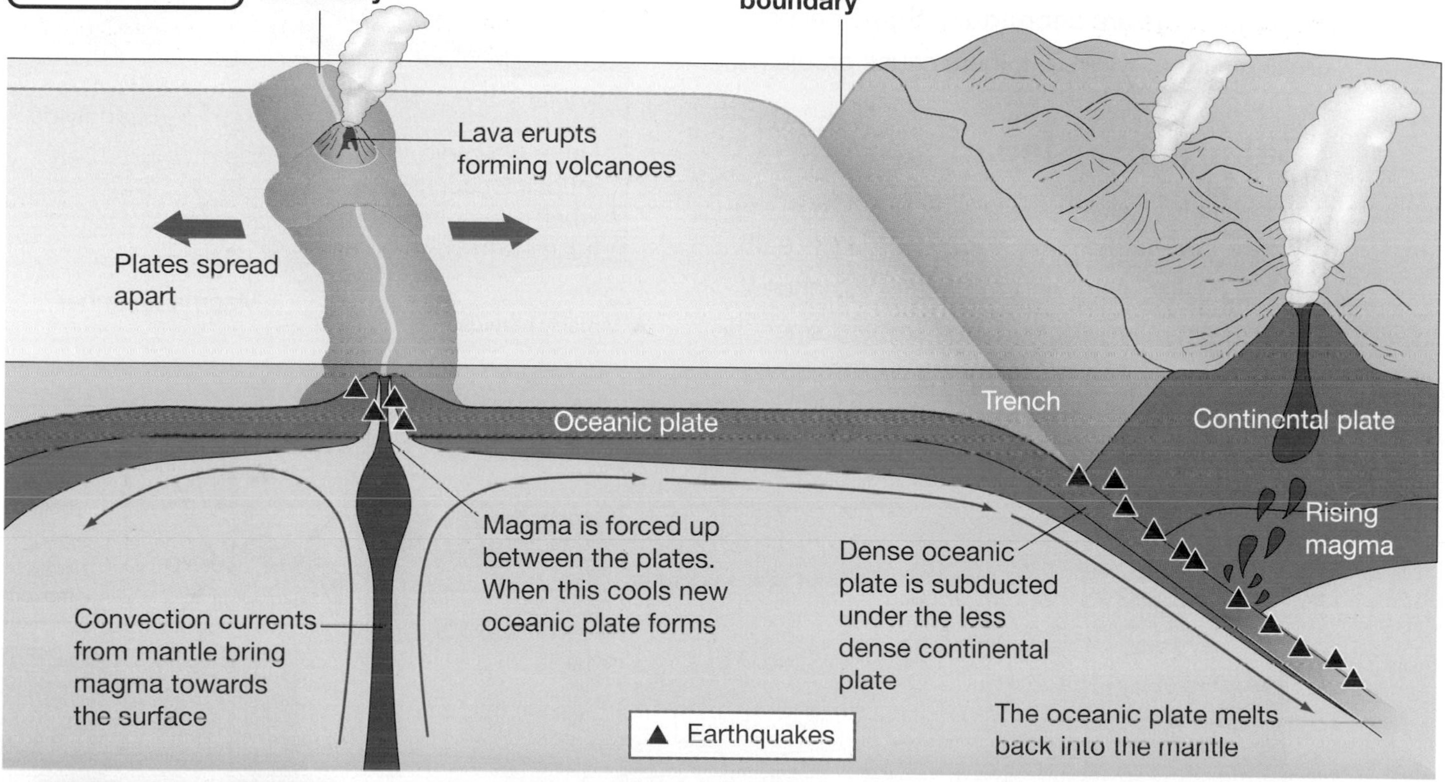

Constructive plate boundary

Destructive plate boundary

Lava erupts forming volcanoes

Plates spread apart

Oceanic plate

Trench

Continental plate

Magma is forced up between the plates. When this cools new oceanic plate forms

Convection currents from mantle bring magma towards the surface

Dense oceanic plate is subducted under the less dense continental plate

Rising magma

The oceanic plate melts back into the mantle

▲ Earthquakes

your questions

1 Which type of plate boundary is most dangerous for humans to live on?

2 Match the words below into pairs:
   constructive   fault   collision zone   fissures   landslides   explosive   destructive   conservative
   Write a brief explanation of your pairs.

3 Look at the diagram of the San Andreas fault. The two plates can 'lock', stopping them from sliding. Why do you think this is?

4 **Exam-style question** Compare the physical features and tectonic hazards of constructive and destructive plate boundaries. (6 marks)

Volcanoes in the developed world

+ In this section you will examine the impact of volcanoes on developed countries.

Destructive power

The most devastating volcanoes are the most explosive ones. The **Volcanic Explosivity Index** (VEI) measures destructive power on a scale from 1 to 8. Mount St Helens, which erupted in May 1980, measured 4. Modern humans have never experienced an eruption measuring 8.

Volcanoes produce many hazards. Some are from the **primary effects** of the volcano, whereas others are **secondary**. Some benefit the areas affected, whereas others cause problems.

Sakurajima, Japan

Japan is on a destructive plate boundary where the Pacific plate is subducted beneath the Eurasian plate, causing active volcanoes. One, Sakurajima (below), has erupted since the 1950s, sometimes 200 times a year. Ash and lava have buried buildings and

Prevailing wind

Ash and gas column

Eruption cloud

Acid rain

Ash fall builds up on roofs, causing buildings to collapse

Lava bombs can kill people close to the crater

Pyroclastic flow

Pyroclastic flows are deadly clouds of hot ash and gas that sweep along at 200 km/h

Lava flow

Lahar (volcanic mudslide) occurs when rain or snow mixes with volcanic ash

Landslide

Magma

Magma chamber

▲ Volcanic hazards

40% of the land is fertile volcanic soil growing tea and rice.

There are lots of urban areas around the bay.

KYUSHU ISLAND

7000 people live at the base of the volcano.

Kajiki   Kokubu

Kagoshima   Sakurajima

Hot springs and lava flows are a popular tourist attraction; the area is a national park.

Kagoshima has a population of 650 000: a big eruption could devastate it with ash, lava bombs and pyroclastic flows.

Tarumizu

The 1914 lava flow joined the island to the mainland.

Kanoya

Kagoshima Bay

Today the volcano hurls volcanic bombs over 3 km from its crater. Pyroclastic flows are 2 km long. 30 km^3 of ash erupt each year.

East China Sea

20 km

Pacific Ocean

The sheltered bay makes a good port and fishing is an important industry.

JAPAN

N

destructive plate boundary

TOKYO

plate movement

Sakurajima

farmland, poisonous gases have caused alerts in local towns and brought acid rain, killing plants.

Volcanic eruptions can be **predicted** – scientists can say when a volcano will erupt. They can then warn people to take shelter or **evacuate**. The diagram below shows how Sakurajima is monitored and also the evacuation procedure there.

Japan is a developed country. It can afford to spend money on monitoring, protection and evacuation. When Sakurajima does erupt, it will probably not cause many deaths. Homes, crops and industries will be destroyed, but most people have insurance and the Government will help to repair the damage. In developed countries, tectonic hazards damage property (economic costs) but cause less harm to people (social costs).

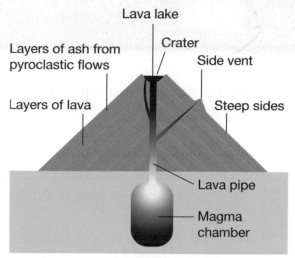

▲ Stratovolcano (or composite cone volcano)

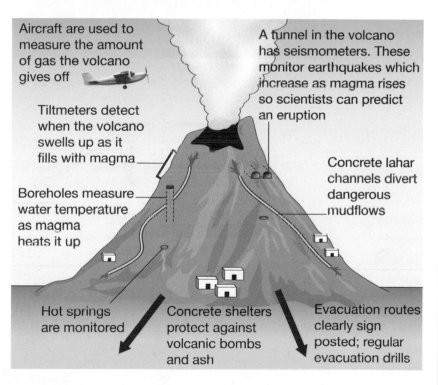

Aircraft are used to measure the amount of gas the volcano gives off

A tunnel in the volcano has seismometers. These monitor earthquakes which increase as magma rises so scientists can predict an eruption

Tiltmeters detect when the volcano swells up as it fills with magma

Boreholes measure water temperature as magma heats it up

Concrete lahar channels divert dangerous mudflows

Hot springs are monitored

Concrete shelters protect against volcanic bombs and ash

Evacuation routes clearly sign posted; regular evacuation drills

▲ Two men take cover in a lava bomb shelter during an eruption of Sakurajima (in the background).

+ **Primary effects** – caused instantly by the eruption. These are directly linked to the volcano e.g. lava, acid rain, gases and earthquakes.

+ **Secondary effects** – in the hours, days, and weeks after the eruption. These are often caused by the volcano e.g. disease, food and water shortages.

your questions

1 Copy and complete the following table to show the effects of Sakurajima:

	Benefits	Problems
Primary effects		
Secondary effects		

2 Make a table like the one on the right:

Protection	Prediction

Use it to list methods used to protect people from Sakurajima, and how scientists predict an eruption.

3 **Exam-style question** Using examples, explain how volcanic eruptions can be predicted. (4 marks)

✛ In this section you will learn how volcanoes can have devastating consequences for people in the developing world.

At risk

In the developing world, people are at greater risk from tectonic hazards than those in developed countries:

- They often live in risky locations, because there is nowhere else to live.
- They can't afford safe, well-built houses, so buildings often collapse.
- They don't have insurance.
- Their governments don't have the money and resources to provide **aid**.
- Communications are poor, so warning and evacuation may not happen.

Most volcanic eruptions with high death tolls are in the developing world.

▲ Lava destroyed 40% of Goma, covered half the airport and destroyed 45 schools. Water and electricity supplies were also cut off by the lava.

Mount Nyiragongo

In January 2002, a fast flowing river of basalt lava, 1000km wide, poured out of Mount Nyiragongo and into the city of Goma (see photo). 100 people died, mostly from poisonous gas and getting trapped in lava. A number of **social impacts** also resulted:

- 12 500 homes were destroyed by lava flows and earthquakes — and as the eruption was predicted, 400 000 people were evacuated. Many people had to move to overcrowded refugee camps.
- Disruption to the mains water supplies caused concern for the spread of diseases.

Many **economic impacts** also resulted:

- Poisonous gases caused acid rain which affected farmland and cattle – many farmers lost income.

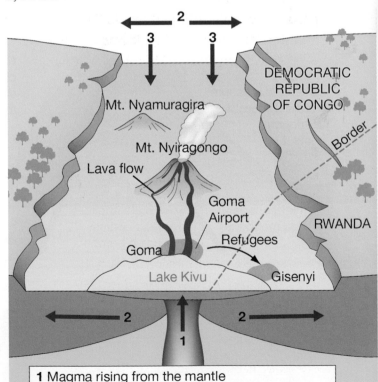

1 Magma rising from the mantle
2 Convection currents pull the two plates apart
3 The valley has formed as some of the crust has sunk downwards

- Due to poverty, most people could not afford to rebuild their homes.

Within days, people began returning to Goma. Over 120 000 were homeless. This was a crisis and people needed help quickly. With little clean water, food and shelter, diseases like cholera could spread. The United Nations and Oxfam began a **relief effort** to help.

- The United Nations sent in 260 tonnes of food in the first week. Families got 26 kg of rations.
- In the UK, a TV appeal asked people to give money to help.
- Governments around the world gave $35 million to get aid to the refugees.
- Emergency measles vaccinations were carried out by the World Health Organisation.

In a developing country, the main problem is poverty. Most people fled from the lava with nothing. It was months before many could start building new homes. By June 2002, however, some roads had been cleared of lava and the water supply repaired.

Future threats

Mount Nyiragongo was active again in 2005. It could erupt at any time. There is also volcanic activity under Lake Kivu. Gases like carbon dioxide and sulfur dioxide rise through the Earth into the lake. They get trapped in mud on the lake bottom. An earthquake could shake these gases free. In 1986 this happened to volcanic Lake Nyos in Cameroon, also in Africa. 1700 people suffocated from breathing in too much carbon dioxide.

	Mt Sakurajima, Japan	Mt Nyiragongo, DRC
Volcano type	Steep-sided stratovolcano (or composite cone) over 1000 meters high	Stratovolcano over 3400 metres, high but not as steep as Mt Sakurajima
Magma type	Andesite. High gas content, high viscosity.	Basalt. Low gas content, very low viscosity.
Explosivity	VEI 4-5	VEI 1
Hazards	Lava flows, volcanic bombs, pyroclastic flows, ash fall. Erupts almost continually, but with major eruptions once every 200-300 years.	Lava flows and gas emissions. Contains a lava lake within its crater, which can drain causing huge, fast moving lava flows.

+ **Social impacts** are the impacts upon people.

+ **Economic impacts** are the impacts upon the wealth of an area.

+ **Refugees** are people who are forced to move due to natural hazards or war.

+ **Aid** is help. It can be short-term such as food given in emergency, or long-term such as training in health care.

+ A **relief effort** is like aid. It is help given by organisations or countries to help those facing an emergency.

your questions

1 Explain in your own words what we mean by aid and relief effort.

2 a Draw an identical table to that for Question 1 on page 17, and complete it to show the effects of Mount Nyiragongo.

   b Which volcano seems to have the greatest effects – Nyiragongo or Sakurajima? Explain your answer.

3 How successful was the relief effort in helping people affected by Mount Nyiragongo's eruption?

4 Why do you think people still live around Mount Nyiragongo and Lake Kivu?

5 Exam-style question Using a named example, explain the economic and social impacts of a volcanic eruption. (6 marks)

➕ In this section you will learn how earthquakes are measured and about their awesome power.

The shaking is worse on the surface if the focus is shallow

Why is the ground shaking?

Earthquakes can't be predicted. They start without warning and can be catastrophic. An earthquake is a sudden release of **energy**. It's a bit like bending a pencil until it suddenly snaps. Underground, tectonic plates try to push past each other – building up pressure. The pressure is suddenly released along **faults** (cracks in the crust), sending out a huge pulse of energy. This travels out in all directions as earthquake waves (see the diagram on the far right).

Magnitude

The power of an earthquake – how much the ground shakes – is its **magnitude**. A **seismometer** measures this using the **Richter scale** (see right) or **Moment magnitude scale**.

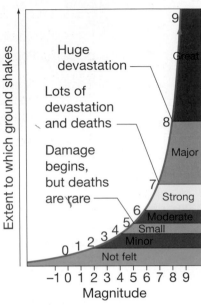

▲ The Richter scale is a logarithmic scale. A magnitude 6.0 earthquake is 10 times more powerful than a magnitude 5.0.

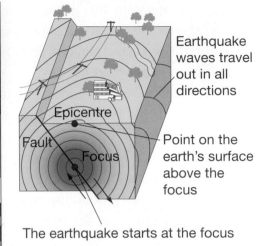

▲ The shallower the focus, the more destructive the earthquake tends to be. The epicentre experiences the most shaking.

The earthquake starts at the focus

| Key | \ Plate movement |

Port-au-Prince, Haiti

On 12 January 2010, a magnitude 7.0 earthquake struck Haiti. This was very similar in size to the 1995, magnitude 7.2 Kobe earthquake in Japan. However, the earthquakes had very different causes and effects. ▶

Kobe, Japan 1995	*Port-au-Prince, Haiti 2010*
• Magnitude – 7.2 (soft ground also made the shaking worse) • Focus – 16km deep on a fault • Epicentre – 20km from Kobe (population 1.5 million)	• Magnitude – 7.0 • Focus – 13km deep on a conservative plate boundary • Epicentre – 25km from Port-au-Prince (population 2.5 million)
Primary effects: • 5000 people died and 26 000 were injured. The population density is very high and people were still in bed when it struck at 5:46am. • Bridges and roads collapsed, train lines were damaged — disrupting transport and communication links. • £100 billion of damage was caused to roads, houses, factories and infrastructure (gas, electric, water and sewage pipes).	**Primary effects:** • 316 000 people died and a further 300 000 were injured. • Many houses were poorly built and collapsed instantly. 1 million people were made homeless. • The port, communication links and major roads were damaged beyond repair. Rubble from collapsed buildings blocked road and rail links.
Secondary effects: • Many fires broke out throughout the city, triggered by broken gas pipes, resulting in further deaths. • Businesses were affected for many weeks due to disruption caused by rebuilding. • Homelessness, disrupted schooling, unemployment and increased stress problems lasted for many months as the authorities struggled to cope with the scale of the damage caused.	**Secondary effects:** • The water supply system was destroyed — a cholera disease outbreak killed over 8000 people. • The port was destroyed — making it hard to get aid to the area. • Haiti's important clothing factories were damaged. These provided over 60% of Haiti's exports. 1 in 5 jobs were lost. • 1 year after the earthquake, 1 million people remained displaced — many still living in refugee camps.

◀ Long-term planning

The secret of survival is long-term planning. Japan is a developed country, so it can afford to do this. There is a 70% **probability** (chance) of a magnitude 7.2 earthquake hitting Toyko in the next 30 years. It could kill 7000 and injure 160 000. There is no way of **predicting** when it might happen.

- Every year Japan has earthquake drills.
- Emergency services practise rescuing people.
- People keep emergency kits at home containing water, food, a torch and radio.

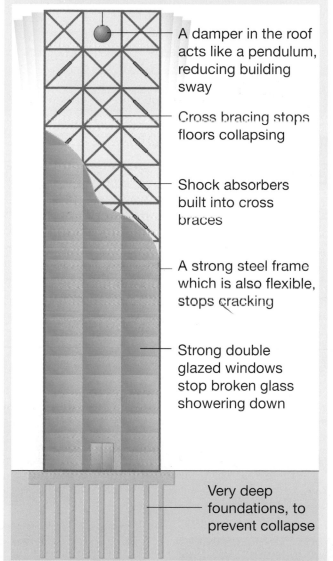

A damper in the roof acts like a pendulum, reducing building sway

Cross bracing stops floors collapsing

Shock absorbers built into cross braces

A strong steel frame which is also flexible, stops cracking

Strong double glazed windows stop broken glass showering down

Very deep foundations, to prevent collapse

▲ Many buildings in Japan are earthquake proof like this one. They can withstand a major earthquake. Gas supplies automatically shut off, reducing the risk of fire.

Tsunami

Earthquakes beneath the sea bed can generate tsunami (see below). Tsunami are waves that travel at up to 900 km/h, with wavelengths of over 200 km. In the open ocean wave height is less than 1m, but as the waves approach the coast they slow down, bunch up and wave height increases up to 30m. When tsunami hit, they cause a very powerful flood, pushing several kilometres inland destroying homes, bridges and infrastructure. Warning systems in the ocean can detect tsunami and set off sirens and alarms but this is only useful if the epicentre is some distance from the coast. None of the countries affected by the 2004 Asian tsunami received any warning because there was no warning system in the Indian Ocean.

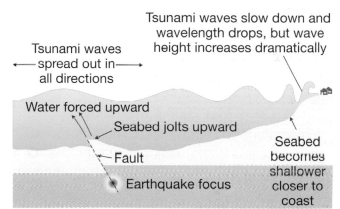

Tsunami waves slow down and wavelength drops, but wave height increases dramatically

Tsunami waves spread out in all directions

Water forced upward

Seabed jolts upward

Fault

Earthquake focus

Seabed becomes shallower closer to coast

your questions

1 Earthquake damage is reduced as you move away from the epicentre. Why is this?

2 **a** Draw a large version of the Richter scale diagram
   b Use a website, like Wikipedia, to find a list of earthquakes since 2000. Mark their magnitudes and death tolls on your diagram.
   c Is there a link between magnitude and death toll?

3 Using the diagram above, list and number the stages in the formation of a tsunami.

4 **a** Classify earthquake impacts in Japan into social and economic.
   b Which are greater? Explain your answer.

5 **Exam-style question** Using named examples, compare the social and economic impacts of two earthquakes. (6 marks)

✚ **In this section you will find out about the impacts of earthquakes on developing countries, and how people respond to them.**

Death and destruction

Earthquakes in the developing world often have very high death tolls compared to volcanoes. Destructive earthquakes happen regularly, as the table shows.

Location	Year	Deaths	Magnitude	Key facts
Sendai, Japan	2011	15 800	9.0	The tsunami (secondary effect) caused the most of the deaths. The economic costs were over $200 billion.
Kashmir, Pakistan	2005	86 000	7.6	One third of the deaths were due to landslides (secondary effect). Many children died in poorly built collapsed schools.
Aceh, Indonesia	2004	280 000	9.3	Most of the deaths were caused by a tsunami (secondary effect) that hit 14 countries around the Indian Ocean.
Bam , Iran	2003	30 000	6.6	Many people were trapped when their poorly built, mud brick homes collapsed in the densely populated city.

Earthquake in Sichuan

Sichuan is a province in central China. On 12 May 2008 it was hit by a magnitude 8.0 earthquake. There was no warning. The social impacts were devastating:

- 70 000 people died
- 375 000 people injured
- 5 million homeless

The economic impacts were large too. Over 1 million people lost their job because their workplace was destroyed. The rebuilding costs were estimated at $75 billion. Many landslides occurred (a secondary effect), some of which dammed rivers creating dangerous 'quake lakes' which can burst, causing flooding. There were up to 200 large aftershocks. On 27 May a magnitude 6.0 aftershock caused 420 000 buildings to collapse. ▷

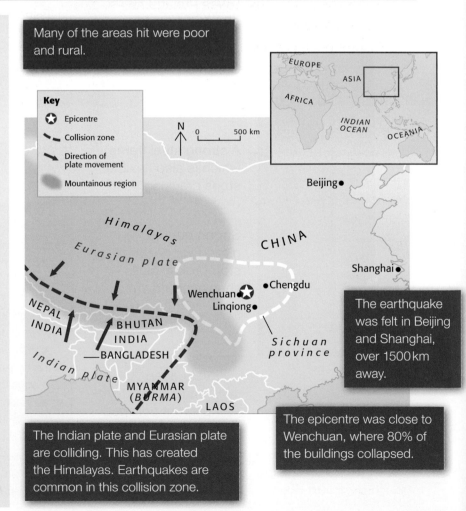

Many of the areas hit were poor and rural.

Key
- ✪ Epicentre
- – – Collision zone
- ↙ Direction of plate movement
- Mountainous region

The Indian plate and Eurasian plate are colliding. This has created the Himalayas. Earthquakes are common in this collision zone.

The earthquake was felt in Beijing and Shanghai, over 1500 km away.

The epicentre was close to Wenchuan, where 80% of the buildings collapsed.

Local responses

Heavy rain, landslides and aftershocks made the rescue effort difficult:

- The Prime Minister, Wen Jiabao, flew to the area very soon after the earthquake.
- 50 000 soldiers were sent to help dig for survivors.
- Helicopters were used to reach the most isolated areas.
- Chinese people donated $1.5 billion in aid.

International responses

China quickly asked the rest of the world to help:

- Some countries sent money. The UK gave $2 million.
- Finland sent 8000 six-person tents, and Indonesia sent 8 tonnes of medicines.
- Rescue teams flew in from Russia, Hong Kong, South Korea and Singapore.

Building for the future?

The Sichuan earthquake caused over 700 schools to collapse. China has strict building rules, so schools should have withstood the shaking. Even in a poorer developing country, buildings can be made cheaply to withstand earthquakes. The diagram on the right shows how.

+ Aftershocks often occur as the fault 'settles' into its new position. They can injure or kill rescuers. In the developing world, aftershocks often destroy buildings that were weakened by the first earthquake.

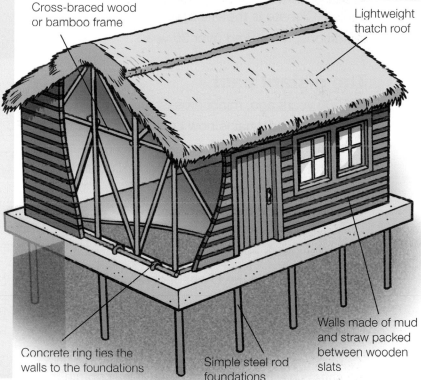

Cross-braced wood or bamboo frame

Lightweight thatch roof

Walls made of mud and straw packed between wooden slats

Concrete ring ties the walls to the foundations

Simple steel rod foundations

your questions

1 Explain why the Sichuan earthquake happened. Remember to include plate names and boundary types in your answer.
2 a Make a list of all the effects of the Sichuan earthquake.
  b Use two colours to circle social effects (effects on people) and economic effects (effects to do with money).
3 Look back on the effects of the two volcanoes and two earthquakes you have studied.
  a Compare impacts each had on people and property.
  b Explain the differences.
4 Exam-style question Explain why earthquakes happen on destructive plate margins. You may draw a diagram to help with your answer. (4 marks)

✚ In this section you will learn how climate was very different from that of today in both the recent, and distant, past.

What is climate?

People say climate is what you expect, but weather is what you get! If you plan a holiday to Majorca in August, you expect it to be hot and sunny (climate). If it rains when you get there, you've got weather!

Weather is short-term day-to-day changes in things like temperature, wind, cloud cover and rainfall. Climate is the *average* of these weather conditions, measured over 30 years.

The distant past

120 000 years ago, rhinoceroses and elephants roamed around what is now London. At other times in the past, huge ice sheets stretched from the North Pole as far south as London. Scientists know that climate was different in the past. They use physical evidence such as:

- fossilised animals, plants and pollen that no longer live in the UK
- landforms, like the U-shaped valleys left by retreating glaciers
- samples from the ice sheets of Greenland (see the photo) and Antarctica.

Ice sheets are like a time capsule. They contain layers of ice, oldest at the bottom, youngest at the top. Each layer is one year of snowfall. Trapped in the ice layers are air bubbles. These preserve air from the time the snow fell. Locked in the air bubble is carbon dioxide. **Climatologists** can reconstruct past temperatures (shown in the graph on the right) by drilling a core through the ice and measuring the amount of trapped carbon dioxide in ice layers.

▲ A lump of ice the size of a house falls off the Greenland ice sheet into the sea.

✚ A **climatologist** is a scientist who is an expert in climate and climate change.

On your planet

✚ The level of carbon dioxide in the atmosphere is higher today than at any time in the past 800 000 years.

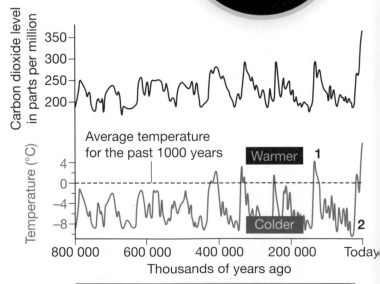

Average temperature for the past 1000 years

Key

1 Warm periods, called **interglacials.**
2 Cold periods, called **glacials.** Some glacial periods became Ice Ages.

During the **Quaternary** (the last 2.6 million years of geological time), warm periods (**interglacials**) lasted for between 10 000 and 15 000 years. Cold periods (**glacials**) lasted about 80 000–100 000 years. During some glacial periods, it became so cold that the Earth plunged into an ice age. Huge ice sheets extended over the continents in the northern hemisphere. There were also vast areas of floating sea ice. The last time this happened was between 30 000 and 10 000 years ago, in the last ice age (see right). The ice sheets were 400–3000 metres thick, and so heavy that they made the Earth's crust sag. So much water was locked up in the ice sheets that sea levels fell by over 100 metres.

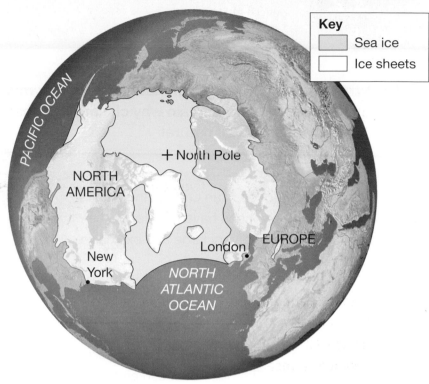

▲ Ice cover in the last ice age.

The recent past

There is also evidence for climate change in more recent times. Evidence comes from:

- old photographs, drawings and paintings of the landscape
- written records, such as diaries, books and newspapers
- the recorded dates of regular events, such as harvests, the arrival of migrating birds and tree blossom.

These sources are often not very accurate, because they were not intended to record climate. However, they can still give us some idea of overall climate trends in the recent past. This type of evidence suggests that climate changes regularly – every few hundred years. Average temperatures over the past 2000 years or so have probably varied between 1–1.5 °C colder or warmer than average temperatures today (see the graph on the right).

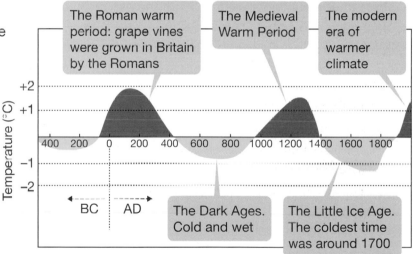

your questions

1 Describe how the northern hemisphere was different during the last Ice Age compared to today.
2 Look at the graph of carbon dioxide levels and temperature over the last 800 000 years.
   a Describe the variations in the temperature graph.
   b Does the Earth warm up faster, or cool down faster?
   c How closely are temperature and carbon dioxide levels linked?
3 Look at the map of ice sheets. Which parts of the land were covered by ice 20 000 years ago, and are now ice-free?
4 **Exam-style question** Describe the evidence that can be used to reconstruct past climates. (4 marks)

The causes of climate change in the past

+ In this section you will learn that there are four main theories that climatologists use to explain why climate has changed in the past. These changes are all natural.

The eruption theory

Big volcanic eruptions can change the Earth's climate. Small eruptions have no effect – the eruption needs to be very large and explosive. Volcanic eruptions produce:

- ash
- sulfur dioxide gas.

If the ash and gas rise high enough, they will be spread around the Earth in the **stratosphere** by high-level winds. The blanket of ash and gas will stop some sunlight reaching the Earth's surface. Instead, the sunlight is reflected off the ash and gas, back into space. This cools the planet and lowers the average temperature.

> + The **stratosphere** is the layer of air 10–50 km above the Earth's surface. It is above the cloudy layer we live in, called the troposphere.

In 1991, Mount Pinatubo in the Philippines erupted, releasing 17 million tonnes of sulfur dioxide (see the top photo). This was enough to reduce global sunlight by 10%, cooling the planet by 0.5 °C for about a year.

Mount Pinatubo was very small-scale compared to the 1815 eruption of Tambora in Indonesia. This was the biggest eruption in human history. In 1816, temperatures around the world were so cold that it was called 'the year without a summer', and up to 200 000 people died in Europe as harvests failed. The effects lasted for four to five years. In general, volcanoes only affect climate for a few years.

Asteroid collisions

Asteroid impacts can alter earth's climate. In 1908 an asteroid with a diameter of 100m exploded in the air 5 km above Tunguska in Russia. The blast flattened 80 million trees but was not large enough to alter the climate. 1 km sized asteroids strike earth once every 500,000 years. An impact of this size would blast millions of tonnes on ash and dust into the atmosphere. This would cool the climate as the dust and ash blocked incoming sunlight. It is similar to a large volcanic eruption and could last 5-10 years.

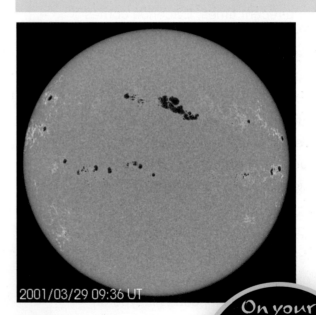

2001/03/29 09:36 UT

▲ Sunspots on the surface of the sun in March 2001.

On your planet

+ In 1947, astronomers observed the Great Sunspot, the largest ever recorded. It was about 30 times bigger than Earth.

Sunspots and solar output

Over 2000 years ago Chinese astronomers started to record sunspots. These are black areas on the surface of the sun (see the bottom photo opposite). Sometimes the sun has lots of these spots. At other times they disappear. Even though the spots are dark, they tell us that the sun is more active than usual. Lots of spots mean more solar energy (solar output) being fired out from the sun towards Earth.

Cooler periods, such as the Little Ice Age, and warmer periods, such as the Medieval Warm Period, may have been caused by changes in sunspot activity. Some people think that, on average, there were more volcanic eruptions during the Little Ice Age, and that this added to the cooling. However, climate change on timescales of a few hundred years, and 1–2 °C, cannot be explained by volcanoes – but it might be explained by sunspot cycles (see the top right diagram).

The orbital theory

Over very long timescales, there have been big changes in climate. Cold glacial periods and ice ages were 5–6 °C colder than today. Some interglacials were 2–3 °C warmer than today. Such big changes need a big cause. Scientists think they know what this is – changes in the way the Earth orbits the sun.

You might think that the Earth's orbit does not change, but over very long periods it does, as the diagrams on the right show.
- The Earth's orbit is sometimes circular, and sometimes more of an ellipse (oval).
- The Earth's axis tilts. Sometimes it is more upright, and sometimes more on its side.
- The Earth's axis wobbles, like a spinning top about to fall over.

These three changes alter the amount of sunlight the Earth receives. They also affect where sunlight falls on the Earth's surface. On timescales of thousands of years, the changes would be enough to start an ice age, or end one. These changes are called **Milankovitch Cycles**.

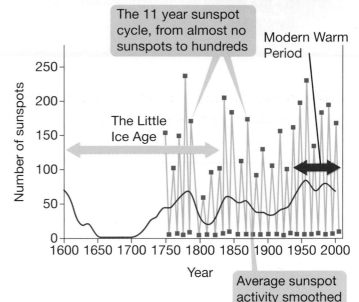

The 11 year sunspot cycle, from almost no sunspots to hundreds

Modern Warm Period

The Little Ice Age

Average sunspot activity smoothed out over time

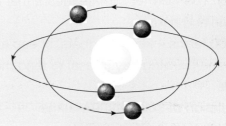

It takes 100 000 years for the Earth's orbit to change from being more circular, to an ellipse, and back again.

It takes 41 000 years for the Earth's axis to tilt, straighten up, and tilt back again.

It takes 26 000 years for the Earth's axis to wobble, straighten up, and wobble again.

your questions

1. Explain how big volcanic eruptions might change our climate.
2. Draw and complete a table to compare the four theories of climate change.
3. **Exam-style question** Explain how solar output and orbital changes can alter earth's climate. (6 marks)

✚ In this section you will examine some of the impacts of past natural climate change and the lessons they can teach us for the future.

Viking Greenland

The Viking saga of 'Erik the Red' tells us that Erik was a Norse Viking, whose parents fled to Iceland. In 982, Erik was banished to Greenland for murder. In western Greenland, he and about 500 other Vikings found a land that was largely free of ice. This was the start of the Medieval Warm Period, when Greenland was very different from today. The warmer climate meant that by 1100 Greenland had:

- over 200 farms – keeping goats, sheep and cows, and growing hay (grass) to feed the animals
- a population of 3000–4000 Vikings
- trade links with Iceland and Norway
- summer hunting expeditions north of the Arctic Circle, for seals and whales.

Life on Greenland was very hard. The Greenland Vikings survived there for over four centuries, but soon after 1410 they died out (see below). What happened? We don't know exactly what killed off the last few Greenland Vikings, but we do know why they declined – the Little Ice Age.

▲ Greenland today.

Learning from the Vikings

The Little Ice Age made life impossible for the Greenland Vikings. They ran out of food and died out. However, they did not help their own survival:

- Deforestation and soil erosion meant that the Viking farms probably produced very little food. The Vikings had damaged the land they depended on.
- They also depended too much on trade with Iceland and Norway. When this stopped, because of sea ice, they had no one else to turn to.
- The Greenland Vikings were not very adaptable. They tried to live as they did in Norway and Iceland.

When the climate became colder, the Greenland Vikings were isolated in a damaged environment with no real way of changing their lifestyle to cope.

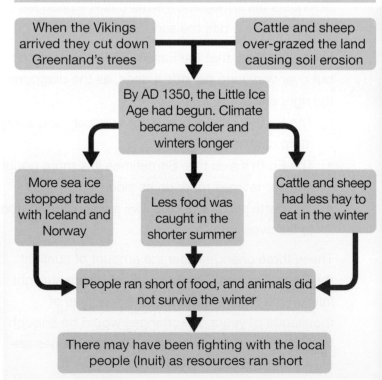

When the Vikings arrived they cut down Greenland's trees

Cattle and sheep over-grazed the land causing soil erosion

By AD 1350, the Little Ice Age had begun. Climate became colder and winters longer

More sea ice stopped trade with Iceland and Norway

Less food was caught in the shorter summer

Cattle and sheep had less hay to eat in the winter

People ran short of food, and animals did not survive the winter

There may have been fighting with the local people (Inuit) as resources ran short

The Little Ice Age

As the world cooled, the Little Ice Age began to have wider impacts. In England, 'Frost Fairs' were held on the River Thames when it froze over in the winter (see right). The first Frost Fair was in 1608 and the last in 1814.

During the Medieval Warm Period, Europe's population grew. New land was given over to farming, often on hillsides and areas that had not been farmed before. The warmer climate made this possible. This growth crashed to a halt in 1315, when the Little Ice Age took hold.

- Cold and rain lashed Europe in the spring and summer of 1315.
- Wheat and oats did not ripen and the harvest failed.
- The cool wet weather continued in 1316 and 1317.
- By 1317 the 'Great Famine' had begun. It lasted until 1325.
- In some areas, 10–20% of the peasant farmers may have died of hunger.

Things did not improve much for Europe. In 1349, it was struck by the Black Death (bubonic plague). This was to kill far more than the Great Famine, but was probably made worse by the colder climate and more difficult farming conditions.

In the Alps, many valley glaciers grew in the colder climate. In the 1820s and 1850s they advanced down valleys, destroying villages and farmland. Many farmers stopped growing wheat, because it needs warm summers. The crop of choice became the cold and wet loving potato, imported from South America in about 1530.

The Little Ice Age caused many problems in Europe. However, people did adapt. They learned to farm new crops, abandoned farms high on hillsides and learned to enjoy fairs on frozen rivers. People can adapt to climate change, but it takes time.

▲ A Frost Fair on the River Thames in London in 1683.

On your planet

+ These examples show that even a 0.5-1°C change in temperature has a big impact on people.

your questions

1 How did the climate at the time affect Erik the Red's prospects when he was banished to Greenland?
2 Imagine you were one of the last farmers in Greenland. Write a brief letter to a relative in Iceland explaining how difficult life has become.
3 Greenland's Vikings did not get on with the local Inuit people. How might befriending them have helped the Vikings to survive?
4 How did people in Europe adapt to the colder climate of the Little Ice Age?
5 **Exam-style question** Using named examples, explain how natural climate change in the past affected people and their lifestyles. (6 marks)

✚ In this section you will discover that, as well as humans, plants and animals are vulnerable to climate change.

Ecosystems

Plants and animals live together in **ecosystems**. They depend on each other, and are linked together in **food chains**. They also depend on the physical environment around them, which includes the climate. Together, plants, animals and the physical environment make up ecosystems like the one in the diagram. Ecosystems can be small, such as a pond, or large, such as the tropical rainforest.

If one part of an ecosystem changes, the other parts will also change. When climate has changed in the past, it has spelled disaster for some plants and animals.

The dinosaurs

65 million years ago many dinosaurs suddenly became **extinct**. Two possible causes are:
- a strike by a massive asteroid in Mexico
- a huge volcanic eruption in Deccan, India, lasting up to 1 million years.

Both of these events are known to have happened at that time. They may have happened together. Both would have thrown up huge amounts of dust, ash and gas into the air, blocking out the sun. Plants would have struggled to grow as the climate cooled. Ecosystems would have broken down as **food chains** collapsed, sealing the dinosaurs' fate.

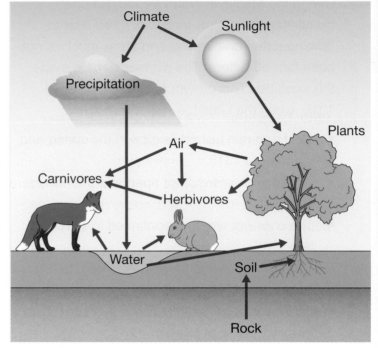

▲ The parts of an ecosystem.

✚ An **ecosystem** is a community of plants and animals (living) which interact with the atmosphere, water and minerals (non-living) as a system. Energy and nutrients move around the ecosystem e.g. a pond, an area of sand dunes.

✚ **Extinction** means a species of plant or animal dying out completely, so none survive.

✚ In a **food chain**, plants provide food for plant-eating animals (herbivores). Herbivores provide food for meat-eating animals (carnivores). Plants and animals are linked together, and depend on each other.

Ice age megafauna extinction

If we fast forward in time from the dinosaurs, we come to a more recent 'mass extinction', when a number of species died out together. The animals were the Quaternary **megafauna**, and the time was 10 000–15 000 years ago.

> + **Megafauna** means 'big animals'. Most weighed over 40 kg and included the woolly mammoth, giant elk, ground sloth, sabre-tooth cat, giant beaver and glyptodon.

What happened to the megafauna? It seems that they were the victims of two new stresses at once – humans, and climate change. As the last major ice age ended, and the climate warmed up by about 6 °C in only 1000 years, many animals had to move.

* They migrated and tried to find new areas to live in, where the climate suited them.
* However, finding the right plants to eat in the new areas would have been difficult.
* This would have disrupted food chains, leaving some animals short of food.
* Climate change stress may also have made the megafauna weaker than normal.

As the climate warmed, humans also migrated into new areas. They hunted some of the megafauna, meaning less prey for carnivores. Some herbivores may have been hunted to extinction, leaving carnivores with nothing to prey on. Humans and climate change seem to have acted together to cause the extinction shown in the graph.

The extinct ice age megafauna provide a lesson for us today. There are many humans on the planet and our climate seems to be changing. Is it surprising that some scientists think that up to 30 000 species are becoming extinct every year?

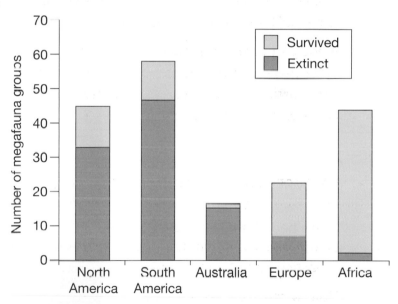

your questions

1 What is an ecosystem?
2 Use the words below to draw a simple food chain:
    Rabbit   Sun   Grass   Wolf   Bacteria
3 Look at the graph of megafauna extinctions.
   a Which regions lost:
      i) the largest number of groups of megafauna?
      ii) the least number of groups of megafauna?
   b Work out the percentage of groups of megafauna that became extinct for each region.
4 **Exam-style question** Using examples, describe how the environment and ecosystems were affected by climate change in the past. (6 marks)

+ In this section you will learn how our **atmosphere** is being changed by human activity.

The greenhouse effect

Earth's atmosphere is vital to life. The gases which make up the atmosphere are important:

- Nitrogen (78.1%) is an important nutrient for plant growth.
- Oxygen (20.9%) is breathed in by animals, which breath out carbon dioxide.
- Carbon dioxide (0.03%) is breathed in by plants, which breath out oxygen.
- Water vapour (about 1%) forms clouds, which are a key part of the water cycle.

Carbon dioxide is a very important gas, even though it makes up only a tiny fraction of the atmosphere. This is because it helps to regulate the temperature on Earth – it is a **greenhouse gas**.

Greenhouse gases

Although carbon dioxide is the most common greenhouse gas, methane is far more potent (see table), and is increasing rapidly. Greenhouse gases make the planet warmer by about 16°C. This keeps the Earth comfortably warm. Without greenhouse gases, most of the planet would be a frozen wasteland. It is important to understand that the **greenhouse effect** is natural (see the diagram).

+ The **atmosphere** is a layer of gases above the Earth's surface.

+ The **greenhouse effect** is the way that gases in the atmosphere trap heat from the sun. The gases act like the glass in a greenhouse. They let heat in, but prevent most of it from getting out.

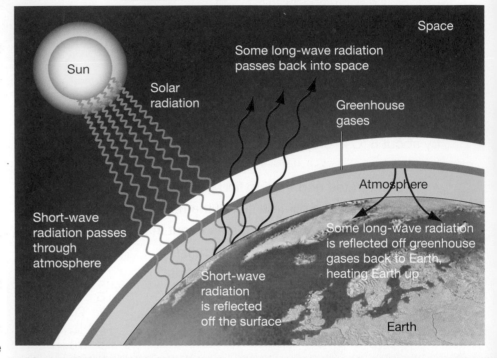

Greenhouse gas	% of greenhouse gases produced	Sources	Warming power compared to carbon dioxide	% increase since 1850
Carbon dioxide	89%	Burning fossil fuels (coal, oil and gas), deforestation which releases carbon dioxide.	1	+30%
Methane	7%	Natural gas extraction, landfill, decomposition of organic matter including manure, emission from rice paddy fields	21 times more powerful	+250%
Nitrous oxide	3%	Jet aircraft engines, cars and lorries, fertilisers and sewage farms.	250 times more powerful	+16%
Halocarbons	1%	Used in industry, solvents and cooling equipment.	3000 times more powerful	Not natural

The extra greenhouse gases which pollute the atmosphere are produced by humans. In the UK we use lots of fossil fuels. Burning these produces carbon dioxide, which ends up in the atmosphere as pollution. The main source of this pollution is power stations that produce our electricity (see below).

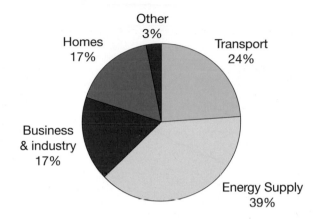

▲ Percentage of carbon dioxide emissions from different sources in the UK (2010).

On a global scale, there are big differences in carbon dioxide production. Most is produced by the developed world.

- The EU, USA and Japan emit 36% of all carbon dioxide.
- China alone emits 24%, with India and Russia both more than 5%.

Most people in the developing world produce 1-3 tonnes of carbon dioxide per person per year, compared to 10–25 tonnes per person in the developed world. The map above right shows the differences.

Many scientists are becoming concerned about greenhouse gas emissions and their effect on our climate. Some issues we need to think about are:

- how to reduce emissions in the developed world, where we use a lot of fossil fuels
- how to persuade big developing countries like China and India to slow down the growth in their carbon dioxide emissions
- how to protect vulnerable people from the future impacts of climate change.

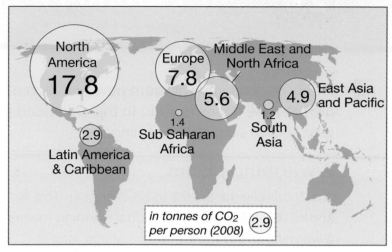

▲ Carbon dioxide emissions around the world.

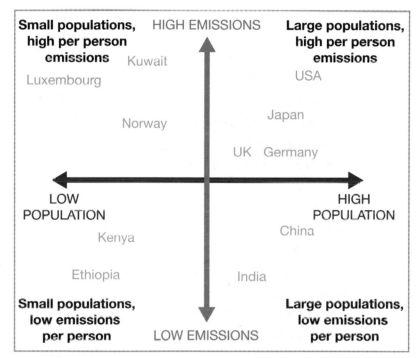

your questions

1 What are the main greenhouse gases?
2 a Draw your own version of the diagram showing the greenhouse effect.
  b Label it to explain how the greenhouse effect works.
3 a Make a list of human activities which add extra greenhouse gases into the atmosphere.
  b Explain why people in the developing world produce only small amounts of greenhouse gases. Think about their lifestyles and activities compared to the developed world.
4 Exam-style question Using examples of countries and human activities, explain why levels of greenhouse gases in the atmosphere are rising. (6 marks)

In this section you will learn how pollution of the atmosphere with greenhouse gases has led to the enhanced greenhouse effect, also known as global warming.

A warming planet

Today our climate seems to be changing. This is known as global warming. Global warming means a warming of the Earth's temperatures, and is caused by the **enhanced greenhouse effect**. 'Enhanced' simply means 'working more strongly'. Because humans have polluted the atmosphere with carbon dioxide and other greenhouse gases, the natural greenhouse effect has been given a boost. The graph shows the increase in carbon dioxide in the atmosphere in Hawaii. More heat is trapped in the atmosphere by the greenhouse gases, and temperatures are rising.

Global warming has been measured:

* Global temperatures rose by 0.75 °C between 1905 and 2005.
* Sea levels rose by 195 mm from 1870 to 2005. They are rising because the sea expands as it warms up. This is called **thermal expansion**. In future, if the glaciers and ice sheets continue to melt, sea levels could rise significantly.

Since 1980, global warming seems to have all been happening more quickly:

* The 20 warmest years ever recorded have been since 1987.
* Summer floating sea ice in the Arctic shrank from 7.6 million square kilometres in 1980 to only 3.6 million in 2012.
* Over 90% of the world's valley glaciers are shrinking. On the right are photos of the Upsala glacier in Argentina in 1928 and 2004.

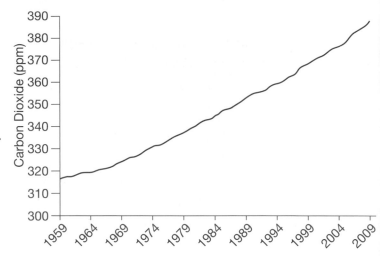

▲ Carbon dioxide concentrations, Hawaii 1959-2009.

▲ The Upsala Glacier, Argentina, in 1928 (top photo) and 2004.

On your planet
+ Worldwide, 2005 was the warmest year on record with 2010 second and 1998 third.

What do scientists think?

Many scientists and politicians are worried about global warming. In 2007, over 2500 scientists and climatologists from 130 countries wrote a report called 'Climate Change 2007'. They work for the Intergovernmental Panel on Climate Change, which is part of the United Nations. Their report said:

'Most of the increase in global average temperatures since the mid-20th century is very likely due to the increase in human greenhouse gases.'

In other words, they blamed humans for the increasing temperatures. But there are some scientists who believe that humans are not the main cause of global warming, or who think most of the warming is natural.

Future climate?

Scientists do not know exactly how global warming might affect our planet. All they can do is try to estimate future changes. Their estimates are that:
- temperatures will rise between 1.1 °C and 6.4 °C by 2100
- sea levels will rise by between 30 cm and 1 metre by 2100.

A 'best guess' might be a warming of 3.5 °C and a sea level rise of 40 cm by 2100. Floods, droughts and heatwaves could become more common, and storms and hurricanes stronger (see the photo). Predicting future global warming is very difficult because we don't know:
- what the world's future population will be.
- if we will continue to use fossil fuels, or change to cleaner energy like wind or solar.
- if people will change their lifestyles and recycle more, or use public transport.

The graph on the right shows how temperatures might increase in three different situations. Also, we don't really know how the climate might react if we continue to pollute the atmosphere. We could be in for some big surprises.

▼ The aftermath of Hurricane Katrina.

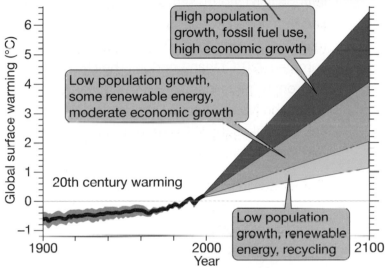

High population growth, fossil fuel use, high economic growth

Low population growth, some renewable energy, moderate economic growth

Low population growth, renewable energy, recycling

20th century warming

your questions

1 Copy and complete the paragraph using the words below:
ENHANCED ATMOSPHERE WARM
CARBON DIOXIDE GLOBAL GAS
POLLUTE

_____ _____ is a greenhouse ____.
These gases in the _____ keep the planet comfortably _____. Human activities are producing more greenhouse gases which _____ the atmosphere. This is causing the _____ greenhouse effect and seems to be leading to _____ warming.

2 Use the graph on the opposite page to describe how carbon dioxide levels have risen since the 1950s.

3 Use data to explain what global warming is.

4 Exam-style question Explain why predictions about future global temperatures and sea level are uncertain. (6 marks)

➕ In this section, you will learn about the climate of the UK today, in order to understand how it might be different in the future.

Ocean influence

The UK has a **temperate maritime** climate, experiencing mild temperatures relative to latitude, with rainfall in all months of the year. Although the UK has seasons, the differences between seasons are relatively small. London's latitude is 51°north, similar to Moscow in Russia, and Winnipeg in Canada. These cities have much colder winters (-10°C or below) but warmer summers (20°C) than London. The UK's climate is strongly influenced by a warm ocean current called the North Atlantic Drift (see map). This current makes the UK warmer than might be expected for its latitude.

Air masses and seasons

The effect of the tilt of earth's axis is to produce seasons. In the UK the seasons are also influenced by the position of the **polar front**. This is the boundary between cold polar air to the north and warm tropical air to the south. During the year the position of the polar front changes (see map). Its exact position influences which air mass sits over the UK (see table above). Air masses have a very strong influence on precipitation, temperature, air pressure, wind, and cloud cover. In winter, the polar front moves south over the UK bringing polar and arctic air masses. In summer it moves north, allowing tropical air masses to move up. The polar front is very variable, and hard to predict. It is often in the 'wrong' place, bringing unseasonable conditions. One reason it rains so much in the UK is because rain forms when cold and warm air meet at the polar front, which usually occurs over, or close to, the UK.

	Source area	Season	Conditions
Polar maritime	Atlantic ocean off Canada and Greenland	Autumn, winter and spring	Cool, wet and windy, 'wet' snow
Polar continental	Russia and eastern Europe	Winter	Cold and dry, 'dry' snow
Arctic maritime	Arctic ocean	Winter	Very cold with northerly winds
Tropical maritime	Atlantic ocean off Spain and west Africa	Autumn, winter and spring	Mild, wet and windy
Tropical continental	North Africa	Summer	Hot and dry

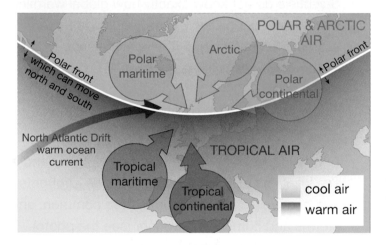

Maritime climates are found close to the sea. Moist air moving in from the sea cools the summer months and warms the winter months, bringing rain all year round.

Continental climates occur far from the sea in the middle of continents. The climate is drier, with hotter summers and colder winters.

Average conditions

Although the climate is similar across the whole of the UK, there are small differences:

- Western and northern locations, like Glasgow, are the wettest. This is because moist air coming in from the Atlantic results in relief rainfall over the mountains in Cumbria, Wales and the Highlands.
- Average precipitation within the UK varies – ranging from 600 mm to 1600 mm depending on location.
- Southern and eastern areas, like Cambridge, are the warmest and driest and tend to have more sunshine hours.

Climate data for Cambridge and Glasgow are shown in the table.

On your planet

+ Average precipitation within the UK varies – most places have between 600 and 1600 mm though there are exceptions!

		J	F	M	A	M	J	J	A	S	O	N	D
Cambridge	Average temperature (°C)	4	5	6	9	12	16	17	18	15	11	7	4
	Average precipitation (mm)	46	35	39	40	47	52	51	54	54	58	55	47
	Average sunshine hours	58	77	110	152	179	176	188	183	140	114	67	49
Glasgow	Average temperature (°C)	3	4	6	8	11	13	15	15	13	10	6	4
	Average precipitation (mm)	142	99	110	60	63	63	68	84	116	132	131	138
	Average sunshine hours	34	59	87	131	179	168	160	145	113	81	52	30

Big shifts?

Most scientists expect the UK's climate to change in the future as a result of global warming (see section 2.8). Over the next 100 years some big changes could take place:

- The polar front may shift north, so the UK experiences tropical air masses more often. This would mean our climate would be warmer – and perhaps drier in the summer, but wetter in winter.
- The North Atlantic Drift could change position or weaken. This could produce a much more variable climate, perhaps even a cooler one.

This is explored fully on pages 38-39.

+ An **air mass** is a huge body of air with uniform temperature and humidity. They form in source areas which could be either; over the sea (maritime) or land (continental), and in cold (polar) or warm areas (tropical).

your questions

1 Use the climate data for Cambridge and Glasgow to plot two climate graphs showing precipitation, temperature and sunshine hours.
2 Explain the differences on the graphs with reference to latitude, seasons, air masses and the polar front's position.
3 Exam-style question Describe the main features of the UK's climate today. (4 marks)

✚ **In this section you will learn how the UK might be affected by global warming.**

There is a lot of debate about what could happen to the UK's climate in future, as the earth warms. Three possibilities are shown here.

1 Getting warmer?

Many assume that because global warming will make the earth warmer, the UK will naturally be warmer. They imagine that warmer air from the south will push the Polar Front (see map on page 36) northwards. Warm tropical air would reach the UK, making summers very hot. Warm air can hold more moisture than cold, so warmer air might also mean reduced precipitation (see maps on the right). In summer, if London was warmer by:

• 2 °C - it would be as warm as Paris
• 4 °C - it would be like the south of France
• 6 °C - it would be like Madrid

Perhaps there would be Mediterranean summers, or maybe more droughts. A warmer UK would have costs as well as benefits, as the table below shows.

2 Getting cooler?

But what if the UK became cooler? Some believe that the Polar Front could actually move south, because melting ice caps around the Arctic would push colder water, and air, southwards — causing the UK to become colder. It might also bring higher rainfall as warm moist air meets cold Polar air, causing large-scale condensation. In this case:

• There would be increased flooding. Existing flood defences — such as the Thames Barrier — might need replacing, at vast expense.
• Summers especially would be colder — perhaps southern England would be more like northern Scotland.

▼ Possible annual temperature change by 2050 if the UK gets warmer.

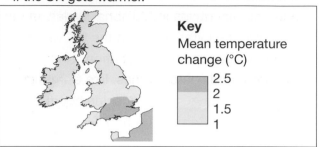

Key
Mean temperature change (°C)
2.5
2
1.5
1

▼ Possible annual precipitation change by 2050 if the UK gets warmer.

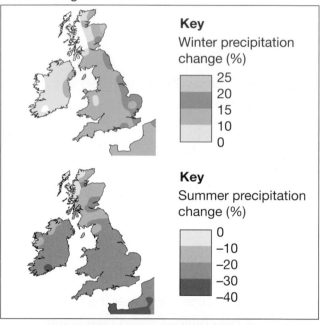

Key
Winter precipitation change (%)
25
20
15
10
0

Key
Summer precipitation change (%)
0
−10
−20
−30
−40

Costs of a warmer UK	Benefits of a warmer UK
In the summer, drought and water shortages could become more common, especially in the south.	In winter, heating costs and road gritting costs could fall.
There could be more illnesses such as heat-stroke and skin cancer.	More people could take their holidays in the UK, which would be good for the tourism industry.
Very hot temperatures can melt road surfaces and buckle railway lines.	There could be fewer deaths of old people in the winter from cold.
Farmers might have to change crops to those that need less water and more sunshine.	New crops might mean new sales opportunities for farmers.
Some plants and animals might die out in the UK if it gets too hot.	More land could be farmed at higher altitudes.

3 Wilder weather?

Some scientists think that global warming will result in more extreme weather in the UK. This could mean more:

- heatwaves, like summer 2003, when temperatures reached 38°C
- flooding, like summer 2007, when parts of the Midlands had a month's rainfall in 1-2 days
- storms, like the 'great gales' in 1987 and 1990, which caused millions of pounds of damage, as shown in the photo.

Extreme weather is hard to predict. Protecting people from extreme weather is an economic challenge. Insurance companies will have to pay out, so the cost of insurance will rise. Billions of pounds will need to be spent on new flood defences to protect homes and businesses.

A Stern warning

In 2005, Sir Nicholas Stern wrote a report on global warming and its impacts. This was called the 'Stern Review'. It warned that we should act now to reduce global warming. Stern said:

- we should spend 2% of our GDP reducing greenhouse gas pollution now
- if we don't do this, the effects of global warming could reduce our GDP by 20%.

He is basically saying 'Spend now, or pay later'.

What can we do?

Could we reduce greenhouse gas pollution? The answer is 'yes'. However, it would only really work if the UK and other countries all acted together. We could;

- reduce our use of fossil fuels and switch to 'green' energy, like wind, solar and tidal power.
- recycle more.
- use cars less and public transport more.

In 1997, the Kyoto Protocol international agreement, set targets for developed countries to reduce greenhouse gas emissions. The UK and Germany have done this, but the USA refused to sign the agreement. In 2009 the Copenhagen Summit involved developing countries like China and Brazil for the first time, but a strong agreement on emissions targets for all countries was not reached.

▲ The aftermath of the great gale in 1987.

On your planet
+ By 2030, heatwaves like the one in 2003 could happen once every three years

your questions

1 Describe how global warming in the UK could affect temperatures and precipitation.
2 How might the UK's weather become more 'wild' due to global warming?
3 What was the Stern Review and what did it say?
4 Write a letter to the Prime Minister saying what you think the UK should do about global warming.
5 **Exam-style question** Using examples, describe how global warming in the UK could have both costs and benefits. (6 marks)

+ In this section you will learn about some of the possible impacts of global warming in the developing world, by exploring the situation in Egypt.

Egypt's contribution

As a developing country, Egypt's greenhouse gas emissions are low. They are 2.6 tonnes of greenhouse gases per person per year. The world average is 6.8 tonnes per person, compared with 11 tonnes in the UK. Egypt's 75 million people produce less than 1% of all greenhouse gases.

+ **Desertification** is the gradual change of land into desert.

Challenges

Egypt is unusual. 99% of its people live on only 5% of the country's land. Much of the country is desert (see the satellite photo).

The River Nile is a very important water supply. It is Egypt's only reliable source of water. Egypt's geography makes it very vulnerable to global warming. It could be on the edge of a climate disaster.

If sea levels rise by only 50 cm, over one third of the city of Alexandria will be under water. The same sea level rise would flood 10% of the Nile delta. This would mean over 7 million people having to find somewhere else to live. The loss of land would also hit farming. Less food would be produced, possibly leading to famine. Rising sea levels, and more frequent storms, are already eroding the delta coastline by over 5 metres per year in some areas.

Global warming could have wider impacts on Egypt, such as:

- temperature rises of 8 °C by 2080, double the global average
- less and more unreliable rainfall
- the spread of the Sahara Desert (**desertification**) onto areas of farmland
- falling crop yields as temperatures rise and water shortages increase
- heatwaves bringing more illness and death
- the spread of diseases like malaria

Mediterranean Sea

Alexandria

Nile Delta

Cairo

Rainfall in Egypt averages under 10 mm per year.

Egypt depends on the River Nile for its water supply.

River Nile

Red Sea

Sahara Desert

Egypt has 6 million acres of farmland, almost all of it is irrigated using River Nile water.

Aswan Dam

Lake Nasser

Egypt grows large amounts of maize, wheat and cotton in the fertile Nile Delta.

Water wars?

Water is already in short supply in Egypt. The amount of water available per person is far below the world average, as the graph shows. Also, the rainfall that feeds the River Nile does not fall on Egypt but in mountainous areas to the south. 86% of the Nile's water starts its journey in Ethiopia.

If climate in Africa changes, water could be in short supply. Countries south of Egypt (see the map) are starting to take more water from the Nile. Many have built large dams and reservoirs;

- Uganda has built the $862 million Bujagali dam near Lake Victoria.
- Sudan has built the $1.8 million Merowe dam near Khartoum.
- Ethiopia has built the $360 million Tekeze dam on a tributary of the Blue Nile.

All of these dams supply hydro-electric power (HEP) and water for irrigation and drinking. However, they also have a serious impact upon the amount of water reaching Egypt which might lead to conflict in the future.

The cost of global warming

As global warming continues, Egypt could face water shortages and the loss of farmland to rising sea levels. In the developed world it might be possible to build sea defences and use water more efficiently. But this costs money. Egypt is a developing country, with debts of $30 billion in 2007 and 44% of its people living on less than $2 per day. Like many developing countries, it may not be able to afford to cope with global warming.

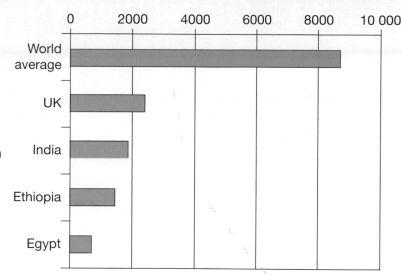

▲ Water resources (m³ per person per year).

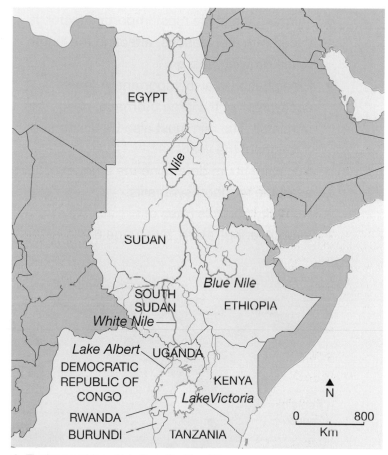

▲ The ten countries which share the River Nile.

---

your questions

1 a Read the text and complete a table to show the economic and environmental impacts of climate change in Egypt.

  b Which impacts are greater - economic or environmental? Explain your answer.

2 How could water cause conflict between Egypt and other countries?

3 **Exam-style question** Using a named example, examine the possible impacts of global warming on a developing country. (6 marks)

✚ **In this section you will learn about the distribution of global biomes.**

The **biosphere** is the part of the Earth's surface inhabited by living things. A **biome** is a world-scale ecosystem. It covers a huge area. The world can be divided into nine major biomes. Each one has its own type of vegetation and wildlife. The location and characteristics of each biome are mainly determined by climate. This is because climatic factors affects the growth of plants.

- Temperature is the most important factor. It varies with the seasons, affecting the length of the growing season.
- Precipitation is also important. A forest ecosystem with a large **biomass**, needs lots of rainfall. The rain must also be distributed throughout the year.
- Sunshine hours determine the amount of light available for photosynthesis.
- Humidity controls rates of **evapotranspiration**. (see page 61)

> ✚ A **biome** is a global scale ecosystem like the tropical rainforest. Tropical forests in Brazil, Indonesia and Nigeria have similar climates and vegetation, but different species of plants and animals.

Mapping biomes

The map below shows the distribution of global biomes. It shows natural vegetation. For instance, you can see that the UK is classified as temperate deciduous forest. There are still forests in the UK, but most of the land is now farmed or built on.

Key

- Tundra
- Coniferous forest
- Temperate deciduous forest
- Temperate grassland
- Mediterranean
- Hot desert
- Tropical rainforest
- Tropical grassland (savanna)
- Other biomes (e.g. ice, mountains)

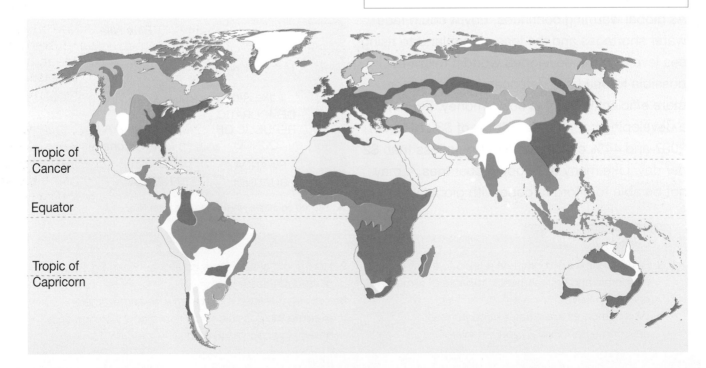

Tropic of Cancer

Equator

Tropic of Capricorn

The diagram below shows a more complex picture. You can see how the biomes gradually change as you move away from the tropics towards the North and South Poles. Take, for example, the tropical region. As you move away from the Equator, tropical rainforest changes to tropical grassland and finally into hot desert.

Altitude and distance from the coast also affect vegetation patterns. At a high altitude few plants will grow. Look again at the tropical region on the diagram. You can see that tropical rainforest develops into coniferous forest and tundra as you gain height and move inland.

On your planet

+ Did you know that, even today, forests cover 30% of the world's land surface? While tropical forests are decreasing, temperate forests are actually increasing.

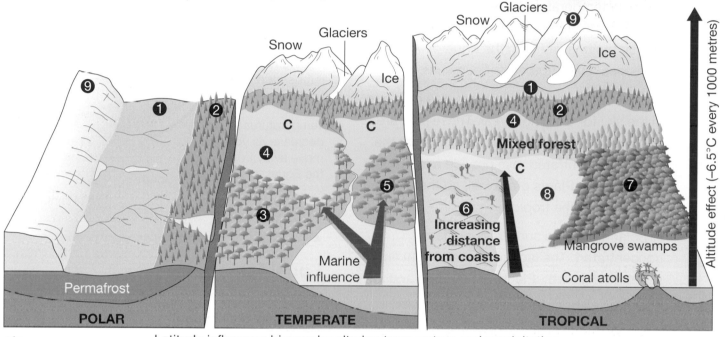

Latitude influences biomes by altering temperature and precipitation

Arctic — Equator

Key

Biomes seasonally lacking in heat and/or water

❷ Coniferous forest

❸ Temperate deciduous forest

❹ Temperate grassland

❺ Mediterranean

❽ Tropical grassland (savanna)

Biomes permanently lacking in heat and/or water

❶ Tundra

❻ Hot desert

❾ Mountain or Alpine

Biomes promoting growth all year round

❼ Tropical rainforest

Marine influence The sea cools nearby land in the hot season and warms it during the cold season. This reduces annual temperature range and increases precipitation.

Continentality C Away from the sea, the land heats up in the hot season and cools quickly in the cold season. This increases the annual temperature range and reduces precipitation.

your questions

1 What is the difference between the biosphere and a biome?

2 a What is the difference between altitude and latitude?
   b How do these affect how plants grow?

3 Draw and complete a table to compare marine and continental climates. Use the diagram above to help you.

4 **Exam-style question** Explain why tropical rainforests are found either side of the equator (4 marks)

In this section you will learn more about the effect of climate and local factors on vegetation.

Temperature

Average temperature is the main factor affecting plant growth. Temperature gradually decreases as you move away from the Equator. As latitude increases, so temperature decreases.

In the tropics, the sun's rays are at a high angle in the sky for the whole year. These rays are concentrated over a smaller area than at the poles. Concentrated rays provide a lot of heat and sunlight. Plants grow well, so there is dense vegetation in the tropics.

In polar areas, the sun's rays are less concentrated. The lack of heat and light limits vegetation growth. Plants are stunted and low growing.

Precipitation

Around the world, precipitation is more likely in some places than others. Precipitation happens in **low-pressure belts**, where air masses **converge** (meet) and air rises. The two main areas of year-round rainfall occur at the Equator and at mid latitudes, such as in the UK. Forests grow in both of these areas.

In polar and desert areas, high-pressure zones occur, causing very dry conditions.

An added complication is that the whole pattern of pressure belts changes with the seasons. Mediterranean and tropical areas sometimes become low-pressure zones and experience rainy seasons for nearly half the year.

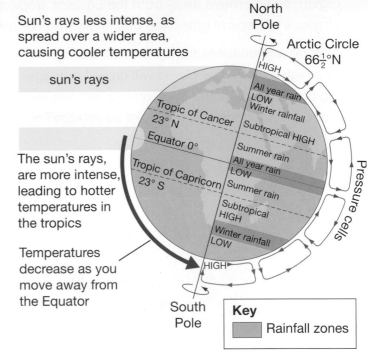

▲ How latitude affects temperature and rainfall patterns.

Local factors

Local factors, as well as global factors also affect ecosystem distribution:

- **Altitude:** temperatures decrease by 6.5°C per 1000 metres in height. So, alpine vegetation and ice at the top of Mount Kilimanjaro give way to tropical savanna at the base.
- **Rainfall:** Altitude affects rainfall – in general, the higher the altitude, the higher rainfall totals tend to be.
- **Drainage:** in large river deltas and areas with impermeable soil and rock, marshes and swamps are present.
- **Soils:** soil type has a smaller influence than other factors. The UK's natural temperate deciduous forest is dominated by oak. Areas of alkaline soil (chalk and limestone) have more ash and beech trees. In waterlogged and acidic soils, birch, willow and alder are more common.

+ **Pressure belts** are regions of the atmosphere which run around the Earth. They are parallel to the Equator. Some are high-pressure areas. Others are low-pressure areas.

your questions

Look at the photos and climate graphs for three biomes.

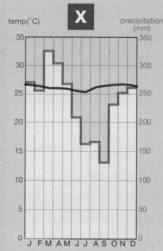

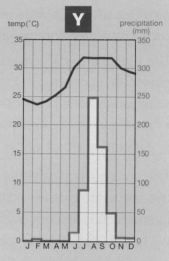

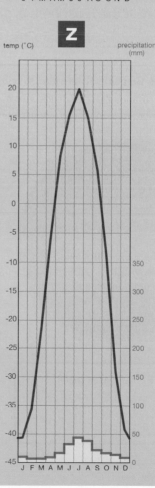

1 Match the photos above (A, B and C) with the climate graphs (X, Y and Z).
2 Which photo is **a** tropical grassland **b** northern coniferous forest **c** tropical rainforest?
3 Design a table to compare the climates, using the 3 graphs. Include maximum and minimum temperatures, total annual rainfall, and the number of months where rainfall is over 50mm.
4 **Exam-style question** Using named examples, examine how local and global factors influence the distribution of biomes. (6 marks)

+ In this section you will learn about the value of the biosphere as a provider of goods and services.

The biosphere is a life-support system. It provides us with a wide range of **goods**, both for survival and for commercial use.

The biosphere also provides many vital **services**, such as:

- regulating the composition of the atmosphere
- maintaining the health of the soil
- regulating water within the hydrological cycle.

The problem is that different people want to use the same biome in different ways. If we overexploit forests or overharvest marine life, we aren't using the biosphere in a **sustainable** way. If the biosphere is damaged, it may fail to provide us with services. This can be disastrous.

> **+ Sustainable** means a process that does no lasting harm to people or the environment.

Green lungs	• Forests remove carbon dioxide from the atmosphere (carbon sinks). This reduces global warming. • Forests give out oxygen – purifying the atmosphere.
Water control	• Forests protect watersheds from soil erosion and intercept precipitation – preventing flash flooding. • By trapping silt, forests keep water pure. • Reefs and mangroves provide protection from coastal storms.
Nutrient cycling	Forests provide leaf litter which forms humus. This makes the soil more fertile for growing crops.
Providing habitats for wildlife/ biodiversity	Rainforests and reefs are very biodiverse. They provide 'homes' for a huge range of organisms, including some very rare animals.
Recreation	Reefs and rainforests provide attractive scenery for tourism.

▲ How the biosphere serves you.

On your planet

+ Sometimes we use up biosphere resources so quickly that they are lost to future generations. For example, the **gene pool** found in rainforests might one day yield cures for cancer. But we won't find this out if we destroy rainforests.

> **+** The **gene pool** is the genetic information contained in living organisms.

▼ The biosphere provides us with goods and services.

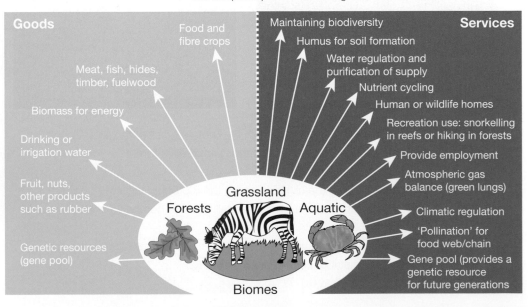

Goods

Services

Food and fibre crops

Meat, fish, hides, timber, fuelwood

Biomass for energy

Drinking or irrigation water

Fruit, nuts, other products such as rubber

Genetic resources (gene pool)

Maintaining biodiversity

Humus for soil formation

Water regulation and purification of supply

Nutrient cycling

Human or wildlife homes

Recreation use: snorkelling in reefs or hiking in forests

Provide employment

Atmospheric gas balance (green lungs)

Climatic regulation

'Pollination' for food web/chain

Gene pool (provides a genetic resource for future generations

Forests   Grassland   Aquatic

Biomes

Delivering the goods

For **indigenous peoples** in the tropical rainforest biome, the biosphere provides almost everything: fuelwood for cooking, timber for building, herbs for medicine, and foods such as nuts, fruit, meat and fish. They can grow subsistence crops, such as yams and millet, using the 'slash and burn' method. Many see this type of farming as being sustainable. The patch of forest is only used for 5-6 years. And once the soil is exhausted, the plot recovers with new forest gradually developing.

The biosphere also provides many goods for commercial use. But there is a problem. Rival commercial users can destroy the rainforest biome for short-term gain. **Transnational companies** exploit the forest by logging for timber or paper manufacture. They deforest the land to grow commercial plantations of rubber, cocoa, and palm oil. Commercial farmers also cut down trees to graze cattle or grow soya beans for biofuels. Drug companies search the forest for plants to provide the ingredients for new medicines. Mining companies search for minerals or oil. And governments may want to develop hydroelectric power.

+ Indigenous peoples are peoples who have originated in and lived in a country for many generations.

+ Transnational companies are giant companies operating in many countries.

	Forested area	Deforested area
Use	Subsistence farming by indigenous local communities.	Ranching, mainly for poor-quality hamburger meat.
Soil	Soil is protected from heavy rain and is nutrient rich. Water moves slowly through the soil, preventing flooding.	Nutrients lost from soil due to heavy rainfall, and surface soil is washed away, blocking rivers and reservoirs. Rapid surface runoff leads to flooding.
Trees	Provide important habitat for wildlife. Tree roots bind the soil, preventing landslides. Wood provides fuel for local communities.	Loss of wildlife because of habitat destruction. Without roots to hold the soil together, landslides can occur. Source of wood for fuel is lost.
Water	Clean river water is fit for drinking.	Water is muddy and unsuitable for drinking.
Economic and environmental gain	Little economic gain, though the forest can support local indigenous communities without suffering permanent damage.	Reasonable economic gain, but only short-term. Deforestation can cause desertification. The natural environment would struggle to recover.

▲ The effect of deforestation on the services offered by the forest.

your questions

1 Write sentences to show how the biosphere keeps
   a air clean   b water protected
   c biodiversity rich   d tourists happy.
2 a Write 200 words explaining why indigenous peoples believe that they maintain the forest, and TNCs cause nothing but damage.
   b Write 200 words to show how a TNC might reply.

3 **Exam-style question** Using a named example of a biome, explain why it is threatened by human activities. (6 marks)
4 **Exam-style question** Explain what is meant by ecosystem goods and services. (4 marks)

Conflicts of interest

✚ **In this section you will learn more about the different demands made on the biosphere.**

As you learnt in section 3.3, the biosphere provides a wide range of goods and services. But different people and organisations (known as **players**) want to use the biosphere in different ways.

Transnational companies and governments often have completely different ideas about the value of the biosphere. Their plans can conflict with the needs of local people. The table shows some of the conflicts which occur between the various players which have an interest in the world's rainforests.

> **On your planet**
>
> ✚ In 100 km² of rainforest, you could find 1500 types of flowering plants, 30 000 insect species, 150 butterfly species, 100 kinds of reptile, and 750 species of tree.

Population pressure in west Africa	Continued population growth in west Africa has led to the removal of all but a small area of rainforest in countries such as Nigeria, Ghana and Sierra Leone.
Rondonia region of Brazil	Brazil's fastest area of deforestation since the 1980s. There have been many groups with conflicting interests: • Forestry operations clashed with ecotourism developments. • Commercial forestry conflicted with native land claims from the local people. • Mining company developments clashed with forest conservation. Therefore areas had to be developed for protection / conservation and production.
The Guyana mountains rainforest	The Guyanan government was short of money due to soaring foreign debts. It wanted to develop the forest for timber and mining. Environmentalists and local people were opposed to this.

▲ Conflicting interests – commercial logging is carried out by transnational companies (left); local people collect fuelwood from the forest (right).

How is the rainforest used?

The diagram here shows an **ecological tree**.
It shows some of the uses of a rainforest.
These include **commercial and industrial**
uses, **ecological** uses or services, providing for
the **subsistence needs** of local people, and
possible **genetic** uses.

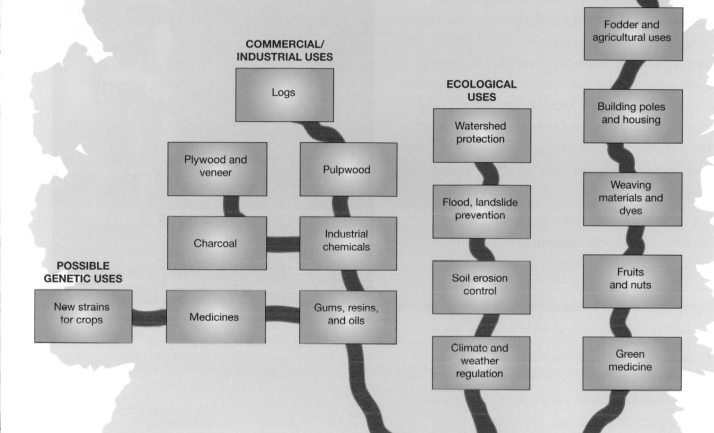

SUBSISTENCE NEEDS

Fuelwood and charcoal

Fodder and agricultural uses

Building poles and housing

Weaving materials and dyes

Fruits and nuts

Green medicine

COMMERCIAL/ INDUSTRIAL USES

Logs

Plywood and veneer

Pulpwood

Charcoal

Industrial chemicals

ECOLOGICAL USES

Watershed protection

Flood, landslide prevention

Soil erosion control

Climate and weather regulation

POSSIBLE GENETIC USES

New strains for crops

Medicines

Gums, resins, and oils

your questions

1 a On a double page draw two copies of the ecological tree above. Include the boxes, but leave them empty.

 b Complete the boxes of the left hand tree with all the things that would happen if TNCs and governments got rid of most of the rainforest.

 c Complete the boxes of the right hand tree with all the things that would happen if the rainforests were well protected.

2 Exam-style question Explain the value of one biome you have studied. (4 marks)

On your planet

+ Chewing gum (as we know it) was first made from the resin of the chicozapote tree, found deep in the tropical rainforest.

On your planet

✚ But it's not all doom and gloom: in 2007 conservationists saved 16 species of beautiful birds from extinction.

✚ In this section you will learn how rainforests are being degraded by human actions, both directly and indirectly.

Species under threat

Norman Myers in his book *The Singing Ark*, published in 1979, wrote: 'By the time you have read this chapter, one species will be extinct. We lose something in the region of 40 000 species every year – 109 a day'. This sounds like a global catastrophe. But put another way, we will only lose 0.7% of all species over the next 50 years. This sounds far more manageable. However, we need to ask ourselves, can we afford to lose these species?

We don't know the true number of species inhabiting the Earth. Estimates range from 2 million to over 30 million, but only 1.4 million have been identified so far. New species are still being discovered in the world's rainforests, which contain huge numbers of species.

▲ The Orang-utan of Borneo and Sumatra are endangered due to human activities.

- Australia's only rainforest is in north Queensland, around the the Daintree River. About 250 000 plant species are currently known in the world, and the Daintree contains 155 000 of these!
- Brazil's Amazon rainforests contain one fifth of all known bird species.
- 90% of all known primates live in rainforests.

Yet rainforests are threatened and it's estimated that 10% of all known species had become extinct by 2000.

▶ Rainforest along the Daintree river, Australia.

Causes of threats to the rainforest

The causes of threats to the world's rainforests can be classified into:

- **Immediate causes**, such as logging, tourist pressures, pollution.
- **Root causes**, such as poverty and debt among nations such as Indonesia, which uses logging as a way of exporting to increase wealth.

A further root cause is economic development, for example in Brazil, which has a huge need for industrial raw materials as its economy grows. Living standards are also improving, which means that people now consume more food and fuel. All this puts a strain on the Amazon's rainforest.

Certain species are particularly under threat, leading to hotspots where biodiversity is endangered. Most hotspots are in developing countries, though in Australia's rainforests of the Daintree, species such as the cassowary bird are endangered as deforestation is destroying its habitat.

> **+ Biodiversity** is the range of animal and plant life found in an area.

Threats to rainforests	Impacts	Examples
Deforestation	Commercial logging destroys forest unless sustainable forestry principles are used. It affects rates of flooding, soil erosion and humus formation.	Logging in the rainforests of the Amazon and Indonesia.
Conversion to farmland or urban use	Commercial intensive farming destroys or alters the ecosystem. Urban sprawl destroys ecosystems and encourages wildfires.	Cattle farming in the Amazon has removed natural rainforest. Soya beans are grown for biofuels. The urban sprawl of Brasilia, the capital city.
Overharvesting/ overfishing	Overharvesting causes wild animals to be hunted to extinction. Overfishing of some species such as krill, needed as food for fish farms, destroys food chains.	Rainforest species such as orang-utan are now endangered. In Uganda's rainforests, only sustained campaigning has prevented gorillas from extinction.
Mining and energy	Mining cuts away whole hillsides. Opencast mining destroys the surface and restoration is only partly successful. Oil drilling in areas of rainforest.	Removal of forests for mining iron ore and bauxite in the Brazilian Amazon. Tin mining in Malaysia. Drilling for oil also began in the Amazon in 2008.
Pollution	Water pollution from oil exploration, agricultural fertilisers and industry. Toxic fumes emitted into the atmosphere.	Alleged exploration by one US oil company in Ecuador has caused skin irritation and cancers for local people, caused by dumping of oil sludge into rivers. This has also killed fish species.
Introduction of alien species	Sometimes we introduce new species deliberately, or they escape (e.g. foxes in Australia). More usually they arrive by accident, or through new housing and gardens. Alien species often breed well and take over.	Invasive plant species brought in as crops or garden plants on edges of the Daintree. Cats and dogs hunt and feed on birds and small rodents in the Daintree communities.
Tourism and recreation	Eco-tourism has little impact. But high-density mass tourism in fragile environments disturbs wildlife.	Tourism threatens the Daintree rainforest, in spite of severe restrictions on housing and development. Traffic levels are high, and native species are often run over.

your questions

1 Draw a table with 3 columns, using the headings Environmental, Economic and Social. Using examples from this chapter, list the impacts of losing so many plant and animal species (a loss of biodiversity). Place them under the correct heading in your table.

2 'The economic gain is worth the environmental loss.' How far do you agree with this statement about the rainforest? Explain your view in 250 words.

3 **Exam-style question** Using a named example, explain how demand for resources threatens biodiversity. (6 marks)

✚ In this section you will learn about direct threats to the biosphere, and how the indirect threat of climate change could be the biggest threat of all.

Direct threats

Amazonia, in South America, is the largest area of tropical rainforest in the world – over half of which is in Brazil. It produces 20% of the world's oxygen and is often called the 'lungs of the planet'. It contains over 40,000 plant species and over 2 million insect species. Deforestation is a huge problem in the area and has many causes (see pie chart). However, the rate of deforestation in Amazonia has slowed since 2004 (see graph).

This is good news for the rainforest. It has happened for several reasons:

- An area the size of France has been protected since 2006.
- The recession and credit crunch since 2008 has reduced demand for resources.
- The government has cracked down on illegal logging and clearance for cattle ranching by seizing land and freezing bank accounts.
- The Forest Code law has been enforced more strictly.
- Brazilians themselves have become more 'green'; 19% voted for a Green Party candidate in the 2010 elections.

The question is, will this reduction continue or will the pressure for more land and resources cause deforestation to increase in the future?

> ✚ **Degradation** is the damage caused to ecosystems and the loss of biodiversity which can eventually lead to their destruction.

Direct threats to Amazonia

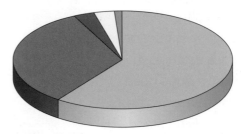

Key

▢ Cattle ranches 60%

▢ Small-scale, subsistence agriculture 33%

▢ Fires, mining, urbanisation, road construction, dams 3%

▢ Logging, legal and illegal 3%

▢ Large-scale commercial agriculture including soybeans 1%

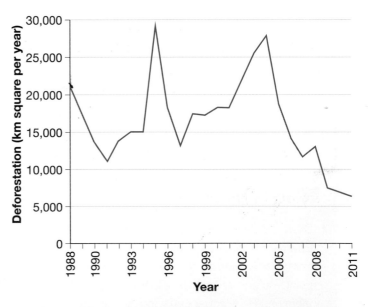

▲ In 2011, deforestation in Amazonia fell to its lowest level since 1970.

Global warming: an indirect threat

Direct threats to biomes can be managed and reduced by countries like Brazil. Indirect threats are much harder to manage. The main indirect threat is global warming.

The rise in population and resource consumption adds greenhouse gases to the atmosphere and this causes the climate to change (see chapter 2). Brazil only emits about 1.5% of global carbon dioxide, so it can't prevent the global warming problem on its own. Some scientists think this will lead to species extinction at an unprecedented rate. Already:

- plants are flowering earlier
- bird migration patterns are changing
- the Arctic tundra is warming rapidly
- vegetation zones are shifting towards the poles by 6km every 10 years.

Global warming is occurring too rapidly for many species to adapt to the changing climate. The table summaries some of the expected changes as temperatures rise. A rise of 3°C could happen by 2060.

Climate stress

The Amazon rainforest suffered from two very severe droughts in 2005 and 2010. Drought is not unusual in the Amazon, but two droughts so close together are. At the time, the 2005 drought was described as a '1 in 100 year' drought.

During the droughts the Amazon switched from absorbing carbon dioxide to emitting it. The fear is that if drought becomes more common, rainforests will suffer permanent damage and die-back. As this happens, they could become sources of carbon dioxide, not carbon dioxide sinks. This could accelerate global warming even more, making the problem worse. Drought also increases the risk of forest fires.

> **On your planet**
> + In 2011, Brazil's exports of beef were worth $5.4 billion.

Temperature rise due to global warming	Impact on species	Impact on biomes
1°C	10% of land species face extinction	Alpine, mountain and tundra biomes shrink as temperatures rise.
2°C	15-40% of land species face extinction	Biomes begin to shift towards the poles and animal migration patterns and breeding time change. Extreme weather at unusual times of the year, such as heatwaves and blizzards, affect pollination and migration.
3°C	20-50% of land species face extinction	Forest biomes are stressed by drought and fire risk increases on grassland. Flooding causes the loss of coastal mangroves. Pests and diseases thrive in the rising temperatures, such as bark beetles which devastate coniferous forests.

▲ The impact of rising temperatures in species and biomes.

your questions

1 Describe the trends on the graph of deforestation and suggest reasons for them.
2 Explain why the speed at which temperatures are rising is such a problem for plant and animal species.
3 **Exam-style question** Using named examples, explain how the biosphere is threatened by direct and indirect means. (6 marks)

+ In this section you will explore strategies for conserving rainforests.

Which parts of rainforests should be conserved?

Sections 3.5 and 3.6 showed that human actions threaten rainforests. But, as the diagram below shows, maintaining the biosphere in general, and reversing its devastation, will require vast sums of money and a massive international effort. However, it's not really a question of can we afford the money, but one of can we afford not to spend it?

How do we decide which parts of rainforests to conserve?

Which habitats or species?

- Do we save the hotspots – the best bits which are most under threat, like parts of the Daintree (see page 50)? Or do we save samples of different plant communities, some from Latin America, some from Australia?

- Is it best to get value for money by conserving areas in developing countries? Money goes further in these countries because of lower costs.

- Is it worth restoring completely devastated areas (like some rainforests in the Amazon), because here the costs would be highest?

- Should we save high-profile animals like orang-utans? Or should we conserve the gene pool and keystone species like insects, which are so important for the whole food web? It's much easier to raise money to save orang-utans than obscure insects.

Global, national and local initiatives are all very important in the battle for rainforests. But there are often tensions over how to conserve it.

> + A **keystone species** is one which has a particularly large effect on other living organisms.

On your planet
+ Should we only protect wildlife in its natural habitat? Or can zoos, seed banks and botanical gardens play an important role?

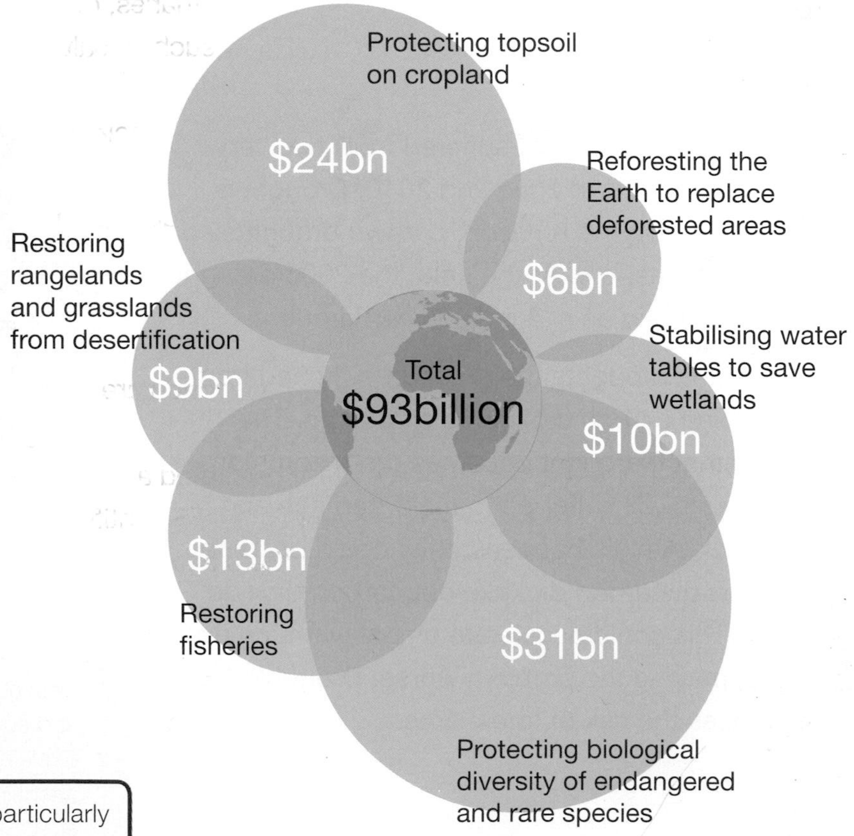

Protecting topsoil on cropland
$24bn

Reforesting the Earth to replace deforested areas
$6bn

Restoring rangelands and grasslands from desertification
$9bn

Total **$93billion**

Stabilising water tables to save wetlands
$10bn

$13bn

Restoring fisheries

$31bn

Protecting biological diversity of endangered and rare species

▲ The price of conserving the biosphere.

Act global

Countries can get together to develop wildlife conservation treaties. Two examples of how this has been achieved include:

The Convention on International Trade in Endangered Species, (CITES)

Signed in 1973 and adopted by 166 countries, the CITES treaty lists the endangered species. The aim is to stop the trade in products such as elephant ivory or handbags made from crocodile skins.

International treaties, such as CITES, are very difficult to manage, there are so many conflicting interests. However, they do provide a useful legal framework for conservation.

RAMSAR wetlands

The 1971 Ramsar convention on Wetlands of International Importance involves 163 countries. They have agreed to protect 196 million hectares of wetland habitat at over 2000 sites worldwide. World Wetlands day (2nd February) raises the profile of wetlands, many of which have very high biodiversity such as coral reefs and river estuaries, or are very important habitats for migrating birds, such as salt marshes.

Individual countries are responsible for protecting their wetlands by following the 3 Ramsar pillars:

- Identify important wetlands and agree to manage them carefully
- Wise use of wetlands – fishing, tourism and other human activities should be sustainable
- Co-operate with other countries when wetlands are shared across a border

The UK has 148 Ramsar sites. Most are designated as Special Protection Areas and /or Sites of Special Scientific Interest (SSSIs) under UK law. This gives them extra protection from damage and development.

Act national

At a national scale, governments can set up protected areas, which help to conserve, manage and restore biodiversity. Three examples from the UK include:

- **Sites of Special Scientific Interest (SSSI)** are areas where rare species are protected from development (e.g. farming or housing developments) by law. Only limited access to sites is allowed, and normally only for scientific research.
- 15 **National Parks** exist in England, Wales and Scotland. These have their own planning authorities to protect the environment from the pressures of tourism (e.g. camp sites, new buildings, or road widening schemes). These authorities manage some of the most attractive, yet fragile, coastal and upland landscapes.
- In 1990, **Community Forests** were set up in England to provide new areas of woodland for leisure and environmental quality near major cities. By 2013, 12 Community Forests had been established, e.g. White Rose Forest around the Pennine areas, surrounding cities such as Sheffield and Leeds.

your questions

1 Look at the diagram opposite. In pairs, imagine you have a budget of US$ 30 billion. What would your priorities be for conserving the biosphere? What would you save, and how might you do it?

2 a Using an outline map of the UK, map the location of National Parks and Community Forests (www. communityforest.org.uk/ and www.nationalparks.gov.uk/)

2 b Design a poster or presentation to explain how National Parks or Community forests help to conserve the biosphere.

3 Exam-style question Using named examples, explain two different ways of conserving ecosystems. (6 marks)

➕ In this section you will learn about sustainable environmental and economic management of ecosystems.

In the 1980s, people realised that 'closing off' great areas of rainforests and reefs to try to protect them was not a success. People still carried out illegal poaching and harvesting. Another form of managing these areas was needed, and this was **sustainable management**.

Sustainable management of ecosystems can be thought of as a middle way between total protection of ecosystems (where no-one has access) and total exploitation where there is no protection.

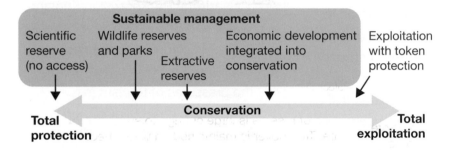

On your planet

➕ In parts of southern Africa local people within reserves can make money from game hunting. Is shooting wildlife really sustainable management?

What is sustainable management?

Sustainable management conserves the ecosystem for future generations, by ensuring that it isn't used faster than it can be renewed. The outcomes of sustainable management can be measured in economic, social and environmental terms:

- **Economic outcomes** may include reducing poverty by helping local people to make a living e.g. from eco-tourism (see diagram opposite).

- **Social outcomes** may involve improving local facilities which benefit the whole community e.g. health clinics, education and training.

- **Environmental outcomes** often protect the natural environment from degradation e.g. trees are planted to replace those which are cut down (see photo).

Sustainable environmental management

Kilum-Ijim forest is an area of mountain rainforest in Cameroon, Africa, home to 35 communities from 3 tribes (the Kom, Nso and Oku). About 250,000 people live a day's walk from the forest and it was under pressure from farming and logging for timber and fuel.

In 1987 the conservation organisation BirdLife International started a project to create a sustainable forest reserve in the area (see diagram). They came together with the Cameroon Ministry of the Environment and Forestry, Kew Gardens in London and funding from DFID and the Dutch Ministry of Agriculture. Working with local communities they:

- Marked out the forest reserve area and made lists of forests resources.
- Developed rules for the sustainable use of the forest.
- Set up a unit to manage and monitor the forest
- Educated communities about replanting trees and safe levels of hunting and logging ▶

The overall aim of the project was to conserve the forest so that future generations could continue to use it, rather than it being destroyed forever. The project has been a success. 50% of the Kilum-Ijim forest was deforested 1958-1988, but the forest area has increased by 8% since the project began.

Kilum-Ijim does face a number of challenges in the future:

- Population growth is bound to increase pressure to deforest areas.
- Urban areas, industry and even roads could encroach on the forest.
- Money and technical support from international donors could end.
- Climate change could begin to degrade the forest

Sustainable economic management

▲ Local people plant new trees to replace the cut down forest.

Key

Core conservation area	
Buffer zone – light use on rotational basis. This surrounds the core	

▼ A sustainable forest reserve, Kilum-Ijim Forest Reserve in Cameroon. The land is divided into zones using the UNESCO biosphere reserve model. The zones are used for different purposes and have different levels of protection.

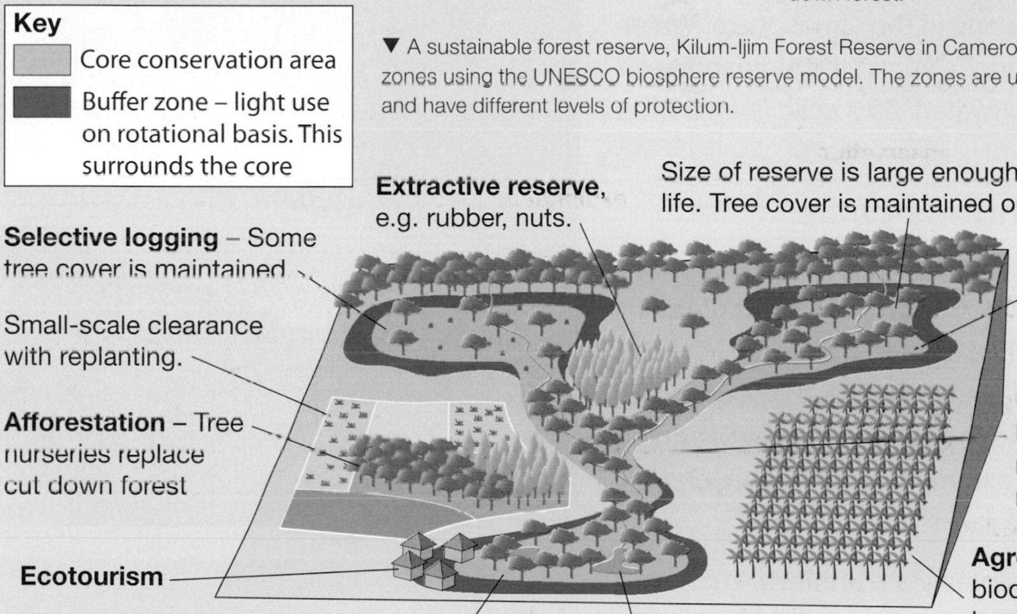

Extractive reserve, e.g. rubber, nuts.

Size of reserve is large enough to support wild life. Tree cover is maintained on watershed.

Selective logging – Some tree cover is maintained

Forest reserve protected area with minimum human interference.

Small-scale clearance with replanting.

Afforestation – Tree nurseries replace cut down forest

Reserves linked by natural corridors for migration.

Ecotourism

Agroforestry – maintains biodiversity of agricultural land. Crops grown beneath the shade of banana trees.

Multiple zoning, e.g. hunting, tourism, conservation.

Tree cover in watersheds reduces flood risk and improves quality and quantity of water.

your questions

1 In pairs, draw a table to show the strengths and weaknesses of sustainable environmental management in Kilum-Ijim.

2 Look at the diagram above. Describe the ways in which the forest is being managed to provide sustainable economic outcomes for the local community.

3 From what you have read here, how far do you think that the battle to conserve the biosphere has been won?

4 Exam-style question Using a named example, explain how sustainable management can help conserve the biosphere. (6 marks)

➕ In this section you'll learn about the hydrosphere and hydrological (water) cycle.

Water and life on Earth

Water is what makes Earth – the **Blue Planet** – unique and different from other planets in the solar system. Without it, life could not exist. It is more important than anything else on Earth. You can live for three weeks without food, but without water you'll be dead in three days!

The **hydrosphere** consists of all the water on the planet – in seas, oceans, rivers and lakes, in rocks and soil, in living things and in the atmosphere. Water exists on the Earth's surface and in the atmosphere in three states: as a liquid (water); as a solid (ice); and as a gas (water vapour).

Water – a continuous cycle

Water flows in a never-ending cycle between the atmosphere, land, and oceans. The **hydrological cycle**, or water cycle, is a **closed system**. The water goes round and round – none is added or lost so the Earth gets neither wetter nor drier. Think of the global water cycle as having a number of **stores**, such as lakes, oceans, soil and rocks underground.

Water flows between stores via **transfers**, such as runoff (overland flow), infiltration, throughflow, and groundwater flow.

Sometimes this involves a change of state.

- Heat energy can change liquid (water) to a gas (water vapour). The process is called **evaporation**.

- Air can only hold so much water vapour before it becomes saturated. As air cools, water vapour turns back into liquid, known as **condensation**.

- Water can freeze to a solid (ice) as the temperature cools, or melt back to liquid as it warms.

The table below shows how important the oceans are as a store, but remember that oceans store salt water which cannot be used by people unless it is converted into fresh water. This can be done but it is hugely expensive. Water stays in the stores for varying amounts of time, from around a day, to thousands of years! Most of the fresh water is stored in ice sheets and glaciers (especially in Antarctica), and these stores are gradually being reduced as a result of global warming. Relatively small amounts of water are stored in rocks as groundwater, and in lakes and rivers, and these are in huge demand as sources of water.

Just think what happens after a heavy storm – water will be dripping off trees, and the grass will be wet for about 12 hours afterwards. The puddles dry up after a day or so but water which finds its way underground can stay there for years. If rainfall is heavy and goes on for a long time, some of the smaller stores fill up and flooding occurs.

Store	Size (km³ x 10 000 000)	% of all water
Oceans	1370.0	97.0
Polar ice and glaciers	29.0	2.0
Groundwater	9.5	0.7
Lakes	0.125	0.01
Soils	0.065	0.005
Atmosphere	0.013	0.001
Rivers	0.0017	0.0001
Living things	0.0006	0.00004

The photos show some water stores:

A Lakes
B Oceans and atmosphere
C Polar ice and glaciers
D Trees and vegetation

your questions

1 a Make a copy of the diagram below.
  b Label A and B to show water changing state.
  c Label the water stores C, D, E and F.

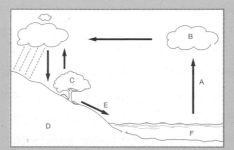

2 **Exam-style question** Explain why the hydrological system on Earth is a 'closed system'. (3 marks)

✚ In this section you'll find out about the role of the biosphere and lithosphere in the hydrological (water) cycle.

The global hydrological cycle

The diagram on the right shows how the global hydrological (water) cycle works. The stores of water are linked by processes which transfer water into and out of them. These processes regulate the water cycle.

- Evaporation from oceans and rivers, and evapotranspiration from trees, condenses to cause precipitation (rainfall).
- This precipitation follows a number of routes:
  - Some runs off over the surface.
  - Some seeps into the soil or rock.
  - Some collects as snow or ice.

The biosphere and lithosphere

The **biosphere** and **lithosphere** play a vital role in the water cycle, and act as sub-cycles. The diagram opposite shows these sub-cycles in a **river basin system** (part of the water cycle which operates on land). In the biosphere, trees intercept precipitation, and over half of it is then evaporated and transpired without ever reaching the ground. This water is known as **green water**. If the storm or rainfall is very heavy, or goes on for a long time, precipitation drips from the leaves and stems and slowly makes its way into the river system. Precipitation infiltrates into the soil, where it flows down hill as **throughflow**, or if the underlying rock is permeable, into the ground, to be stored as **groundwater** – a vital supply of water. Only after many hours is the water released into the river basin. Both the biosphere and the lithosphere help to regulate the water cycle.

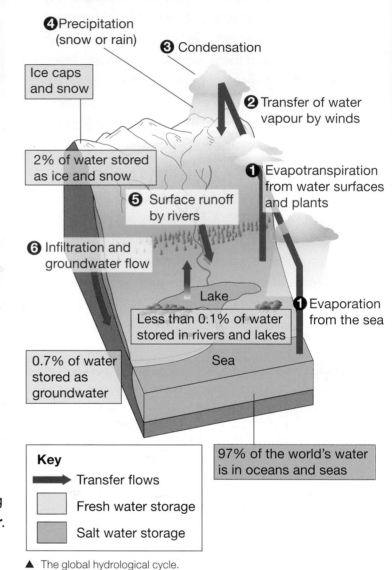

❹ Precipitation (snow or rain)

❸ Condensation

Ice caps and snow

❷ Transfer of water vapour by winds

2% of water stored as ice and snow

❶ Evapotranspiration from water surfaces and plants

❺ Surface runoff by rivers

❻ Infiltration and groundwater flow

Lake

❶ Evaporation from the sea

Less than 0.1% of water stored in rivers and lakes

0.7% of water stored as groundwater

Sea

Key

➡ Transfer flows

☐ Fresh water storage

☐ Salt water storage

97% of the world's water is in oceans and seas

▲ The global hydrological cycle.

✚ The **biosphere** is the part of the Earth and atmosphere in which living organisms exist.

✚ The **lithosphere** is the outer layers of the Earth's surface (the crust and upper mantle).

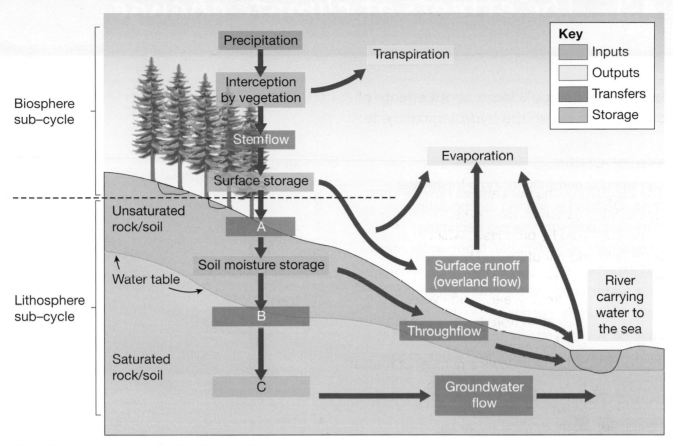

▲ A river basin system showing the biosphere and lithosphere.

Infiltration – movement of water into the soil from the surface
Percolation – movement of water into underlying rocks
Groundwater storage – water stored in rocks following percolation
Saturation – when soil is full of moisture
Water table – the level at which saturation occurs in the ground or soil
Inputs – things which enter the system
Outputs – things which leave the system
Transfers or flows – movements within the system
Stores – held within the system

+ Evapotranspiration is the combined process of; Evaporation – the changing of water from a liquid to a gas (water vapour) due to the heat of the sun, and, **Transpiration** – the movement of water through a plant (from roots to leaves) and its loss into the atmosphere as water vapour.

your questions

1 Make a copy of the river basin system above, and complete the terms in boxes A, B and C.
2 Define these words: evaporation, evapotranspiration
3 Using the diagram and definitions, explain the passage of water from the time it falls as precipitation, to the time it reaches the river.
4 **Exam-style question** Describe how the hydrological cycle links the biosphere, atmosphere and lithosphere. (4 marks)

On your planet
+ Did you know that some of the groundwater deep down in the Earth has been stored there for over 10 000 years?

➕ **In this section you'll learn about effects of climate change on the hydrological cycle.**

Water crisis

The world is currently facing a freshwater crisis. Demand is soaring as population increases, and supplies are becoming increasingly unpredictable. Many economists and experts predict that in the future we will have **water wars**, where countries fight over water resources – especially in the Middle East. As you can see from the graph, our use of water is increasing, and much of the increasing demand is from agriculture. Modern farming often requires **irrigation**, which uses vast quantities of water.

As if that was not enough, the world now faces major change caused by climate change. In 2007, Oxfam published a report 'Africa – Up in smoke' which showed the kinds of threats faced by Africa as a result of climate change. How can this be? Most global authorities are agreed – that climate change will affect water vapour, cloud formation, precipitation patterns, surface run-off, and river flow.

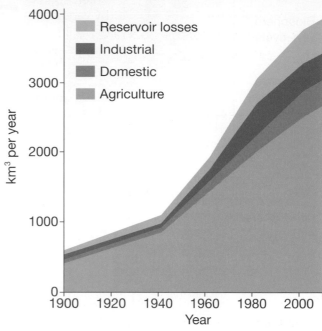

▲ Increasing global water use.

▼ The increased likelihood of storms (shown in blue)

Change in Precipitation Intensity (standard deviations)

–1.5	–1	–.5	0	.5	1	1.5

The impacts on precipitation

It is difficult to predict whether precipitation amounts will increase or decrease as a result of climate change. However, some changes are already occurring:

- Warmer temperatures are likely to lead to more precipitation falling as rain, instead of snow.
- If there is more water vapour in the atmosphere, then there should be an increase in the amount of precipitation.
- However, the most likely impact of climate change is an increase in precipitation intensity, that is, a larger proportion of rain will fall in a shorter amount of time. There are more likely to be more intense storms over land areas, as the map shows.

The impacts on evaporation

As the atmosphere gets warmer, evaporation rates increase, leading to an increase in the amount of moisture in the atmosphere. Warmer temperatures would also lead to greater evaporation from soil surfaces in some areas, making drought more likely.

The impacts on drought

With a combination of increased temperatures, increased evaporation, and reduced river flow, one very likely outcome of climate change is increased drought. This is likely to vary, as the map from NASA shows. The map shows a drought severity index, where -4 is most likely and +4 least likely. It is most likely that areas already suffering drought will suffer worse droughts than now; areas with little likelihood of drought at present are likely to be much less affected.

The impacts on river flow

The Northern Hemisphere has more land surface, and so most changes in river flow will occur there. A warming climate would lead to an earlier arrival of spring. Mountain snows would melt earlier, increasing river flow in the spring, and therefore reducing river flow in the summer. NASA, the US Earth Observatory, fears that the impact will be the reduced availability of fresh water in the hot, summer months, when water demands are highest.

▼ The increased likelihood of drought, shown in orange and brown

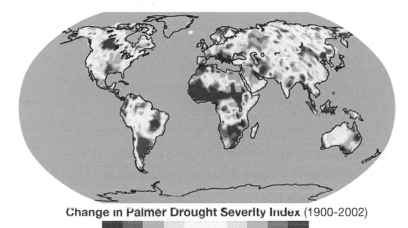

Change in Palmer Drought Severity Index (1900-2002)

-4   -2   0   2   4

your questions

1 Make a sketch of the hydrological cycle from page 60, and change the labels to show which parts of the hydrological cycle will increase, decrease, or stay the same.

2 Write sentences beneath the diagram to explain why these will change in the ways that you have shown.

3 **Exam-style question** Use the map on the page opposite to describe those parts of the world where rainfall will become more intense. (4 marks).

4 **Exam-style question** Using examples, explain how climate change is likely to alter the hydrological cycle. (6 marks)

In this section you'll find out about chronic water shortage in the Sahel, and how climate change may affect water supplies in other parts of the world.

Drought in the Sahel

The Sahel is a narrow belt of semi-arid land immediately south of the Sahara desert, (see map). Rain falls in only 1 or 2 months of the year, and both the total amount of rainfall (usually between 250 and 450 mm) and the length of the rainy season are very variable. Since 1970, rainfall has more often than not been below average – and in some cases up to 25% below average. Sometimes the rain comes in torrential downpours and is then lost as surface runoff, causing flooding. In other years the rains fail completely – recent years have seen several lengthy droughts.

- Drought causes seasonal rivers and water holes to dry up and the water table to fall.
- Drought spells disaster for the **nomads** who graze animals, and for **subsistence farmers**, who rely on rain to grow millet and maize.
- Grasses die, and overgrazing by animals causes soil erosion and desertification.

Many of the countries in the Sahel are developing countries, such as Chad, Niger and Sudan and are among the poorest in the world. They have rapidly growing populations, which puts pressure in drought years on failing food supplies. These semi-arid lands are very fragile – water stress soon causes humanitarian crises and regular famines.

About 780 million people worldwide lack a reliable and sufficient water supply. This has many impacts:

- Waterborne diseases like cholera, typhoid, and dysentery kill one child every second.
- Women and children spend up to 200 million hours worldwide, each day, just collecting water.
- High risk of crop failure and animal death.
- Everyday tasks like cooking and washing are very difficult.
- Businesses, schools and clinics all require reliable and safe water supplies, without which they may be unable to open.
- In slums, in developing cities, people have to pay high prices for clean water – so have less to spend on food, education and health.

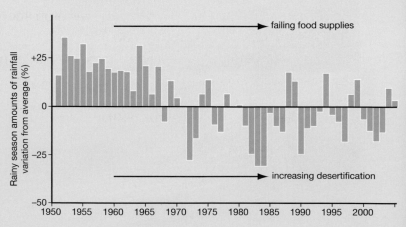

▲ The Sahel, and variations in rainfall.

The climate change threat

The World Bank estimates that by 2025, 50% of the world's population will face water shortages. Global warming will lead to:

- Some regions such as the Sahel, Australia and southern Europe becoming drier.
- Both drought and flood becoming more common, and more severe — possibly in the same location.
- Melting of some water stores, especially mountain glaciers in the Himalayas, Alps and Andes.

The top map on the right shows the impact of climate change on water scarcity by 2025. The lower map shows the combined effects of population growth and climate change. Population growth to 8 billion by 2025 (increasing water demand), combined with global warming (disrupting water supply), could be a devastating combination.

Global warming is likely to have different impacts on people:

- In richer countries the cost of water will rise. New reservoirs, pipes and desalination plants will keep water flowing, but it will be increasingly expensive.
- Subsistence farmers in the developing world, who rely on rainfall, will become more vulnerable to drought and floods, increasing their food insecurity as crops fail.
- If average precipitation falls, river flow will drop and groundwater will not be recharged — wells might run dry
- In south Asia many people rely on seasonal glacier melt for their summer water supply — as the glaciers shrink year to year, so will water supply.

▼ Climate change will influence water scarcity ...

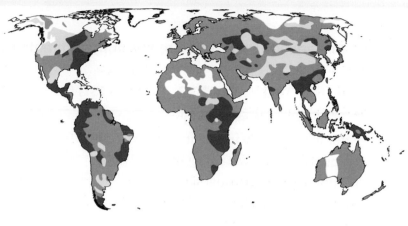

▼ ... but population growth with climate change could be devastating.

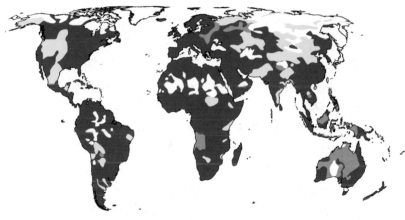

Key
Water demand (as a percentage of the available water supply)

☐ Less than 80

▨ 81–119

▪ 120 or more (i.e. demand is at least 20% more than the available supply)

your questions

1 Explain what the graph opposite shows about rainfall in the Sahel since 1950.
2 Place these in order to show how drought leads to soil erosion: drought, roots can't hold the soil together, wind blows, grass dies, soil blows away, soil dries out.
3 Explain how and why farmers and cattle can make this worse.
4 **Exam-style question** Using a named example, explain how an unreliable water supply affects people in a vulnerable region. (6 marks)

+ **In this section you'll explore the human threats to water quality.**

Quality not quantity

Not only do we need enough water for everyday use, but the quality of our water is just as important as the quantity. People can suffer from water stress if their water isn't safe, or is contaminated.

The diagram below shows a wide range of sources of pollution – the worst can be classified as:

- Industrial pollution
- Sewage disposal
- Intensive agriculture
- Other

Industrial Pollution

The highest levels of water pollution are often found in countries with rapid rates of economic growth, such as India and China (see graph) — which tend to put economic growth before environmental protection. Three main types of water pollution from industrial activity are:

- Chemicals (e.g. from the manufacture of plastics, oil, pesticides, and PCBs) and inorganic solid compounds (such as toxic materials and cyanide from mines) — these can poison stretches of river completely, killing all life.
- Radioactive substances (e.g. from nuclear waste treatment) — these can lead to cancers such as leukaemia.
- Thermal (heat) pollution (e.g. from power stations pumping hot waste water into rivers) — increasing the rate of decomposition of bio-degradable waste, which reduces the river's ability to retain oxygen.

However, as countries develop, they often control pollution by law. In the late 1960s, Japan's lakes, rivers and seas were all badly polluted causing major health problems and damage to ecosystems. As a result of recent pollution laws, the rivers entering Tokyo and Osaka Bay – two of Japan's industrial areas – are now almost free of pollutants.

▼ Sources of water pollution

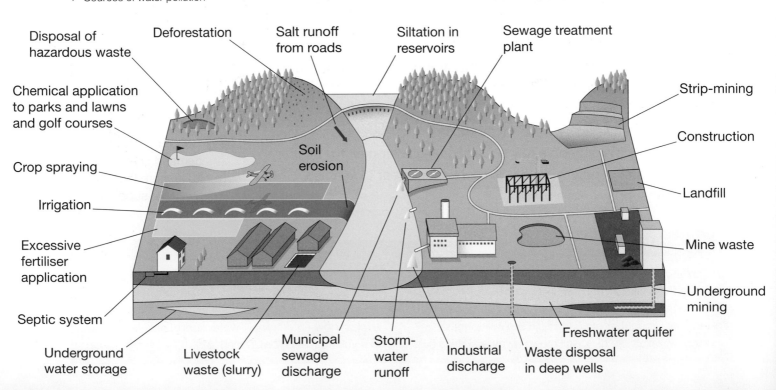

Disposal of hazardous waste · Chemical application to parks and lawns and golf courses · Crop spraying · Irrigation · Excessive fertiliser application · Septic system · Underground water storage · Livestock waste (slurry) · Municipal sewage discharge · Storm-water runoff · Industrial discharge · Waste disposal in deep wells · Freshwater aquifer · Deforestation · Salt runoff from roads · Siltation in reservoirs · Sewage treatment plant · Soil erosion · Strip-mining · Construction · Landfill · Mine waste · Underground mining

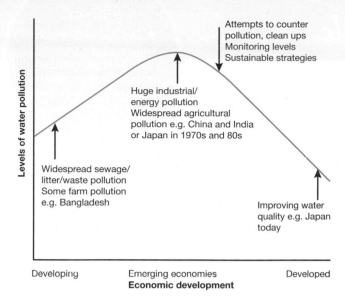

Attempts to counter pollution, clean ups
Monitoring levels
Sustainable strategies

Huge industrial/ energy pollution
Widespread agricultural pollution e.g. China and India or Japan in 1970s and 80s

Widespread sewage/ litter/waste pollution
Some farm pollution e.g. Bangladesh

Improving water quality e.g. Japan today

Levels of water pollution

Developing — Emerging economies — Developed
Economic development

Intensive agriculture

There are two main types of water pollution caused by intensive agriculture:

* Chemical — Modern commercial agriculture relies on pesticides and fertilisers to increase crop yields. The run-off of those chemicals increases water pollution as plant nutrients from fertilisers (e.g. nitrates and phosphates) seep into rivers, causing **eutrophication**. In this case, river water is polluted by rich nutrients, leading to the rapid growth of algae. These starve the river water of oxygen.

* Solid — While chemical pollution exists widely, it tends to be less concentrated. Sometimes, animal manure escapes into streams as **slurry**, polluting streams with raw sewage. Like eutrophication, slurry dumping deprives river water of oxygen, killing many organisms.

Sewage disposal

Many of the world's mega-cities such as Mumbai are experiencing rapid urban growth, faster than piped water and waste disposal systems can be installed. As a result, streams flowing through the slums and shanty towns of megacities are badly affected by pollution (see the photo). People use the rivers for washing clothes, boiling water for cooking, or swimming. For people living there, contact with this water presents three potential impacts:

* Diseases — such as cholera, typhoid, dysentery, or hepatitis — caused by harmful organisms (e.g. bacteria, viruses, river worm, rats).

* Domestic sewage consumes oxygen in the water and kills many organisms through oxygen depletion.

* Suspended solids affect the colour of the water and kills fish / shellfish.

your questions

1 a Draw a table to classify all of the pollutants from the diagram on the opposite page. Use the headings industrial pollution, sewage disposal, intensive agriculture, and others.

b Which category has the worst impacts? Explain your answer.

2 Draw a spider diagram to show why industries and farmers may resist laws to control and reduce pollution.

3 Why do you think countries such as India and China do not have standards for water quality, as Japan now does?

4 Exam-style question Using examples, explain how pollution threatens water quality. (6 marks)

Interfering in the hydrological cycle

+ In this section you'll look at some of the impacts of human interference in the hydrological (water) cycle.

Human interference

Water isn't just for drinking and washing. It is used for industry, farming (irrigation), hydro-electric power generation (HEP), and also for waste disposal – plus we use it for recreation. Water creates wetland habitats which are important environmentally for their **biodiversity**. Many wetlands are protected as **Ramsar Sites**.

The wide range of uses reflects the processes and links within the water cycle. The diagram on the right shows how people intervene in the water cycle. While some of these have positive impacts, others have negative impacts on water supplies. And the downside of the links within the cycle means that over-use for one purpose, or disruption of the cycle, can have serious knock-on effects elsewhere in the system.

Overabstraction

In recent years, overabstraction of water in the Thames Valley in Southern England has led to a dramatic drop in river flow. Some tributary streams have dried up completely – damaging the river ecosystem, home to many plants and animals. Droughts in Southern England, and rising demand from increasing numbers of homes, has led to increased use of groundwater supplies. This has lowered the water table across the Thames Basin so the aquifer (underground water store) is not being used sustainably.

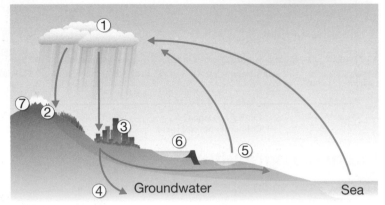

Groundwater   Sea

Interventions

① Cloud seeding to make rain

② Deforestation and changes in land use, leads to loss of interception capability and possible flooding (Himalayas)

③ Widespread urbanisation (cuts evapotranspiration)

④ Overabstraction of groundwater leads to falling water table

⑤ Overabstraction from rivers and lakes leads to conflict between users and loss of evaporation

⑥ Building of dams and larger reservoirs

⑦ Impacts of global warming melts glaciers

+ **Overabstraction** means too much water is being taken from the river, lake or other water source.

Most water companies now have strict policies called CAMS (Catchment Abstraction Management Strategies) for managing local water resources. Water levels are managed to keep the competing demands of the area in balance – sufficiently high for all the users, but not so high that there is an increased flood risk.

▼ Water abstraction in the Test-Itchen CAMS area in Hampshire.

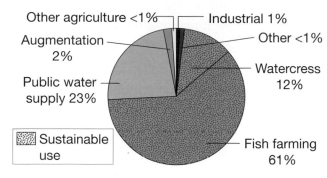

Reservoir building

In some parts of the world, natural lakes are drying up, such as Lake Chad in Africa, but in other areas artificial reservoirs can add new stores to the water cycle.

There are several types of reservoir, including those that are used for HEP generation (which usually have high dams in order to get sufficient water to generate power) and those that are used for water storage (which usually have a much less spectacular earth dam). Reservoirs can be very useful, but they can also bring problems:

* Loss of land. In the UK, nearly 275 km² are covered by reservoirs, some of which have drowned whole villages and large areas of valuable farmland.
* In some countries they can be a source of disease, because they are home to insects such as mosquitoes.
* Vegetation drowned by the lakes decays and releases methane and carbon dioxide (greenhouse gases).

Reservoirs are not all bad news. Sometimes they are designed to be multi-purpose – like Grafham Water in the diagram on the right.

Deforestation

Deforestation affects the water cycle:

* Removing the trees reduces evapotranspiration, so less green water is recycled. This can lead to a reduction in rainfall and the possibility of desertification.
* It exposes the soil surface to intense heat, which hardens the ground – making it impermeable and increasing runoff.
* It leads to a loss of soil nutrients (because of a loss of biomass). Nutrients are quickly flushed out of the system.
* Raindrop splash washes out the finer soil particles, leaving behind a coarser, heavier sand surface.
* It cuts out the process of interception. If trees are removed from watershed areas, this can increase the siltation in rivers and lead to increased flood risk.

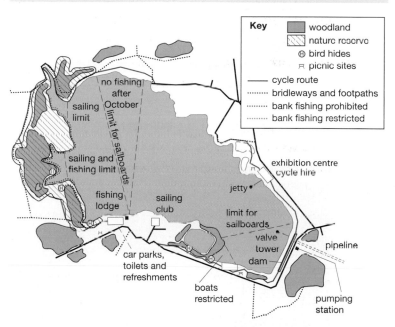

Grafham Water is a water supply reservoir covering 600 hectares near Huntingdon in the UK. It has been zoned to allow wildlife conservation and different leisure activities, such as windsurfing, sailing, fly-fishing, cycling and birdwatching.

your questions

1 Draw a flow diagram for each of the following, to show the knock-on effects: deforestation, urbanisation, dam building, over-abstraction.
2 Which of these has the worst effects? Explain your reasons.
3 **Exam-style question** Using named examples, examine how human activities can disrupt water quantity and quality. (6 marks)

✚ In this section you'll evaluate large-scale solutions to managing water supplies.

> **✚ Blue water** is the water contained in lakes, reservoirs, rivers and aquifers – accessible for human use.

Big dams – costs and benefits

Our demand for water is constantly increasing — yet supplies can be unpredictable. If we use water faster than it can be replenished, its use is not sustainable. Not only that, water may not be available where it is needed most, so it has to be stored which can be expensive.

Large-scale water management schemes tend to involve big dams. They are complex, multi-purpose projects, with many **costs** (or losses) and **benefits** (or gains). These may be economic, social and environmental (see below).

Loss of farmland and villages

Habitat for water birds

Scenic asset – recreational use

Increase in humidity

Fish stocking

Water for domestic use and irrigation – can reduce quality

Sedimentation in lake affects marine life

Dam interferes with logging, navigation and fish migration

Hydroelectric power attracts industry

Dam acts as a knickpoint – energy is reduced and deposition results

Less sediment means more energy, leading to 'clear water' erosion

Regulated flow – floods are controlled

▲ The possible effects (both positive and negative) of dam building on the environment.

Around 45 000 dams worldwide, affect 6 out of 10 major rivers. They currently supply:

- 40% of the world's irrigated water
- 20% of the world's electricity
- 15% of all blue water

The Hoover Dam, USA

The Hoover Dam on the Colorado River, USA opened in 1936. It was the most expensive engineering project in U.S. history – costing $49 million. The dam created Lake Mead, a 180km reservoir, which supplies water to 8 million people in Nevada, California and Arizona — including the city of Las Vegas. This is an arid region that depends on the water for farming, homes and recreation. But the Hoover Dam has problems:

- Virtually all of the Colorado River's water is now used, so almost none reaches the sea.
- The estuary of the river is basically dead and because seasonal flooding no longer occurs bird and fish species numbers have declined.
- Lake Mead is running out of water, as demand increases, but drought reduces inputs. In 2010, Lake Mead was only 40% full and it could fall below useable levels as early as 2020.

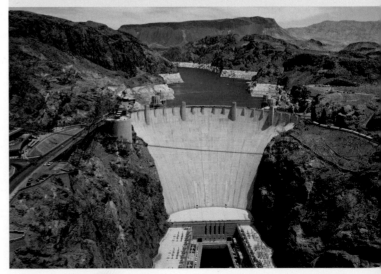

China's big schemes

China's centralised government enables it to develop huge schemes. Water is plentiful in the south of China, but scarce on the parched Northern Plains, and likely to get scarcer. China's answer to its water problems has been to develop major schemes such as the South-to-North Water Diversion Project and the Three Gorges Project (see right).

South-to-North Water Diversion Project

This huge water diversion project will transfer water to the drier North of China. Planned for completion in 2020, the scheme will eventually divert 44.8 billion m³ of water annually. The work will link China's main rivers and requires the construction of three diversion routes. The complete project is expected to cost twice as much as the Three Gorges Dam and has caused many environmental concerns, especially:

* the loss of ancient sites
* the displacement of people
* the destruction of pasture land
* an increased risk of water pollution from new industry along the diversion routes

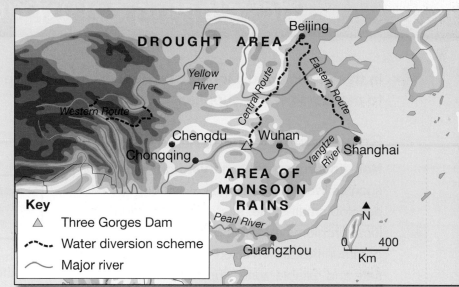

Key
△ Three Gorges Dam
-·-·- Water diversion scheme
~~~ Major river

Three Gorges Dam

The $26 billion Three Gorges Dam was completed in 2012. It was designed to reduce seasonal flooding on the Yangtze River, improve water supply by regulating the river flow, generate electricity, and make the river easier for ships to navigate.

The 175m concrete dam has created a reservoir 600km long. The project has 34 hydro-electric generators that produce 80 billion kWh of electricity each year. During China's 2011 drought the reservoir provided farms downstream with millions of cubic meters of water, but in 2012 it held back flood waters. On the downside, 632km² of land was inundated by the reservoir, and river wildlife has suffered. 1.3 million people had to be relocated from 1350 villages and 150 towns. Water quality in the reservoir is low, because of upstream industry, sewage, and farm waste entering the water.

your questions

1 In pairs, draw a large table with social, environmental and economic down the left hand column, and benefits and problems across the top. Complete the table to show the advantages and disadvantages of a big dam.

2 How far do big dams produce economic benefits, but social and environmental problems? Write 300 words to explain your opinion.

3 Discuss as a class – are big dams sustainable?

4 **Exam-style question** Using named examples, explain the costs and benefits of large scale water management schemes (6 marks)

+ In this section you'll look at small-scale sustainable solutions to managing water supplies.

Small-scale solutions

Non-governmental organisations (NGOs), such as WaterAid or Practical Action, often develop small-scale, sustainable solutions to local problems in developing countries. Local communities are involved in projects to develop safe and reliable water supplies. NGOs set up low-cost projects using **appropriate** or **intermediate technology**. This means that it is appropriate to the geographical conditions of the local area, and within the technical ability of the local community, so that they can operate and maintain it themselves. Local people are trained to take responsibility for the development and management of the schemes. The aim is, that when the NGO ends its involvement, local people can continue with the projects, and even replicate them elsewhere.

The schemes include rainwater harvesting (see the diagram on the right), protecting springs from contamination, developing gravity-fed piped schemes, and building hand-dug or **tube wells** for villages. Water can be obtained from hand-dug wells using buckets or hand/treddle pumps. Recently, scientists at Bristol University have developed a simple hand-held device for communities to check the safety of their water supplies. Other NGO schemes aim to purify water supplies.

> + **Tube wells** are built where the water table is too deep to be reached by a hand-dug well.

> + **Appropriate** or **intermediate technology** – development schemes that meet the needs of local people and the environment in which they live.

Gutters collect rainwater

Tank is made from clay covering a simple bamboo frame

Local people make the taps and dig the collection pit

Each tank takes 3 days to make and can be maintained by the family

▲ A rainwater harvester for use in rural villages.

▶ A handpump at a hand-dug well.

Many of the NGO projects are in rural areas but others have been developed to supply fresh water for the shanty towns in urban areas. They have also developed projects to help with sanitation problems, such as low-cost pit or composting toilets, which will prevent water supply contamination.

Dhaka, Bangladesh

Old Zhimkhana, a slum community built on the site of a disused railway station in Dhaka, Bangladesh, had no safe water or toilets. But with the help of an organisation called Prodiplan (one of WaterAid's partners) things are changing. Prodiplan is working with the community to help deliver water, sanitation, and hygiene education.

Six deep tube wells have been constructed, saving people time and energy in collecting water, and two new sanitation blocks provide toilets and water for washing.

People in Old Zimkhana are no longer continually ill. People run the facilities themselves and the project has helped them begin to move out of poverty.

Large scale or small scale?

Are there any snags to small-scale intermediate technology schemes? Some don't succeed because they can be inefficient. Others may not be suited to the specific climate, geology, resources available, or even the cultures in different locations. But, in general, intermediate technology works well and is often more sustainable than the large-scale schemes. Access to clean water can make a huge difference to people's lives. In the past, people (usually women) often had to walk long distances to get water. Now they can have safe drinking water, and water for cooking, washing and personal hygiene.

> **What do you think?**
> ✚ Large-scale schemes or small-scale projects – which is the best way to get sustainable water supplies?

your questions

1 Explain how a hand-dug well works. Draw a diagram to help your answer.

2 **a** In one column of a table, write out the main benefits and problems with hand-dug wells. In a second column, compare the main advantages and disadvantages of big dams. (Look back to section 4.7.)

   b Which is more sustainable? Explain your answer.

3 Look at the pie chart below showing domestic uses of water in England and Wales.

   a Add up percentages to show total water used for
     i personal hygiene
     ii sanitation
     iii food and cooking
     iv leisure.

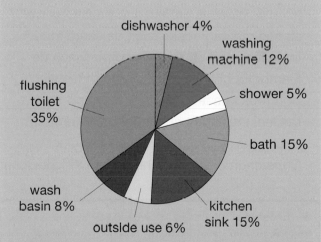

dishwasher 4%
washing machine 12%
shower 5%
bath 15%
kitchen sink 15%
outside use 6%
wash basin 8%
flushing toilet 35%

   b For each of these, suggest ways of reducing the amount of water used.

4 **Exam-style question** Using named examples, explain how small-scale intermediate technology can improve water supplies (6 marks).

In this section you will learn about the coastal zone, and investigate how rock type influences coastal landforms.

The coastal zone

The coastal zone is the transition zone between the land and the sea. Coasts are very dynamic places, constantly changing. Coasts are very popular places for people because they allow access to the sea for fishing, trade and resources like oil and gas. They are popular residential locations and are used for recreation and tourism.

▲ Hard, resistant rocks produce vertical cliff profiles — caves, arches and stack (left). Soft, less-resistant, rocks produce less steep cliffs.

Geology and rock type

The most important feature of a coast is the type of rock in the area. Some rocks are hard and resistant to erosion, whereas other rocks are soft and easily eroded:

- Hard rock coasts consist of resistant rocks such as igneous granite and basalt, as well as sedimentary rocks such as sandstone, limestone and chalk — Examples include Flamborough Head and Lulworth Cove.
- Soft rock coasts consist of clay and shale – the least resistant and most easily eroded. Examples include the Holderness Coast, Christchurch Bay and the North Norfolk Coast

> ✚ **Erosion** is the process of wearing away and breaking down rocks. There are three main types of coastal erosion: abrasion, attrition and hydraulic action.

▼ The three main types of coastal erosion.

1. Water is forced into cracks in the rock. This compresses the air. When the wave retreats the compressed air blasts out. This can force the rock apart. This is called **hydraulic action**.

2. Loose rocks, called sediment, are thrown against the cliff by waves. This wears the cliff away and chips bits of rock off the cliff. This is called **abrasion**.

Cliff

Waves crashing against cliff

3. Loose sediment, knocked off the cliff by hydraulic action and abrasion, is swirled around by waves. It constantly collides with other sediment, and gradually gets worn down into smaller, and rounder sediment. This is called **attrition**.

Coastal landforms

On coasts with hard, resistant rocks, erosion is slow. It may only be a few millimetres or centimetres a year. Most erosion happens during big storms, when waves are very powerful. Gradually, erosion produces certain characteristic landforms that give the coast its shape.

- Wave power is concentrated at the base of a cliff, where abrasion forms a **wave-cut notch**.
- Above this notch there is a cliff **overhang**.
- As the notch grows, the overhanging cliff becomes unstable and eventually collapses.
- The resulting pile of rock debris at the base of the cliff protects the cliff from further erosion.
- Over time, the loose rock is eroded by attrition, exposing the cliff to erosion again.

Over thousands of years, a succession of wave-cut notches form, and the cliff collapses again and again. Gradually the cliff retreats inland. You can tell where the cliff line once was, because a level area of smooth rock is left (called a **wave-cut platform**), which stretches out into the sea.

Hydraulic action

Hydraulic action produces coastal landforms by eroding weaknesses in the rock – cracks, joints and fissures. This produces the classic sequence of landforms shown in the diagram below. It starts with a large crack, which grows into a cave, then forms an arch, then a stack and finally a stump.

Hard rock coasts erode slowly. But, from time to time, a slab of cliff, or an arch, collapses spectacularly.

▼ The hard rock chalk cliffs at Flamborough Head in Yorkshire are famous for their arches, stacks and caves.

On your planet

✚ In 1990, the London Bridge Arch in Victoria, Australia suddenly collapsed. Two people were left stranded on the newly formed stack and had to be rescued by helicopter!

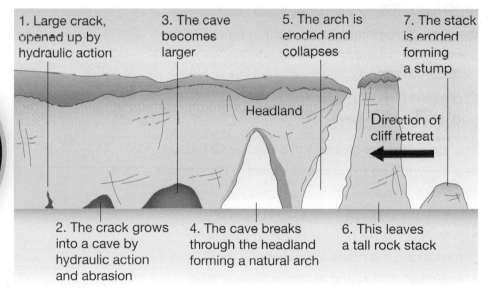

1. Large crack, opened up by hydraulic action
3. The cave becomes larger
5. The arch is eroded and collapses
7. The stack is eroded forming a stump

Headland

Direction of cliff retreat

2. The crack grows into a cave by hydraulic action and abrasion
4. The cave breaks through the headland forming a natural arch
6. This leaves a tall rock stack

your questions

1 Make a list of the advantages of living on a coast.
2 What does erosion mean?
3 Look at the diagram showing how arches, stacks and caves form. Imagine you could visit the coast in 50 years' time. Explain:
  a how the large crack might have changed
  b what could have happened to the arch.
  c what might have happened to the stump.
4 **Exam-style question**
  a Describe the difference between abrasion and attrition. (2 marks)
  b Explain how cliffs are eroded. (6 marks)

✚ In this section you will explore how rock structure influences the shape of the coast.

Rock structure

Rock structure simply means the way different rock types are arranged. Rocks are generally found in layers, called strata. This means there may be several types of rock in one cliff (see right). The cliff will only be as resistant as its weakest layers. Rock strata can be arranged in two ways along coastlines:

- If the layers are parallel to the coastline, the coast is **concordant**.
- If the layers are perpendicular (at 90°) to the coast, the coast is **discordant**.

Concordant coasts have the same type of rock all along the coastline. Discordant coasts have lots of different rock types. When these two types of coast erode, different landforms are produced, as the diagram shows.

Concordant coasts: coves and cliffs

The most famous example of a concordant coast is at Lulworth on the Jurassic Coast in Dorset (see the photo opposite). This is a World Heritage Site, because the geology here is so important.

- A resistant layer of hard Portland limestone runs along the coast at Lulworth.
- Hydraulic action and abrasion have eroded this and 'punched' through, exposing the less resistant rock behind.
- Where the waves have been able to reach the softer rock, a **cove** has quickly widened.
- The erosion at Lulworth slows when the waves reach the more resistant chalk at the back of the cove.
- A steep chalk cliff has formed at Lulworth.

This resistant sandstone layer forms a very large overhang

Coal, a very weak layer. A Wave cut notch has formed.

Ironstone, a resistant layer

Shale, a weak layer

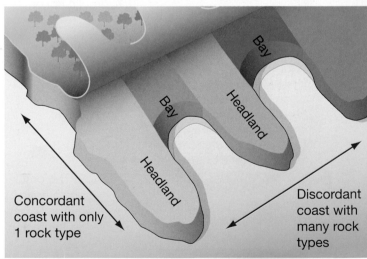

Bay

Bay

Headland

Headland

Headland

Concordant coast with only 1 rock type

Discordant coast with many rock types

✚ A **cove** is an oval-shaped bay with a narrow opening to the sea.

On your planet

✚ Because coves have narrow entrances from the sea, but sheltered beaches hidden by steep cliffs, they were often used by smugglers in the past.

Close to Lulworth Cove is Stair Hole. This is a cove that is beginning to form. The sea has punched an arch in the resistant limestone, and is widening the cove behind.

Another factor which influences erosion is weakness in the rocks forming the cliffs. There are two types of weakness:

- Joints are small, natural cracks, found in many rocks.
- Faults are larger cracks caused in the past by tectonic movements.

The more joints and faults there are in a cliff, the weaker the cliff will be. Hydraulic action attacks faults and joints, causing erosion.

Discordant coasts: headlands and bays

In south west Ireland there is an unusual coastline, with very long headlands and bays. This is a discordant coast (see below). Layers of resistant sandstones and less resistant limestones are found along the coast. Waves have eroded the limestone to form bays, leaving the harder sandstone as headlands.

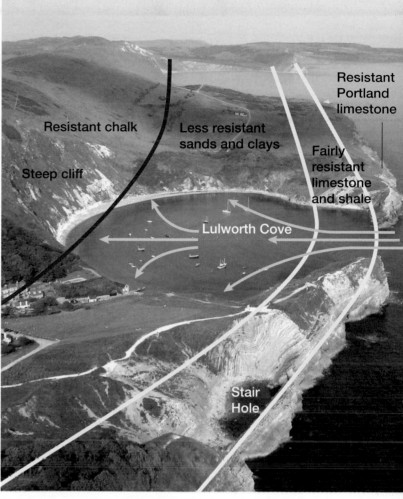

Resistant chalk

Less resistant sands and clays

Resistant Portland limestone

Fairly resistant limestone and shale

Steep cliff

Lulworth Cove

Stair Hole

▲ Concordant coast – Lulworth Cove, Dorest

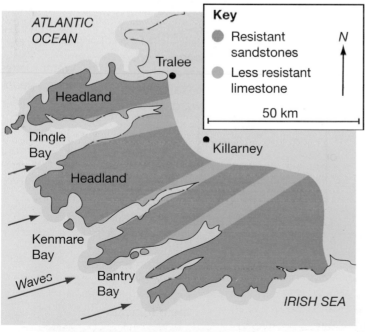

ATLANTIC OCEAN

Tralee

Headland

Dingle Bay

Killarney

Headland

Kenmare Bay

Waves

Bantry Bay

IRISH SEA

Key

- Resistant sandstones
- Less resistant limestone

N

50 km

▲ Disconcordant coast – South West Ireland

your questions

1 Draw a sketch of the cliff photo on page 76 and label it to show the most and least resistant rock.
2 Explain in a sequence of diagrams how it got to look like this.
3 Study the photo of Lulworth Cove (above) and put these changes into a correct sequence:
   - erosion by sea
   - less resistant sand and clays are eroded
   - cuts through resistant limestone
   - cove forms
   - cliffs of resistant limestone
   - sea can't erode resistant chalk so widens cove
   - forms a break in the cliff
   - sea reaches resistant chalk
4 **Exam-style question** Using named examples, explain the differences between concordant and discordant coasts (8 marks)

Waves

+ In this section you will learn how waves form, and how different types of waves affect beaches.

What causes waves?

If you blow across the surface of a glass of water, ripples will form. The same process forms waves in the sea. When wind blows across the sea, friction between the wind and water surface causes waves (see below). The size of the waves depends on:

• the strength of the wind
• how long the wind blows for
• the length of water the wind blows over – called the **fetch**.

Some waves have a fetch of thousands of miles. Waves can start near Florida and travel right across the Atlantic Ocean before hitting Cornwall, a fetch of about 6000 kilometres.

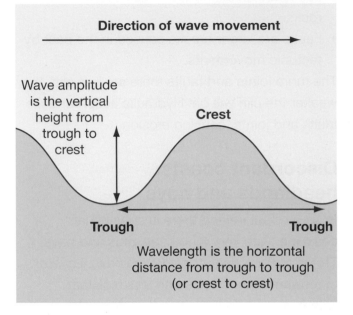

Direction of wave movement

Wave amplitude is the vertical height from trough to crest

Crest

Trough   Trough

Wavelength is the horizontal distance from trough to trough (or crest to crest)

▲ Wavelength and amplitude

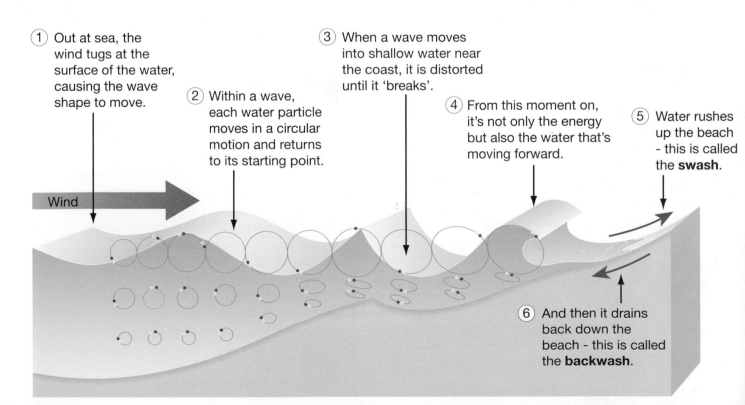

① Out at sea, the wind tugs at the surface of the water, causing the wave shape to move.

② Within a wave, each water particle moves in a circular motion and returns to its starting point.

③ When a wave moves into shallow water near the coast, it is distorted until it 'breaks'.

④ From this moment on, it's not only the energy but also the water that's moving forward.

⑤ Water rushes up the beach - this is called the **swash**.

⑥ And then it drains back down the beach - this is called the **backwash**.

Wind

Summer waves and winter waves

Not all waves are the same. The shape of a beach, or **beach profile**, is a result of how waves break on the beach.

In the summer, waves are small. They are called spilling waves. They have long wavelengths and low amplitudes. These are the type of waves everyone runs away from on summer holiday – when they break they spill up the beach!

- They have a strong swash.
- This transports sand up the beach.
- The sand is deposited as a bank or beach berm.

Because these waves build up the beach by depositing sand on it, they are called constructive waves (see top right).

In the winter, when storms and strong winds are more common, waves are different. They are taller (larger amplitude) and closer together (shorter wavelength). These are called plunging waves.

- They have a strong backwash.
- This erodes sand from the beach.
- This creates a steep beach profile.
- The sand is carried offshore by an underwater rip current.
- The sand is deposited out at sea, forming an offshore bar.

Plunging waves are dangerous, because they arrive one after another, very quickly. After a wave breaks, the next incoming wave arrives so quickly that the backwash has to flow under the incoming wave. This flow is called a rip current (see bottom right). These currents can be very strong and can drag weak swimmers out to sea, under the water.

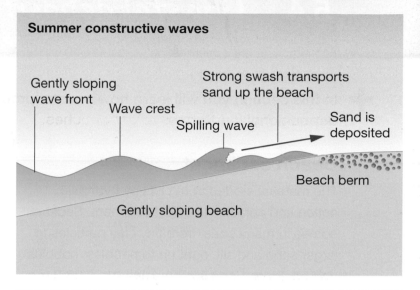

Summer constructive waves

Gently sloping wave front

Wave crest

Spilling wave

Strong swash transports sand up the beach

Sand is deposited

Beach berm

Gently sloping beach

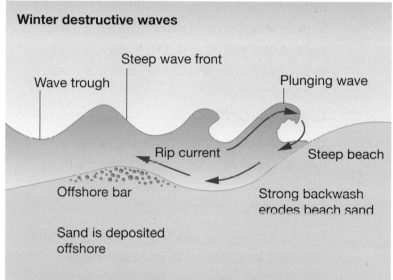

Winter destructive waves

Steep wave front

Wave trough

Plunging wave

Rip current

Steep beach

Offshore bar

Strong backwash erodes beach sand

Sand is deposited offshore

your questions

1 a Using an atlas, measure the fetch in kilometres:
   - from Calais to Dover
   - from Denmark to the Yorkshire coast
   - from Greenland to the west coast of Scotland.
  b Which of the three places in the UK will get the biggest waves?
  c Explain why this is so.
2 Explain why:
  a beaches in the summer are flat
  b beaches in the winter are steep.
3 **Exam-style question**
  a Describe the difference between swash and backwash. (2 marks)
  b Explain how beach formation can depend on different kinds of waves. (6 marks)

✚ In this section you will learn how coastal processes can create depositional landforms.

Beach sediment

The material eroded from cliffs by hydraulic action and abrasion is called sediment. Sediment comes in many sizes, from tiny clay particles to larger sand and silt, right up to pebbles, cobbles and boulders. Sand is material that is 0.06-2 mm in size. Over time, attrition will make sediment smaller and rounder. Beach sediment is also transported from where it was eroded to new locations.

Get the drift?

The main way sediment is transported is by **longshore drift**. This happens when waves break at an angle to the coast, rather than parallel to it.

Because prevailing winds are mostly from one direction, longshore drift is usually in one direction too. Longshore drift transports sediment along coastlines, as the top right diagram shows – sometimes for hundreds of kilometres before it is eventually deposited.

Depositional landforms

When rocks are eroded, creating sediment, it will first be deposited very close to where it eroded. In a partly sheltered area such as a cove or bay, a beach will form because the sediment is trapped in the bay.

Sediment transported by longshore drift will create new landforms where it is deposited – as the bottom right diagram shows.

Many **beaches** are simply rivers of sand and shingle (pebbles) slowly moving along the coast,

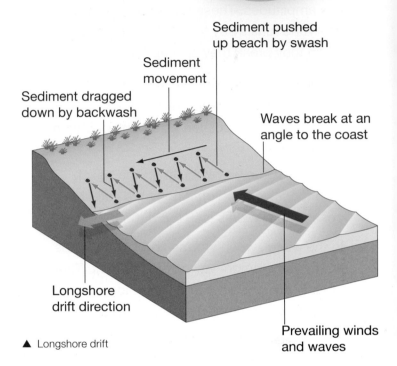

Sediment dragged down by backwash

Sediment movement

Sediment pushed up beach by swash

Waves break at an angle to the coast

Longshore drift direction

Prevailing winds and waves

▲ Longshore drift

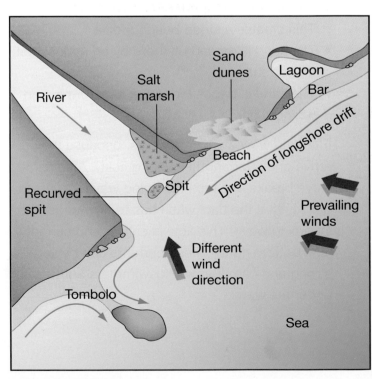

River

Salt marsh

Sand dunes

Lagoon

Bar

Beach

Direction of longshore drift

Spit

Recurved spit

Prevailing winds

Different wind direction

Tombolo

Sea

▲ Depositional landforms

as sediment is transported by longshore drift.

- Strong onshore winds can blow sand inland, forming **sand dunes** parallel to the shoreline.
- Small bays on the coast can sometimes be blocked by a **bar** of sand which grows across the mouth of the bay due to longshore drift. Behind the bar, a shallow **lagoon** forms. These are often important habitats for birds.
- Longshore drift carries sand along the shore until it reaches a river estuary, where it gets pushed out into the river channel. The river flow halts the drift, so sand is deposited, forming a long sandy neck, called a **spit**. The spit stops growing when deposition of sand, by longshore drift, is balanced by erosion from the river.
- The river moves out to sea at low tide, whilst at high tide, the sea flows inland. Each tide erodes the spit, and causes it to curve back on itself; many spits have a hooked or **recurved** end. The water behind a spit is protected from storms and tides, so remains calm, allowing salt marshes to form (see bottom photo).

Depositional landforms are made of loose sediment, so they are not very stable. Sand dunes form when plants grow on the sediment helping to stabilise it. Plants that grow on beach sand need to be tough.

- They have long roots to hold them in place in the strong winds e.g. Marram grass.
- They have tough, waxy leaves to stop them getting sandblasted.
- They can survive being sprayed by salt water.

lagoon

Slapton Ley bar

Salt marshes and mudflats

Hurst Castle spit

recurved end

your questions

1 Why is shingle usually round?
2 Why do spits not continue to grow right across the mouths of rivers?
3 Explain how vegetation stabilises depositional landforms at the coast.

4 **Exam-style question** Explain the process of longshore drift. You may use a diagram to help with your answer. (6 marks)

✚ In this section, you will examine how climate change might increase the risk of erosion and flooding on coastlines.

Rising sea levels

Many scientists fear that global warming will cause sea levels to rise. How much they will rise by is not known, but there are estimates of between 30 cm and 1 metre by the year 2100. Sea levels are rising today, as the sea is warming up and expanding. Melting ice sheets are likely to speed up the rise.

For people who live on very low-lying land next to the sea, this could spell trouble. There are many areas around the world at risk:

- In Bangladesh, if sea levels rose by 1 metre, up to 15% of the country might be flooded.
- In the UK, London and Essex are at risk, because they are low-lying.
- Many small coral islands in the Pacific and Indian Oceans, like the Maldives and Tuvalu, could disappear underwater.

Flood risk

Sea levels are constantly changing. Twice a day, due to the gravity of the moon, high tides cause raised sea levels. A few times every year there are exceptionally high tides, called spring tides. During spring tides the flood risk rises.

If spring tides occur when there are large waves, the sea will be even higher. Worse, if spring tides and waves combine with low air pressure, then a **storm surge** can form. Storm surges are caused by hurricanes and depressions, which are both low-pressure weather systems.

It is possible that global warming could make hurricanes and depressions more powerful. They might also become more frequent (see page 35-39). This would mean storm surges would happen more often. If melting ice sheets raise sea levels as well, the combined results could be very serious indeed.

On your planet

✚ In 1953, when there was a 5 metre storm surge in the North Sea, over 300 people were killed in the UK and 1800 in the Netherlands in devastating flooding.

✚ During a **storm surge**, sea level rises. This is because the air pressure falls. Sea level rises by 10 mm for every 1 millibar drop in air pressure.

Flood wall

Spring tide, storm surge and global warming?

Spring tide and storm surge

Spring tide

High tide

Low tide

▲ People will become increasingly vulnerable to spring tides and storm surges as sea levels rise due to global warming.

Storm Sandy

As the newspaper headlines show, coastal flooding is a real risk. In October 2012 storm Sandy struck New York and New Jersey in the USA, unusually far north for a tropical storm. High winds and a 4m storm surge caused massive flooding, knocked out power for 5 million people, killed over 100 and caused around $50 billion in damage. Scientists estimate that '1 in 100-year' events like Sandy, could be '1 in 20-year' events by 2050.

▲ Some scientists warn that events like storm Sandy could become more common as climate change raises sea temperatures.

Erosion and deposition

With higher sea levels in the future, and possibly increased storms, the balance of erosion and deposition may change. Beaches, spits, barrier islands and river deltas may erode faster, and in some cases, become submerged.

For the south and east of the UK, this is a problem — as land is low-lying and rocks are easily eroded. A sea level rise of only 50cm would make existing sea defences useless. The only choice would be to build new, higher defences or abandon some areas to the sea.

"Thousands flee to higher ground as stormy seas gather momentum"
Newspaper extract, Nov 2007

"There is a threat we could become the world's first global warming refugees"
Spokesperson for the Maldives Government, 2006

"US President Barack Obama has declared a "major disaster" in New York state, after storm Sandy smashed into the US East Coast, causing flooding and cutting power to millions"
Newspaper extract, Oct 2012

your questions

1 What are the estimates of sea level rise by 2100?
2 Use an atlas to find 10 major cities in the world that are on the coast, and that could be at risk from rising sea levels.

3 Explain how a storm surge forms.
4 **Exam-style question** Using examples, explain how sea level rise could threaten people and their property. (6 marks)

In this section you will learn why coasts vary in their rate of erosion, and the impacts that erosion has on people.

A complex problem

Some coastlines are eroding quickly, by over 2 metres a year. Others hardly change at all.

- In the UK, the Holderness coast in Yorkshire, the North Norfolk coast, and some coastal areas of Hampshire and Dorset have very high erosion rates due to weak rocks which collapse easily. They all suffer from erosion (hydraulic action and abrasion), but **weathering** and **mass movement** make the problem much worse. Weathering and mass movement are called **sub-aerial processes**. Water movement is often important in causing weak cliffs to collapse.

- Other coasts remain stable with low rates of erosion. The coast of Cornwall has many cliffs of granite and slate which are resistant rocks. They are able to withstand the energy of even the biggest Atlantic waves for long periods.

- As well as geology, much depends on wave energy. This in turn depends on the length of fetch over which the wave has travelled (see page 78).

> + **Weathering** is the breakdown of rocks *in situ*. This means it happens where the rock is. Rocks are weakened by being chemically attacked and mechanically broken down.

> + **Mass movement** is the movement of materials downslope, such as rock falls, landslides or cliff collapse.

Why cliffs collapse

Marine processes

1 The base of the cliff is eroded by hydraulic action and abrasion (cliff foot erosion), making the cliff face steeper.

Sub-aerial processes

2 Weathering weakens the rock. This can be mechanical weathering, like freeze-thaw action, or chemical weathering, like solution.

3 Heavy rain saturates the permeable rock at the top of the cliff. Rainwater may also erode the cliff as it runs down it or emerges from the cliff at a spring (cliff face erosion).

4 The water flows through the permeable rock, adding weight to the cliff.

Human actions

5 Building on top of the cliff adds a heavy load, which pushes down on the weak cliff.

During a big storm, heavy rain saturates the permeable rock, and erosion by the sea undermines it. Eventually a large chunk of cliff gives way and slides down the cliff – a rotational slide (see the photo opposite).

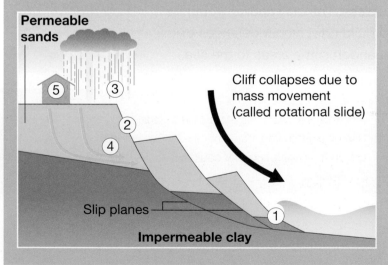

Permeable sands

Cliff collapses due to mass movement (called rotational slide)

Slip planes

Impermeable clay

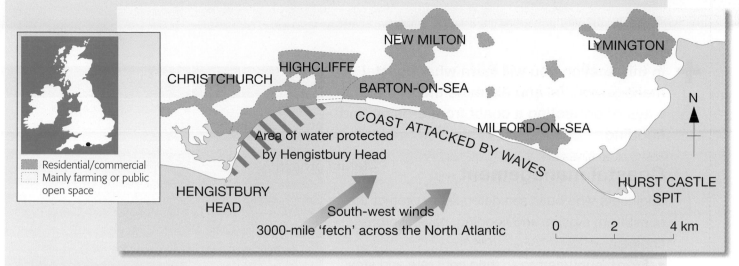

Residential/commercial
Mainly farming or public open space

NEW MILTON
LYMINGTON
HIGHCLIFFE
CHRISTCHURCH
BARTON-ON-SEA
COAST ATTACKED BY WAVES
Area of water protected by Hengistbury Head
MILFORD-ON-SEA
HENGISTBURY HEAD
HURST CASTLE SPIT
South-west winds
3000-mile 'fetch' across the North Atlantic
N
0   2   4 km

Christchurch Bay

Christchurch Bay on the UK's south coast has a severe problem. Without management, the cliffs erode by over 2 metres a year; threatening residential areas along the coast, such as Barton-on-Sea (see the map). At Barton-on-Sea, mass movement is the major problem, caused by weathering and water movement. But, cliff foot erosion plays a part, as the Atlantic fetch brings big waves. This makes managing this type of erosion very difficult.

Impacts

Erosion in Christchurch Bay affects many people:

- Homeowners lose their homes to the sea. House values fall, and insurance is impossible to get.
- Rapid cliff collapses are dangerous for people on the cliff top, and on the beach.
- Roads and other infrastructure are destroyed.
- Erosion makes the area unattractive.

People living in Christchurch Bay argue that they need sea defences to protect their coast. These are expensive, and there is no agreement about which defences work best.

your questions

1 What does the term mass movement mean?
2 Look at the map of Christchurch Bay (above). Make a list of the type of people who might live and work here.
3 Read the section on why cliffs collapse, and draw a spider diagram to show all the different factors.
4 **Exam-style question**
   a Describe the difference between weathering and erosion. (2 marks)
   b Using a named example, explain why coastlines experience rapid erosion. (6 marks)

+ In this section you will learn what coastal management is, and examine the traditional ways of protecting a coast from erosion and flooding.

Coastal management

Engineers who build sea defences to protect the coast from erosion and flooding, have two basic choices:

- **'Hard' engineering** – using concrete and steel structures, such as sea walls, to stop waves in their tracks.
- **'Soft' engineering** – using smaller structures, sometimes built from natural materials, to reduce the energy in waves.

The 'hard' method is the traditional method of coastal management. However, it has two major problems:

- It is very costly.
- It usually makes the coast look unnatural, and often ugly.

On the positive side, large, very strong sea defences may be the best way to stop erosion. Engineers have many different types of sea defences they can use, as the table shows. Often they use several types of defences together.

Type of defence	Cost	About the defence
Sea wall	£2000 per metre	• Reflects waves back out to sea. • Can prevent easy access to the beach. • Suffers from wave scour, where plunging waves erode the beach and attack the wall's foundations.
Sea wall with steps and bullnose	£5000 per metre	• Steps **dissipate** wave energy, the bullnose throws waves up and back out to sea.
Revetments	£1000 per metre	• Break up incoming waves. • Restrict beach access and look ugly. • Can be destroyed by big storms.
Gabions	£100 per metre	• A cheap type of sea wall. • Absorb wave energy as they are permeable. • Not very strong.
Rock armour (rip-rap)	£300 per metre	• Easy to build. • More expensive if built in the sea. • Dissipate wave energy and look 'natural'.
Groynes	£2000 per metre	• Prevent longshore drift, trapping sand and shingle. • Larger beach dissipates wave energy, reducing erosion. • May increase erosion downdrift.

+ **Dissipate** means to reduce wave energy, as some of it is absorbed as waves pass through, or over, sea defences.

Flood gates are closed during very high tides and storms

Concrete bullnose sea wall deflects waves back out to sea

Wooden groynes prevent longshore drift and build up the beach

Rip-rap dissipates wave energy before it can hit the sea wall

▲ Coastal management at Hornsea, East Riding of Yorkshire.

Coastal change and conflict

Conflict in Christchurch Bay

People who live on rapidly eroding, soft rock coasts are often in favour of hard engineering. It looks like something 'serious' is being done to protect their coast. However, not everyone agrees (see table).

Groynes

Beaches are nature's sea defences. When a wave breaks on a wide beach, the energy of the wave is dissipated by the beach before it hits the cliff. On coasts with rapid erosion, there are often very narrow beaches or no beach at all.

The traditional way of solving this problem is to 'grow' a beach. This is done by building wooden or stone groynes across the beach at right-angles to the coast. Each groyne costs £200 000-£250 000. Groynes stop longshore drift by trapping sediment, building up the beach. But building groynes in one place stops sediment reaching other places further downdrift, so beaches elsewhere can disappear.

Sea walls

It is likely that sea walls built today will have to be made higher in future to cope with rising sea levels. On low-lying coasts, settlements could end up in a race to build higher defences in the future. This will be very expensive.

▲ Erosion near Naish Holiday Village in Christchurch Bay showing how groynes have protected one part of the coast but caused erosion elsewhere. This is called 'terminal groyne syndrome'.

Stakeholders	Views on coastal protection.
Coastal residents and business owners in Christchurch, Barton and Milford	They are in favour of a 'hold the line' policy to protect their homes, businesses and tourist sites. Most favour hard engineering, but this is expensive. Ugly sea defences would be an eyesore and deter visitors.
Local politicians and the council	Need the support of all local people, so have to be careful not to favour one group more than another; they want effective coastal protection, but not at any price.
Local people living further inland	They are not affected by the erosion, but are concerned local taxes will rise to pay for coastal protection; they prefer low cost options and don't want low value land, such as farmland, protected.
Environmentalists	Fear that habitats and sensitive ecosystems will be affected by the construction of sea defences; they prefer a 'do nothing' approach and soft engineering.
Residents and businesses downdrift	They worry that defences built updrift will reduce their beach size and protection; they want an integrated approach to management that includes everyone affected.
Fisherman and boat users	Their priority is access to the sea, so they want harbours, marinas and some beaches protected to maintain this.

▲ About £28 million has been spent protecting Christchurch Bay since 1964, and the costs will only rise in the future.

your questions

1 List all of the types of defences mentioned, in order of cost (highest to lowest).
2 How can building sea defences in one place increase erosion in another place nearby?
3 **Exam-style question** Using named examples, examine the costs and benefits of hard and soft engineering coastal defences (8 marks)

87

In this section you will learn how coasts today are often managed in a more holistic, sustainable way.

Managing the whole coast

Coastal management is expensive. Now, holistic coastal management is adopted; meaning that some places are protected from flooding or erosion if they are worth it, whereas others aren't. This is because:

- the value of land and buildings does not justify the cost
- building defences might actually cause more erosion elsewhere
- climate change is likely to bring rising sea levels anyway
- it might be better for the environment e.g. creating new areas of marsh.

This involves making different choices which impact upon a whole stretch of coast, such as Christchurch Bay — and not just on one area, such as Barton-on-Sea.

Holistic management takes into account the:

- needs of different groups of people
- economic costs and benefits of different strategies today, and in the future
- environment, both on land and in the sea.

Integrated Coastal Zone Management (ICZM) is the name for this approach to managing coasts. For long stretches of coast, a plan is

The choices

In the UK, local councils pay for sea defences. They may get some money from the Government, or the Environment Agency if there is a flood risk. There are four choices that councils can make about how to manage the coast:

1. **Hold the line** – use sea defences to stop erosion, and keep the coast where it is today. This is expensive.
2. **Advance the line** – use sea defences to move the coast further into the sea. This is very expensive.
3. **Strategic realignment (strategic retreat)** – gradually let the coast erode, and move people and businesses away from at-risk areas. This may involve financial compensation for people when their homes are lost.
4. **Do nothing** – take no action at all, and let nature take its course.

These choices can cause conflict. Choices 3 and 4 may mean some people lose land, businesses or homes. The diagram below shows the choices that have been made along part of the North Norfolk coast.

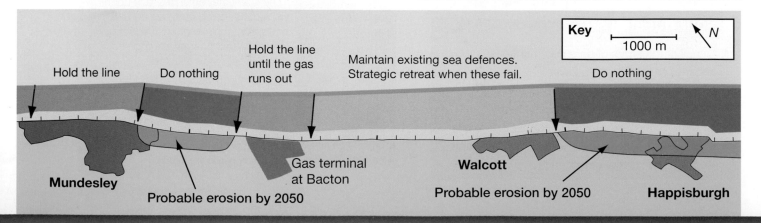

Key — 1000 m — N

Hold the line | Do nothing | Hold the line until the gas runs out | Maintain existing sea defences. Strategic retreat when these fail. | Do nothing

Mundesley — Probable erosion by 2050 — Gas terminal at Bacton — Probable erosion by 2050 — **Walcott** — **Happisburgh**

drawn up, called a **Shoreline Management Plan (SMP)**, which sets out how the coast will be managed. This should prevent one place building groynes, if this will then cause more erosion downdrift.

Soft engineering

On many coasts, soft engineering is replacing hard engineering. This works with natural processes, and tries to stop erosion by stabilising beaches and cliffs and reducing wave energy. Soft engineering solutions can be cheaper than hard, and are often less intrusive.

The diagram on the right shows some soft engineering solutions:

- Planting vegetation: £20 50 per square metre.
- Beach nourishment: £500-1000 per square metre.
- Offshore breakwaters: £2000 per metre.

Soft engineering may not work in all places. Where rocks are very weak, hard engineering or 'do nothing' may be the only choices.

Here comes the sea

Holistic management and soft engineering solutions are seen as more sustainable. In some places in the UK, sea defences have been abandoned and nature is taking its course.

In the next few decades, the UK will face many difficult decisions about how best to protect the coast.

- At the moment the Government thinks it is too expensive to protect farmland and isolated houses.
- Residents, councils and businesses often disagree.
- It is very hard to persuade people who have lived by the coast all their lives that protecting their property is not sustainable.
- We don't know exactly what the impact of rising sea levels will be, so planning new defences is difficult.

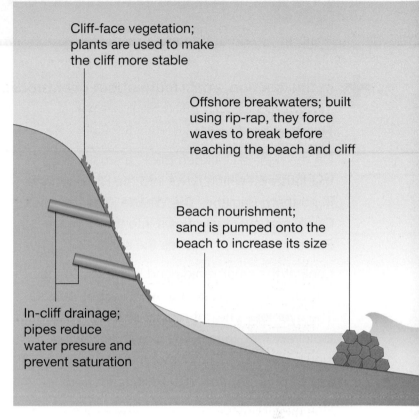

Cliff-face vegetation; plants are used to make the cliff more stable

Offshore breakwaters; built using rip-rap, they force waves to break before reaching the beach and cliff

Beach nourishment; sand is pumped onto the beach to increase its size

In-cliff drainage; pipes reduce water presure and prevent saturation

▲ Soft engineering solutions.

your questions

1 What does 'holistic' coastal management mean?
2 Who pays for most sea defences in the UK?
3 a Look at the map of the North Norfolk coastline (opposite). Make a list of land uses and features that will be lost to the sea by 2050.
   b Make a list of people this could affect.
   c Why is the gas terminal at Bacton being protected?
   d Explain how people in this area might have very different views about the shoreline management plan.
4 **Exam-style question** Using named examples, explain how coastal management decisions can lead to conflict at the coast (6 marks)

➕ In this section, you'll learn about river processes in upland areas.

Buckden Beck

On the right is Buckden Beck, a small stream – or tributary – which flows into the River Wharfe in northern England. The Wharfe joins the River Ouse, and eventually flows into the Humber Estuary – where it reaches the sea.

Look at the small rapids and waterfalls in the photo. This is typical of a river's **upper course**. The river loses height rapidly, making the stream look very fast. In fact, the stream is flowing slowly, because so much of the river's energy is lost through **friction** with the stream bed.

However, rainstorms increase the river's energy a lot, and allow it to carry large boulders and stones, which wear away – or erode – the **channel**. Therefore, most **erosion** is done during periods of wet weather.

> + **Channel** refers to the bed and banks of the river.
> + **Erosion** means wearing away the landscape.

▲ Buckden Beck – the **gradient** (slope) of the river course is steep.

How a river carries its load

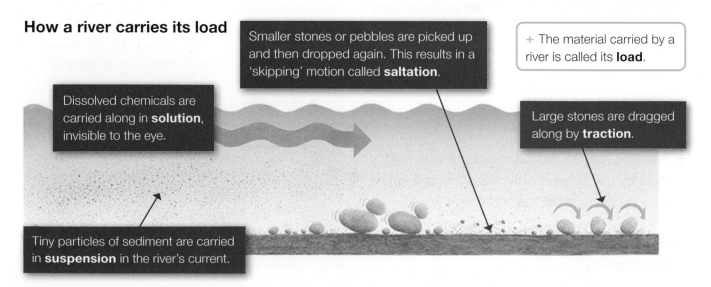

Smaller stones or pebbles are picked up and then dropped again. This results in a 'skipping' motion called **saltation**.

> + The material carried by a river is called its **load**.

Dissolved chemicals are carried along in **solution**, invisible to the eye.

Large stones are dragged along by **traction**.

Tiny particles of sediment are carried in **suspension** in the river's current.

How a river erodes its channel

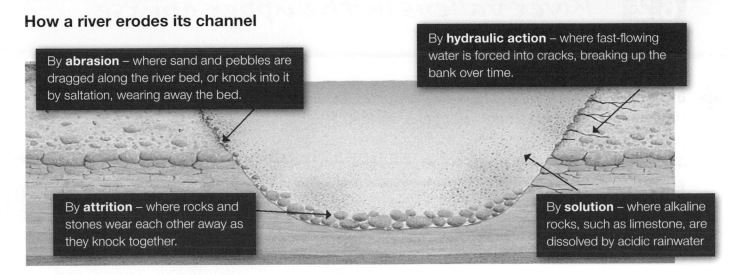

By **abrasion** – where sand and pebbles are dragged along the river bed, or knock into it by saltation, wearing away the bed.

By **hydraulic action** – where fast-flowing water is forced into cracks, breaking up the bank over time.

By **attrition** – where rocks and stones wear each other away as they knock together.

By **solution** – where alkaline rocks, such as limestone, are dissolved by acidic rainwater

In the upper course, most erosion is vertical (downward). Whenever the river meets a hard – or **resistant** – rock, a step is formed. This can eventually become a waterfall. The river gains a great deal of energy where it falls over the lip of the waterfall, allowing it to erode rapidly.

How waterfalls are formed

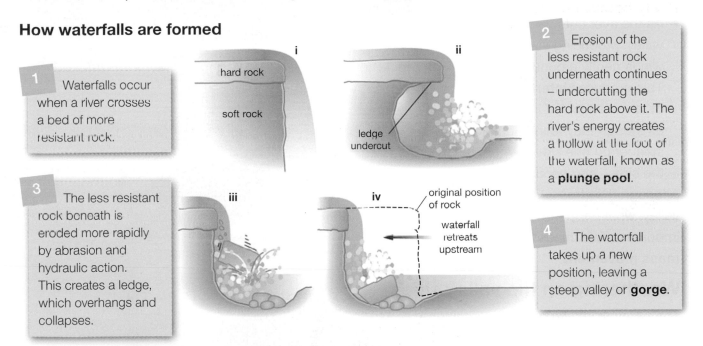

1 Waterfalls occur when a river crosses a bed of more resistant rock.

hard rock

soft rock

ledge undercut

2 Erosion of the less resistant rock underneath continues – undercutting the hard rock above it. The river's energy creates a hollow at the foot of the waterfall, known as a **plunge pool**.

3 The less resistant rock beneath is eroded more rapidly by abrasion and hydraulic action. This creates a ledge, which overhangs and collapses.

original position of rock

waterfall retreats upstream

4 The waterfall takes up a new position, leaving a steep valley or **gorge**.

your questions

1 For each set of words below, **a** say which is the odd one out, **b** explain your choice.
  • stream, tributary, river, waterfall
  • solution, attrition, plunge pool, abrasion
  • suspension, gorge, traction, load
  • hydraulic action, saltation, suspension, traction

2 Explain how the following might change during wet weather:
  • The amount of water in the stream, and its energy.
  • The amount of erosion that a stream can do.
  • A waterfall.

3 **Exam-style question** Explain the processes that lead to the formation of a waterfall. You may want to use a diagram to help with your answer (6 marks)

In this section, you'll learn how rivers and their valleys develop in upland areas and what causes this.

▼ Buckden Beck is typical of a valley in its upper course – it forms a V-shape, with interlocking spurs.

The valley of Buckden Beck

Look at the steep valley sides in the photo. This is typical of a river's upper course in upland areas. It has two main features:

- It has steep sides and the valley bottom is narrow (you can see why valleys like this are called **V-shaped**).

- As the river cuts vertically into the valley, producing steep sides, it also winds gently around areas of resistant rock. This produces **interlocking spurs** – or ridges of land – which jut into the river valley, looking as though they're interlocked. They are steeper on one side – where the river cuts into the side – and gentler on the other.

Although the stream is vital in eroding the valley, what happens on the valley sides is also important. Two things happen: **weathering** and **mass movement**.

Weathering on the valley sides

The valley sides are pale grey limestone, and have been attacked and broken down by weathering. The cliffs of rock are known as **rock outcrops**, where rock comes to the surface. Below them, the valley is covered in limestone rubble, called **scree**. The scree has broken away from the cliffs above because of weathering. There are different types of weathering – physical, chemical and biological, as shown in the photo.

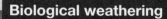

Biological weathering

Although rocks look solid, small cracks allow plant roots to penetrate in search of water and nutrients. As they grow, root cells force the cracks apart, widening them and breaking the rock into pieces.

Physical weathering

Physical weathering occurs when physical force breaks rock into pieces. In winter, cracks in the limestone rock fill with rain. This freezes, expanding in volume by 10% and widening cracks so that more water gets in. This process is known as **freeze-thaw**. If repeated often enough, pieces of rock break away, becoming scree at the base of the cliff.

Chemical weathering

Chemical weathering is any chemical change or decay of solid rock. Rainwater mixes with atmospheric gases, e.g. CO_2, to form weak acids which dissolve alkaline rocks such as limestone.

+ **Weathering** is the breakdown of rocks *in situ*. This means it happens where the rock is. Rocks are weakened by being chemically attacked, and mechanically broken down.

Mass movement

Once rock is broken up, the fragments move down the slope towards the stream. Some move quickly, others slowly. This is known as mass movement, but there are different processes:

- Rapid – such as **landslides** and mudflows. These are less common in the UK, although they do occur on railway cuttings and along cliff coastlines.
- Slow – the most common of which is **soil creep**. Soil creep occurs really slowly – perhaps 2 cm a year – but over decades it has many effects, like those in the top diagram.

The shape of the valley

Valley shape is affected by three things, as shown in the bottom diagram:

- The speed of weathering. If scree piles up, weathering is taking place rapidly.
- The speed of mass movement.
- How quickly the river can remove the material brought by mass movement.

If the river has plenty of energy, it takes the material away and uses it to help erode the valley, making it steeper. However, if it is slow and cannot cope with all the material, weathered rock collects at the bottom of the slope, making the valley gentler and flatter. You'll see in the middle course how this alters the valley shape.

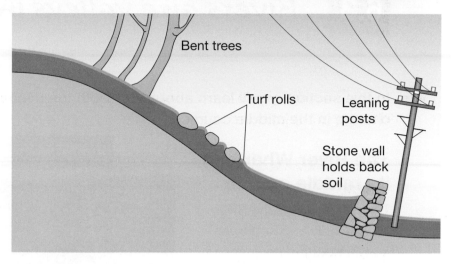

▲ Different ways in which soil creep can be recognised in the landscape. Although it is slow, soil creep can cause walls or telegraph poles to lean.

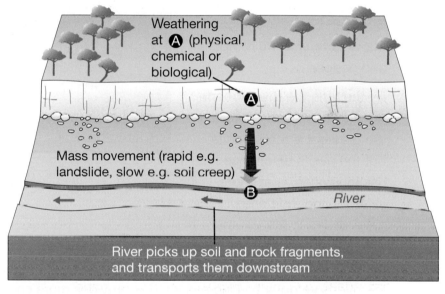

▲ Weathering takes place at A on the valley sides to break down solid rock into fragments. These move down the slope to the river at B, where they are removed. The river then uses these to wear away – or erode – the river bed.

your questions

1 Draw labelled diagrams to explain these features of a valley in the upper course:
   - How scree is formed.
   - Why valley slopes get covered in scree.
   - Why hedges and stone walls can fall over.
   - How trees can grow out of solid rock cliffs.
   - Why stream beds contain large rock fragments.
2 **Exam-style question** Explain how mass movement and weathering affect the shape of river valleys (6 marks)

✚ In this section, you'll learn about how both the river and its valley change in the middle course.

The River Wharfe in its middle course

The photo shows the River Wharfe between Kettlewell and Starbotton, downstream from Buckden Beck. By now, the river is in its **middle course**. Several streams like Buckden Beck have joined the Wharfe, making it wider and deeper.

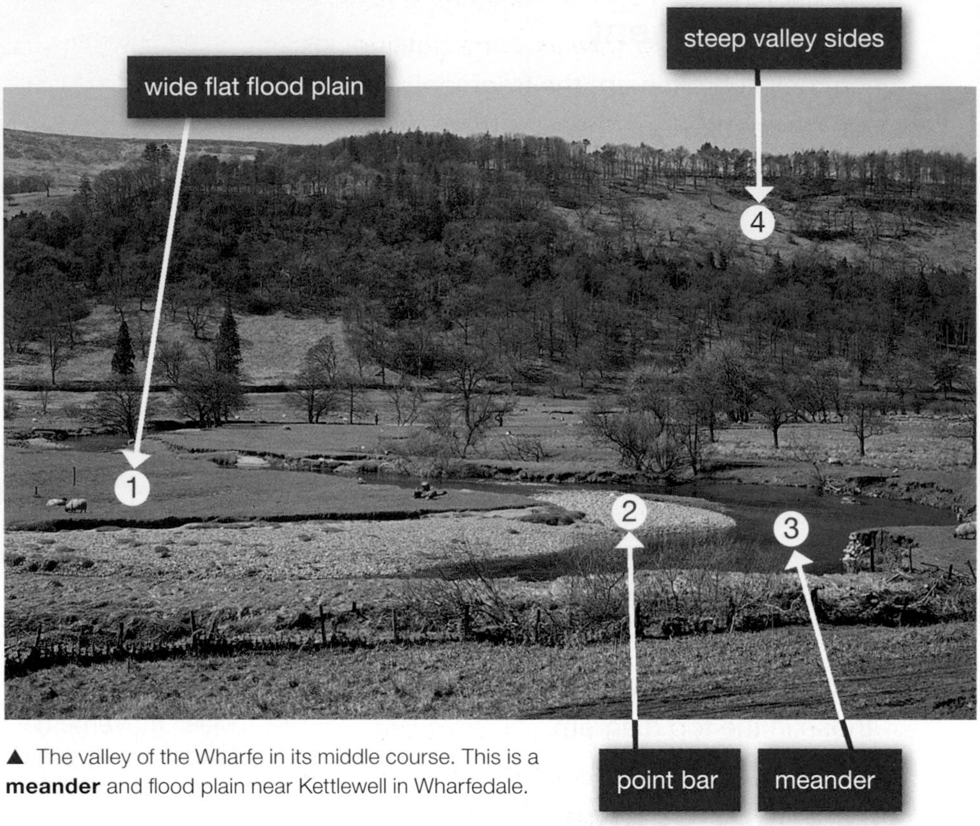

By the time the River Wharfe reaches its middle course, the gradient is more gentle, but the river's **discharge** has increased — as more streams and tributaries have added to the volume of water in the river. The river channel is also smoother — creating less friction to slow down the river flow. As a result, the river's **velocity** has increased. This provides more energy for the river to erode laterally (i.e. side to side, rather than downwards), creating **meanders**, **point bars** and **flood plains** (see photo). The valley shape changes in the middle course from a V-shape to a **U-shape**, as the flood plain has broadened the valley floor.

▲ The valley of the Wharfe in its middle course. This is a **meander** and flood plain near Kettlewell in Wharfedale.

wide flat flood plain

steep valley sides

point bar

meander

+ **Velocity** is the speed the river water flows at, usually measured in metres per second.

+ **Discharge** is the volume of water flowing in a river, measured in cubic metres per second.

During wet weather, the volume of water can be so great that the river sometimes floods over the **flood plain**. During this time, fine sands and clays brought down by the river from upstream, settle over the flood plain and form layers of **alluvium**. Flood plains are at risk of flooding, but alluvium is fertile and attracts farmers.

How meanders change the valley

1

The river bends in its middle course; each sharp bend is called a meander. Meanders are natural; rivers almost never flow in a straight line. Water flows naturally in a corkscrew pattern. This is called **helical flow**.

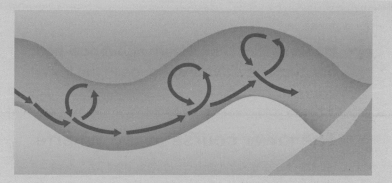

2

Helical flow sends the river's energy laterally – i.e. to the sides. The fastest current (called the **thalweg**) is forced to the outer bend (A), where it undercuts the bank. This produces a steep edge, or **river cliff**, which eventually collapses. Gradually, the river undercuts more and more, so that the channel moves to a new position. Helical flow shifts sediment across the channel to the inner bank (B). Here the sediment is deposited by the slower moving water, to form a point bar. As the river channel shifts, the point bar gets bigger and bigger, forming a flat area known as a **flood plain** (C).

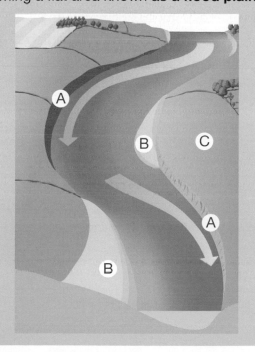

3

Continued erosion can create a narrow neck between two meanders (X). Eventually, the neck will be breached at (Y), cutting off the meander to create an **ox-bow lake** (Z).

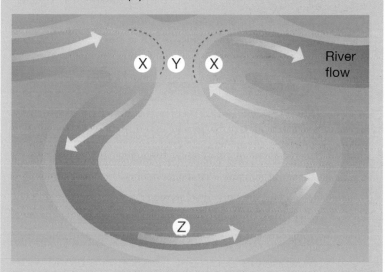

your questions

1 Classify each of these features by whether they are part of **a** the **river** in its middle course, **b** its **valley** in its middle course:

flood plain	alluvium	U-shape
lateral erosion	meander	helical flow
thalweg	river cliff	point bar
meander neck	ox-bow lake	

2 Draw a sketch of the photo on page 94, adding the following labels: flood plain, alluvium, lateral erosion, meander, thalweg, undercutting, point bar.

3 **Exam-style question** Explain the processes that lead to the formation of an ox-bow lake. You may want to use a diagram to help with your answer (6 marks)

✚ In this section, you'll learn about how both the river and its valley change in the lower course.

The lower course of the Wharfe

By the time the Wharfe has reached its lower course, the differences with its upper course are really obvious! The river – once narrow, with waterfalls – is now wide and deep, flowing over a gentle gradient. Its meanders are large, the volume of water is also larger, and there is a wide, flat flood plain. This is low-lying and floods easily.

Beside the river, there are embankments. These can be either natural or artificial, and are known as **levées** (see right).

Towards the sea

The Ouse River (joined by the Wharfe) eventually joins the Humber and forms an estuary – where the river meets the sea. Two directions of flow take place here – **outwards** by the river, taking water out to sea, and **inwards** by incoming high tides. Twice a day, incoming tides meet the outgoing river and the flow stops, forcing the river to deposit its sediment. This forms a broad wide area where the river deposits mud – hence the name **mudflats** (see below).

▲ The lower part of the Wharfe at Tadcaster. By this stage, the river is flowing over an almost flat gradient.

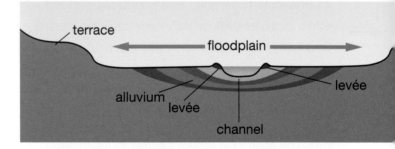

Natural levées form beside the river's bank where it first floods. As a river reaches **bankful** – i.e. before it spills on to the flood plain – it deposits sand and clay particles where the flow is slower. These build up beside the river as a bank.

Artificial levées are built by engineers to protect farms or towns from flooding. These are common in the UK along rivers like the Severn and the Ouse, where flooding is common, and along some of the world's giant rivers, e.g. the Mississippi in the USA.

◄ Mudflats and salt marshes on the River Humber, near its estuary. Note the artificial levée in the top right of the photo.

Because the estuary is tidal, submerged by the sea twice a day, **salt marsh** plants have to be able to stand salt water as well as fresh. Salt marshes are valuable for wildlife; migrating birds use them to shelter during stormy weather, and the mud is rich in shellfish and worms. But salt marshes are under threat from ports and industry.

Conclusion – spot the changes!

Two kinds of changes occur during the upper, middle and lower courses of a river.

- Changes in the **long profile** of the river. As the first diagram shows, the long profile is the way that the gradient of the river changes from its upper to lower course. Put simply, it's steep in upland areas, and gentle in the lowlands.
- Changes in the **cross profile**, or valley shape. Put simply, the valley is V-shaped in the upper course and almost flat by the lower course.

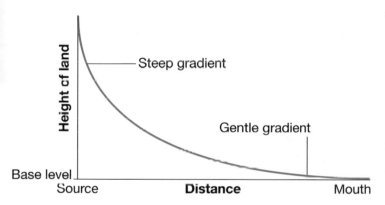

▲ The long profile of a river – showing how river gradient changes as the river moves from its upper to its lower course.

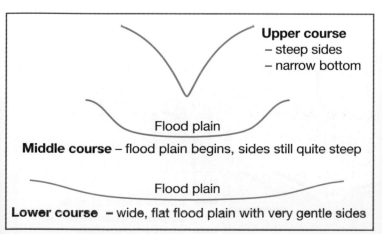

▲ The cross profile of a river valley – showing how valley shape changes as the river moves from its upper to its lower course.

Upper course	Mid course	Lower course	
			River discharge
			Channel width
			Channel depth
			Velocity
			Sediment load volume
Sediment particle size			
Channel bed roughness			
Slope angle			

▲ The Bradshaw model summarises the changes to river characteristics from source to mouth down the long profile. Five features increase downstream, and three decrease. See Question 1 below.

your questions

1 Study the Bradshaw model above. Copy and complete the table to show what happens to each feature of a river as it moves from the upper to lower courses, and give reasons for these.

Feature	Increase or decrease downstream?	Reasons for the change
River discharge		
Channel width		
Channel depth		
Velocity		
Sediment load volume		
Sediment particle size		
Channel bed roughness		
Slope angle		

2 Describe what natural levees are, and explain how they form.

3 Exam-style question

  a Describe the difference between velocity and discharge (2 marks)

  b Explain how channel shape and characteristics change along a river long profile (6 marks)

Why does flooding occur?

+ In this section, you'll learn about storm hydrographs, and how they can help to explain the Sheffield floods.

Stage 1 – Precipitation and runoff

When it rains, very little falls directly into rivers; most falls elsewhere. As the diagram below shows, leaves and branches of plants trap a lot of the rain that falls. This is known as the **interception zone**. The amount intercepted depends on the vegetation, and also the season. For example, deciduous plants (those which lose their leaves in winter) intercept more rain in summer, when they're in leaf.

Some intercepted water is **evaporated** into the atmosphere. The rest drips from leaves to the soil and soaks in – this is called **infiltration**.

Eventually, the soil becomes **saturated**, and cannot take any more. Any extra rain flows overground – called **surface runoff**. How quickly this happens depends on three factors:

1. How much rain has fallen recently – known as **antecedent rainfall**. If the weather has been wet, the soil may already be saturated.
2. How permeable the soil is. Sandy soils are **permeable** – they absorb water easily, and surface runoff rarely occurs. Clay soils are more closely packed, so are **impermeable** – water can't infiltrate easily.
3. How heavily the rain falls. Heavy storms cause rapid surface runoff; the greater the runoff, the greater the flood risk.

Stage 2 – Throughflow and groundwater flow

Once water enters the soil:

- some is taken up by plants and **transpired** through leaves into the atmosphere
- some seeps into the river through soil air spaces – known as **throughflow**
- some continues into solid rock, and saturates it. The upper limit of saturated rock is known as the **water table**. From here, water seeps slowly towards the river as **groundwater flow**, which keeps a river flowing even when there is no rain.

▼ A cross-section of a river valley to show how water moves down valley sides to a river.

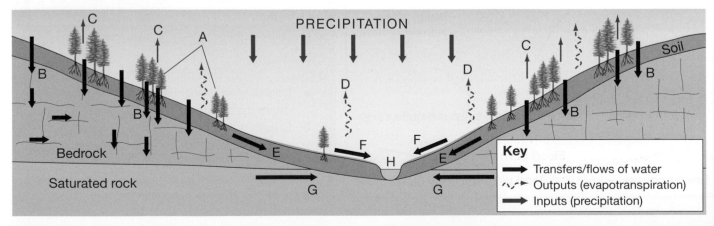

Key
→ Transfers/flows of water
〰️ Outputs (evapotranspiration)
➡️ Inputs (precipitation)

Understanding storm hydrographs

A **storm hydrograph** (see the graph) shows how a river reacts to a rainfall event. Rainfall takes time to reach the river, because most falls on the valley sides and flows to the river as throughflow and surface runoff. Once rainfall reaches the river, the discharge rises from the baseflow level (the rising limb) and reaches peak discharge. Discharge then falls back (the falling limb) as the water flows out of the river system.

> + A **storm hydrograph** is a graph which shows how a river changes as a result of rainfall. It shows two things: rainfall and discharge.

Hydrograph shape

- **Hydrograph 'A'** (see graph) reacts to the rainfall very quickly. This happens in urban areas where impermeable surfaces like roads, roofs and car parks allow lots of surface runoff and very little infiltration. Drains force the water into rivers very rapidly so river levels rise quickly, and flooding is a risk.
- **Hydrograph 'B'** is the shape we would expect in a natural landscape – perhaps a forested area with permeable rocks like sandstone. This increases interception and infiltration so water reaches the river more slowly. The lag time is longer and peak discharge lower.

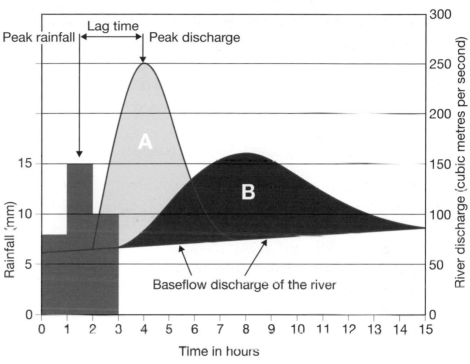

Natural factors influence the shape of the hydrograph. These include: the size and shape of the river basin, temperature, type of precipitation (e.g. snowfall), and geology. However, human activity can also affect the shape of the hydrograph:

- Urbanisation replaces fields and forests, with buildings and roads. This increases run-off, allowing the rainfall to reach the rivers more quickly, reducing the lag time.
- Deforestation would also change the hydrograph shape from 'B' to more like 'A', by reducing interception and infiltration. Afforestation (planting trees) would have the opposite effect.

your questions

1 Read the text, then match the letters on the diagram of the valley opposite with these terms: groundwater flow, evaporation, infiltration, interception zone, river channel, surface runoff, throughflow, transpiration.

2 Compare the lag times and peak discharges of hydrographs A and B on the diagram above.

3 Explain how previous rainfall, vegetation and soil / rock type could affect the shape of the storm hydrograph.

4 Exam-style question
   a Describe what is meant by 'lag time' on a storm hydrograph (1 mark)
   b Using examples, explain how human activity can increase flood risk (6 marks)

✚ In this section, you'll learn how sudden floods affected Sheffield in the summer of 2007.

A June to remember

In June and July 2007, several periods of extreme rainfall gave rise to widespread flooding in parts of England and Wales. It was the wettest May – July period since 1766. Nationally, 49 000 households and 7000 businesses were flooded. Major transport links, schools, power and water supplies were disrupted. South Yorkshire suffered record-breaking floods.

What caused the Sheffield floods?

1 Prolonged rain

Sheffield's flood story began when extreme wet weather hit the north of England. Heavy and prolonged rain fell across South Yorkshire because of a depression over northern England.

- On 15 June, 90 mm of rain fell over Sheffield; more than one month's normal rainfall!
- This was exceeded by even heavier rain on 25 June, when almost 100 mm fell in just 24 hours – the most rain Sheffield had ever had in one day.

With almost 190 mm in those two days, June was the wettest month in Yorkshire since 1882. The result was a hydrograph shape exactly like 'Hydrograph A' on page 99.

2 Soil saturation

After the rains of 15 and 16 June, the soil was **saturated**, causing localised flooding. The ground had not dried out by 25 June, when more extreme rain overwhelmed the rivers and drains in Sheffield. All the extra rain just ran straight off into rivers as **surface runoff**.

Two swept to death in Sheffield

Hundreds of families in South Yorkshire have been moved to safety amid severe flooding, and a man and a teenage boy have been drowned in Sheffield.

About 900 people in Sheffield are using emergency shelters, and about 700 have left villages near Rotherham, fearing the nearby Ulley dam could collapse.

Police closed the M1 completely between junctions 32 to 34, because of the risks posed by the dam, and rail companies announced cancellations to services.

The village of Catcliffe near Sheffield was under water.

London Fire Brigade sent two 'high volume' pumps to help with the flooding problems.

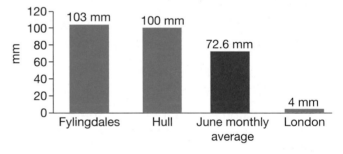

▲ Rainfall in Yorkshire on 25 June, compared to a) June's average monthly rainfall total and b) the London area.

▼ Club Mill Bridge in Sheffield was washed away.

3 The confluence of several rivers

The map shows where the worst Sheffield flooding occurred, in the Hillsborough area near the football ground. This was because of a combination of circumstances.

- The River Rivelin joins the River Loxley at confluence X. The volume of water increased hugely at this point, and flooding became a risk straight away.
- Only a short distance away, confluence Y occurs – this time with the River Don. At this stage, the Don was already almost up to its banks.
- A 'backlog' of water therefore occurred along both rivers, causing water to back up and overflow the banks.

4 The physical landscape

Sheffield lies at the foot of the Pennines, at the point where three rivers meet. Several storage reservoirs at the heads of these rivers provide water for Sheffield, and normally help to store water in rainy periods. This time, however, they filled rapidly and overflowed. The slopes near Sheffield are steep (see below), so water runs off rapidly during wet spells, quickly filling the rivers. The heavy rains raised river levels, and soon blocked drains in parts of the city.

▼ The slopes of the Don valley on the outskirts of Sheffield.

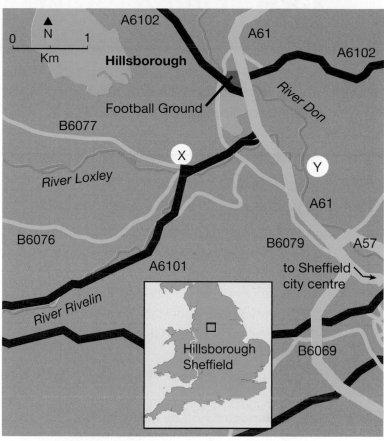

▲ The area around Hillsborough, north-west Sheffield, where some of the worst flooding occurred. Also see section 6.7.

your questions

1 Make a copy of 'Hydrograph A' on page 99. Label it with details about; **a** the rainfall, and **b** the rivers — to show why flooding occurred in Sheffield in June 2007.

2 **a** Copy the diagram below. On it, label how each of the factors shown helped to produce the floods in Sheffield.

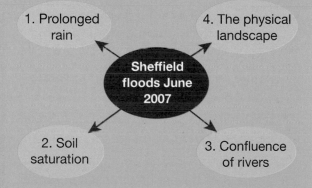

2 **b** Now add other points to show how human factors helped to cause the flood event.

3 **Exam-style question** Using an example, explain the factors that contributed to a flood event on a named river. (8 marks)

✚ In this section, you'll learn about the impacts of Sheffield's floods in 2007.

The impacts of the Sheffield floods

Most of Sheffield's flooding was caused by drains, river channels and flood defences being overwhelmed by extreme flows of water (see section 6.6). Sudden downpours of rain made everything happen so quickly, and made it difficult to predict where flooding might occur.

- Fallen trees along the rivers Sheaf, Loxley and Don quickly blocked river channels.
- As surface runoff raced down Sheffield's hills, drains were blocked and overflowed.

A Hillsborough and north-west Sheffield

- Hillsborough Football Stadium (on the right) was flooded up to 8 metres deep. It cost several million pounds to repair the damage.
- Of 300 homes on one estate, 128 (43%) were flooded. All council tenants returned home within 9 months; some owner-occupiers had to wait much longer as builders were so inundated with work.
- Displaced people suffered stress. Some families were moved into caravans for the winter.
- There were severe health risks from raw sewage escaping into the floodwaters.

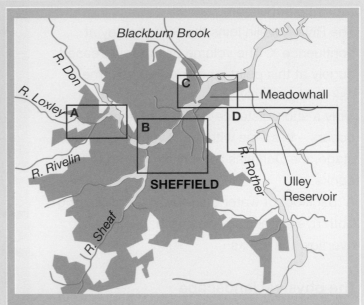

The floods had a devastating impact on people living and working in Sheffield.

- Two people drowned in the floodwaters.
- Over 1200 homes were flooded across the city, and more than 1000 businesses were affected.
- Roads were damaged and a bridge collapsed, blocking roads and affecting travel for several days.
- 13 000 people were without power for two days.

B The city centre and the River Don flood plain

In Sheffield city centre, on 25 June, motorists abandoned their vehicles as roads flooded and traffic gridlocked.

- As the floodwaters rose, many people were caught unaware and had to be evacuated from flooded buildings.
- So many trains and buses were cancelled that many people were unable to get home. Some were trapped overnight; over 900 people spent the night in their offices.
- 20 people were airlifted to safety from one building, a further 3 from the roof of another.
- 200 people were stranded on the first floor of a Royal Mail distribution centre.

C The Lower Don Valley

- The deepest flooding occurred in areas near the River Don, where some industries were badly affected.
- The tool-making company Clarkson Osborn suffered £15 million in flood damage, with equal damage to Sheffield Forgemasters International and Cadbury Trebor Bassett.
- Meadowhall Shopping Centre was flooded (see below), causing millions of pounds' worth of stock losses and damage, and closing the centre for a week. Some shops were closed for three months. Meadowhall's flood defences, built to withstand a '1 in 100-years' flood event, were overwhelmed.

D The Ulley Reservoir area

On 26 June, there were fears that the Ulley Dam might collapse, following the rains and damage to its structure (see above).

- 700 residents were evacuated from the villages of Whiston, Canklow, Catcliffe and Treeton. Some were allowed back 2 days later; others were away for 2 weeks.
- 100 people took shelter at Dinnington Comprehensive School, where the Salvation Army provided food, clothing and bedding.
- About 100 people had to be rehoused for up to a year where flood damage was severe.
- The M1 motorway was closed for two days between junctions 32 and 36 because of the risk of the dam bursting.

your questions

1 Classify the effects of the flooding in Sheffield in a copy of the table below. Use different colour pens to write in the effects for the four areas of Sheffield, A–D.

	Short-term	Medium-term	Long-term
Social			
Economic			
Environmental			

2 Which were the greatest effects – social, economic or environmental? Explain your answer.

3 **Exam-style question** Using named examples, examine the impacts of flooding on people and the environment (8 marks)

In this section, you'll learn how Sheffield has attempted to manage flooding using hard engineering.

Since 2007, Sheffield has debated what it can do to avoid flooding again. Engineers have to choose between 'hard' and 'soft' solutions.

- 'Hard' solutions are structures built to defend areas from floodwater.
- 'Soft' solutions adapt to flood risks, and allow natural processes to deal with rainwater.

Hard engineering in Sheffield

Drains and culverts

Sheffield's drains and culverts (which carry away rainwater) were designed to deal with rainfall amounts that might only occur once in 30 years. But the 2007 floods were a 1 in 400 year event, and people wonder whether anything could have protected them. The drains had two problems.

- They could not cope with all the rain, so streets flooded.
- Where drains met the rivers, water could not escape as the rivers were already so high.

The River Sheaf

The River Sheaf has an enlarged concrete-lined channel where it joins the River Don in central Sheffield. It aims to speed up the flow of water away from the city. In June 2007, it worked, and there was no flooding in this part of Sheffield.

Meadowhall shopping centre

Meadowhall has its own flood defences, constructed when the centre was built, consisting of an embankment around it to hold off water. The photo on the right shows that the bank simply wasn't high enough to contain all the floodwater!

Sheffield City Council denies that the city's drainage system was to blame for the extensive flooding. The council says the rain was so severe that no amount of planning could have prevented it. They said: 'The rain was the worst for 35 years. The rivers, drains and streams were already full, due to heavy rainfall for many days, and couldn't cope. Most floods were caused by rivers overflowing. Pipes and drains simply weren't big enough to cope with the deluge.

▲ Adapted from a newspaper extract, 19 June 2007.

What do you think?

+ The UK's Environment Agency believes that:
- hard defences are not the solution
- buildings at risk must be flood-proofed (e.g. on raised sites or protected by walls), or companies relocated to safe locations
- councils should increase maintenance – streams and drains block easily with debris or tree branches. Are they right?

▼ Flood defences at Meadowhall shopping centre just weren't high enough.

Flood storage reservoirs

In Rother Valley Country Park, east of Sheffield, lakes have been shaped from old quarries along the River Rother to take floodwater. They hold it until it is released into the River Don a few miles away. In 2007, they prevented flooding in this area. The lakes are now also used for boating within the park.

How effective are Sheffield's flood defences?

The table below shows some hard engineering methods that could be considered for Sheffield. But how effective are they?

▲ Creating a river diversion in Rotherham.

▼ Some hard engineering methods.

Method	How it works	How effective is it?
Build flood banks	Raise the banks of a river to increase its capacity.	These are fairly cheap, one-off costs. However, they disperse water quickly and increase flood risk downstream.
Increase the size of the river channel	Dredge the river to increase its capacity, or line it with concrete to speed up the flow of water.	Lining with concrete is expensive, and dredging needs to be done every year. Speeding up the water can increase the flood risk downstream.
Divert the river away from the city centre	Create a diversion for excess water to avoid flooding the city centre.	This was done in 2008 in Rotherham, 10km away (see above photo). It costs £14 million for a 1km stretch. It protects the city-centre but could cause flooding elsewhere.
Increase the size of the drains	Dig up every major road into Sheffield and enlarge the major drains.	Gets runoff away from the city, but it is only as good as the capacity of the river to take all the water from the drains.
Increase the maintenance budget	Clear rivers, drains and sewers to remove debris or vegetation.	This needs to be done every year, so it can be very costly.

What do you think?
+ Should cities just build bigger drains?

your questions

1 In pairs, copy the table below to show the costs and benefits of 'hard' flood protection. Award up to 5 points for each benefit, and up to 5 minus points for each cost. Add up the total for each method. Which is best?

Method of 'hard' protection	Costs	Benefits
1 Build flood banks		
2		
3		
4		

2 **Exam-style question**
a Describe what is meant by 'hard engineering' solutions to flooding. (2 marks)
b Explain the costs and benefits of using hard engineering to reduce flood risk (6 marks)

In this section you will learn about flood management through soft engineering.

Is soft engineering the solution?

After 2007, the Environment Agency considered the situation in Sheffield.

- Increased building in Sheffield since the 1980s had increased surface run-off, but nothing more was spent on flood protection.
- The reservoirs in the upper Don valley outside the city were originally designed to store water for Sheffield – it was an added advantage that, by storing water, they helped to prevent flooding. However, in 2007, they were so full that they were of no use when heavy rains fell.

The UK's Environment Agency now believes that hard defences are not the solution. They cost a great deal and can rarely be made big enough to cope with the largest floods. Instead they suggest:

- upstream, upland areas should be planted with trees to reduce surface run-off
- buildings at risk would be better protected using flood-proofing (e.g. protecting with walls, or building on raised land)
- planning permission should not be given for building near rivers
- companies should relocate if flood proofing is not possible.

So far, Sheffield has almost no 'soft' methods, but the Environment Agency believes that these would make flood management more sustainable. They would also help to produce a river's response more like 'Hydrograph B' on page 99 every time it rained.

Method	How it works	How effective is it?
Flood abatement	Change land – use upstream, e.g. by planting trees.	This delays the passage of water into rivers by increasing interception and transpiration.
Flood proofing	Design new buildings or alter existing ones to reduce flood risk.	Only affects new buildings; can be expensive to alter existing buildings.
Flood plain zoning	Refuse planning permission where flood risk is high.	It phases out development in risk areas, or restricts permission to certain uses, e.g. leisure centres.
Flood prediction and warning	The Environment Agency monitors rivers (see photo), and uses forecasts from the Met Office.	They provide accurate predictions help to reduce flooddamage and evacuate people.

▲ Some soft engineering methods.

▼ The Environment Agency monitors rivers, to help them predict possible floods.

Sustainable management along the River Skerne

The Environment Agency is responsible for the environment around rivers as well as the rivers themselves. In Darlington, the River Skerne has recently been improved to create an amenity for local people AND to prevent flooding.

In the 19th century, its meanders were straightened to allow industries to build on the flood plain. This decreased its length by 13% which increased the flood risk. Therefore the flood plain was raised with industrial waste. This made the river environment poor, especially when industries closed and left derelict land.

So could it be restored, without increasing the flood risk in Darlington? Meanders could not be restored to their former length because pipes containing gas, electricity and sewage followed the new course.

In the end, the Environment Agency restored 2km of river, creating a riverside park as an outdoor space.
- Some meanders were rebuilt, lengthening the river and slowing water down.
- Banks were lowered to make the river flood the park instead of Darlington.
- The flood plain was lowered to increase its ability to store floodwater.

Has it worked?

The River Skerne did flood in June 2007, but rainfall that year was exceptional. Despite this the damage caused was still less than in the floods of 2000, before the work was done. 50 homes were flooded, all due to the backlog of rainwater entering the city's drains.

Another result has been large increases in species with a 30% increase in birds and insects, within one year. Locals also liked the changes – in a survey, 82% of people liked it either 'mostly' or 'strongly'.

your questions

1 a Make a copy of Hydrograph B on page 99. Label it to show how managing the River Skerne should produce hydrographs like this in future.
  b Why is Hydrograph B on page 99 more desirable than Hydrograph A?
2 a In pairs, copy the table below to show the costs and benefits of 'soft' flood protection. Award up to 5 points for each benefit, and up to 5 minus points for each cost. Add up the total for each method.
  b Compare the totals above with those for hard engineering (Question 1, page 105). Which strategies give i the greatest costs? ii the greatest benefits?
3 Exam-style question
  a What is meant by 'soft engineering' solutions to flooding (2 marks)
  b Using examples from named rivers, explain why 'soft engineering' solutions are increasing used to manage flooding (8 marks)

Method of 'soft' protection	Costs	Benefits
1 Flood abatement		
2		
3		
4		

7.1 Bad news

+ In this section you will learn how and why the oceans are threatened with destruction.

Global patterns

In 2008, scientists completed the environmental damage map below. It shows the combined impact of 17 types of human activities on 20 types of ocean ecosystems.

The map shows that many of the world's oceans were experiencing serious environmental damage in 2008, with only a small percentage in good condition. Enclosed seas in densely populated, industrialised areas (North Sea, Mediterranean, South and East China Seas) are most at risk. Polar areas are, at present, the least damaged — but for how long? As global warming melts polar ice, especially in the Arctic, future exploitation could threaten even these ocean environments.

The increasing number of **dead zones** is very worrying. In dead zones, ecosystems have collapsed completely. The biggest one, in the Baltic Sea, is half the size of the UK.

Think about this:

- Our supplies of seafood (fish, crabs, lobsters, shellfish) could die out by 2050 unless we farm them ourselves.
- Every year, hundreds of dolphins are killed by deep sea trawling.
- Vast areas of coral reefs and mangrove forests are at risk.
- The depletion of individual species could upset **food chains** and destroy entire ocean **ecosystems**. This is especially likely near crowded coastal areas, where nearly 60% of people live.

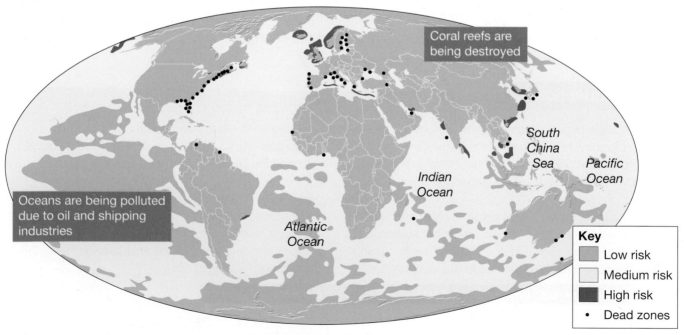

▲ Environmental damage map (2008): the distribution of ocean habitats at risk, and the location of dead zones.

The value of coral reefs

Coral reefs are the 'rainforests of the oceans', because of their high levels of **biodiversity** (biological richness of species and habitat).

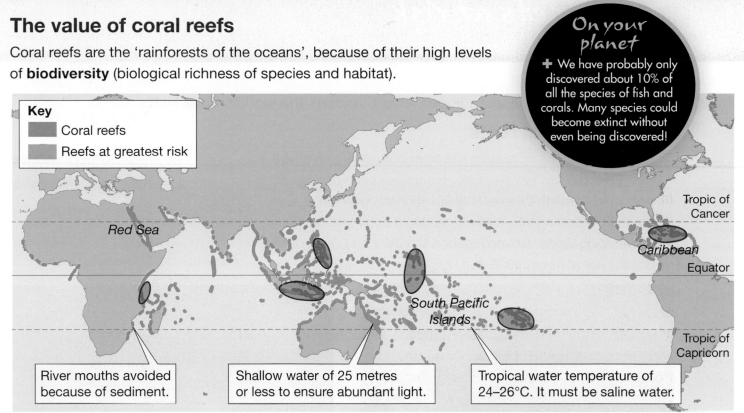

Key
- Coral reefs
- Reefs at greatest risk

Red Sea

Tropic of Cancer

Caribbean

Equator

South Pacific Islands

Tropic of Capricorn

River mouths avoided because of sediment.

Shallow water of 25 metres or less to ensure abundant light.

Tropical water temperature of 24–26°C. It must be saline water.

▲ The distribution of the world's coral reefs.

Coral reefs are of great value, as this table shows.

Exploitation for fish	4000 species of reef-living fish (25% of all known marine fish) provide food for local communities. 25% of the world's total commercial fish catch comes from coral reefs.
Shoreline protection	Reefs provide shoreline protection from storms, tsunami and wave erosion. Reefs can grow with rising sea levels and protect against the impacts of climate change.
Aquarium trade	Reefs supply tropical fish, sea horses, and 'plants' for the aquarium trade.
Tourism	Reefs are a magnet for the world's tourists. Many countries in the Caribbean receive over half of their income from reef tourism.
Education and research	Reefs can be visited easily to learn about marine life.
Other uses	As a medicine source (some drugs originate from reef organisms). To make decorative objects, such as jewellery. A source of lime, for cement and building.

Trends in coral reefs:

In the Indian and Pacific Oceans, living coral cover on reefs fell from 48% in 1980 to 27% in 1990. It has been fairly stable since, but only 4% of reefs now have coral covering 50% or more of the reef, compared with over 60% of reefs in 1980.

In 2008 it was estimated that 75% of coral reefs were threatened by human activity. This is an increase from 58% in 1998. Overfishing is the biggest threat with 55% of all coral reefs affected. In southeast Asia, the region with the most reefs, 95% are threatened.

your questions

1 Make a list of the different uses of the coral reefs from the table above. In groups, rank these in order of importance. Be prepared to justify your ranking to the class.

2 Why are enclosed seas in densely populated areas the most at risk?

3 In pairs, find 3-4 pictures that show the value of the coral reef. Produce an A3 colour poster that illustrates the reef's value.

4 **Exam-style question** Describe and explain the distribution of coral reefs. (6 marks)

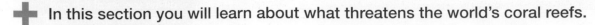

➕ In this section you will learn about what threatens the world's coral reefs.

Threats to coral reefs

Coral reefs are under increasing risk from both local and global threats. Over the last few decades, reefs have been damaged by population growth in coastal areas, tourism, overfishing and river pollution flowing into the oceans. Most coral reefs are in the developing world. Isolated reefs in the Java and South China seas were not at risk in 1998, but now are.

Coral reefs under threat		
	1998	**2008**
Indian Ocean	56%	81%
Pacific Ocean	40%	67%

▲ A coral reef can support a fantastic range of life.

▲ Coral bleached due to global warming is unable to sustain life.

Global factors

- Global warming will devastate reefs as ocean temperatures rise. An increase of only 1-2°C will cause coral bleaching.
- Warming, caused by **El Niño** events, has the same effect as global warming, and can affect huge areas of ocean.
- Ocean acidification will prevent corals growing. As more carbon dioxide is dissolved in the oceans the water's pH is expected to fall from its natural 8.2 to 7.9 by 2050

+ El Niño events occur every 3-7 years in the Pacific Ocean and have been blamed for causing coral bleaching. Winds and ocean currents reverse for 6-18 months. This has a drastic impact on the weather, but also warms up the ocean which bleaches coral. It is possible that El Niño events could become more common due to global warming.

Local factors

- Fishing—blast fishing (using dynamite), cyanide fishing (see photo opposite), and trawling, all damage reefs.
- Coral mining—sand, rubble and lime are used for building materials.
- Biological threats include diseases such as black band coral disease, unwelcome predators such as Parrot fish that graze the algae on coral reefs, and crown of thorns starfish (COTS) which eat the coral.
- Pollution from sewage, oil spills, and toxic industrial chemicals, all damage reefs.
- Hurricanes can produce huge waves, which break up reefs, and heavy rain, which washes pollution from land into the sea.
- Siltation - reefs can become smothered with a blanket of fine silt (see page 113); this is eroded from farmland and washes into oceans via rivers. Silt reduces light levels in the sea and prevents corals feeding and plants growing.

Tourism

Tourism benefits the economy and gives people jobs. But the diagram below shows that tourism can have a wide range of direct and indirect impacts on the very reefs the tourists come to see.

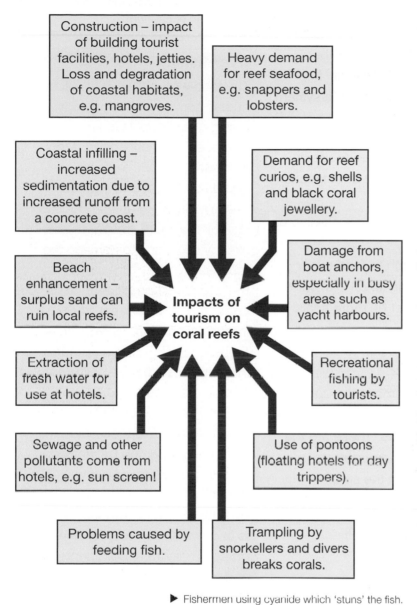

Construction – impact of building tourist facilities, hotels, jetties. Loss and degradation of coastal habitats, e.g. mangroves.

Heavy demand for reef seafood, e.g. snappers and lobsters.

Coastal infilling – increased sedimentation due to increased runoff from a concrete coast.

Demand for reef curios, e.g. shells and black coral jewellery.

Beach enhancement – surplus sand can ruin local reefs.

Damage from boat anchors, especially in busy areas such as yacht harbours.

Impacts of tourism on coral reefs

Extraction of fresh water for use at hotels.

Recreational fishing by tourists.

Sewage and other pollutants come from hotels, e.g. sun screen!

Use of pontoons (floating hotels for day trippers).

Problems caused by feeding fish.

Trampling by snorkellers and divers breaks corals.

▶ Fishermen using cyanide which 'stuns' the fish. They can then be sold for use in aquariums.

Fishing

Fishing has damaged reef ecosystems for a number of reasons:

- Overfishing of reef species, especially to meet huge demand in the Far East, is widespread.
- People living on the coast near reefs are often very poor. They need fish as a source of protein. They fish at a subsistence level using spears and traps, but some use blast fishing with dynamite. This damages the coral and kills many species – not just the fish being hunted.
- Many breeding grounds for reef fish are being destroyed.

your questions

1 'The real threats to the oceans are activities that happen on land'. How far to you agree with this statement?
2 Research a coral reef tourist destination such as the Great Barrier Reef. Use a table to consider the **economic**, **social**, and **environmental**, costs and benefits of tourism.

3 Explain why fishing is such a threat to ocean ecosystems.
4 **Exam-style question** Using named examples, explain how tourism can damage marine ecosystems. (6 marks)

In this section you'll explore physical processes in marine ecosystems and the threats of overfishing and pollution.

Marine ecosystems are the same as any other – they consist of plants and animals living in co-existence with each other. However, this is a dynamic relationship; any increase or decrease in species number has an impact on the whole system.

Marine food webs

In any ecosystem, there is a balance. One species feeds on – and in most cases is fed on – by another. Their populations depend on the balance.

- In the Arctic food web below, the polar bears are carnivores and feed on species such as seal.
- Each polar bear needs 3-4 seals on which to feed every summer, so that they can hibernate in the long Polar winter. Therefore, for them to survive, there must be more seals than polar bears.
- Seals feed on fish, so there must be more fish to survive, and so on.
- Smaller fish feed on microscopic organisms in the upper sunlit layer of the ocean – known as phyto-plankton – which rely on the sun's energy for photosynthesis. These are the **primary producers**. Everything else is a **consumer**.
- Everything therefore relies on energy from the sun.

The relationships between these species is called a **food web**, and the links between each one form a **food chain**.

> **+ food web** – the links between animals and plants feeding on each other in an ecosystem

Over-fishing

The food web is always at risk from human activity and technology. Rising global population and increased prosperity has led to increased demand for fish. On-board fish processing and fast-freeze technology enable ships to take ever bigger catches. This is already beginning to damage the food web.

Krill are tiny shrimplike animals, under increasing threat from intensive 'suction' harvesting – a method which gathers up huge quantities of the tiny creatures. They are a keystone species, feeding on algae and plankton at the bottom of the food chain. In turn, they are eaten by whales, penguins and fish. However, the krill population has declined by 80% since the 1980s. Krill is in growing demand for use in omega 3 health supplements, and as food in fish farms. Until intensive harvesting by humans began, numbers of krill had been increasing. Many whale species eat vast amounts of krill. However, when whale numbers decreased due to overhunting by whalers, krill numbers rose. It's a fine balance.

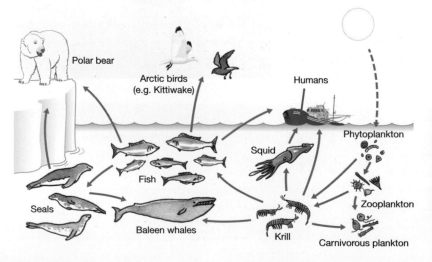

Polar bear
Arctic birds (e.g. Kittiwake)
Humans
Squid
Phytoplankton
Fish
Zooplankton
Seals
Baleen whales
Krill
Carnivorous plankton

◀ A marine food web from the northern Atlantic and Arctic oceans

Nutrient cycles

As well as food, ecosystems also rely on other systems. The whole ecosystem relies on the movement from one species to another of nutrients – known as the **nutrient cycle**. This works simply, though the chemistry is more complex. It's based on nitrogen. When fish consume algae or plant matter, they take in nitrates. As they excrete, bacteria converts waste into ammonia (NH_4) and into nitrates. Plants and algae absorb nitrates, and complete the cycle. Again, it's a balance – as the diagram shows.

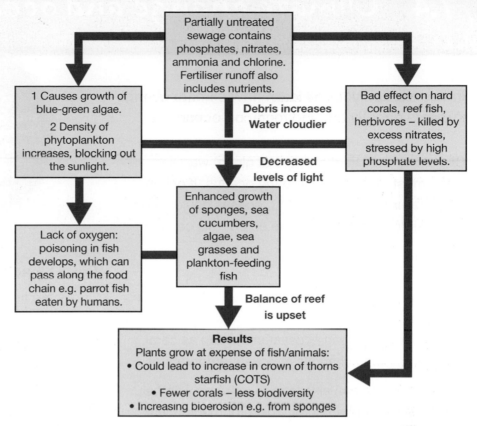

▲ The impacts of eutrophication

Eutrophication and siltation

One of the world's biggest threats is pollution. It leads to the creation of dead zones, ocean areas which are critically low in oxygen, and cannot support life. Once rare, they are now common in many areas under threat from development of industry and tourism.

- Nutrient cycles can be disrupted by chemical pollution, especially from nitrate fertilisers (see diagram). Nutrient overloading is called **eutrophication**, and it's happening on a huge scale.

- **Siltation** – the increased cloudiness of water by sediment – is also a threat. Siltation occurs in areas where deforestation (see page 99) causes run-off of sediment into rivers and then into the sea. It clouds sea water, which prevents sunlight reaching as far into the ocean. The impacts on the food web can be disastrous.

your questions

1 Explain the difference between 'food web' and 'nutrient cycle'.

2 Draw a food chain to show the species which link the sun's energy and polar bears.

3 Using the diagram of the food web, explain:

  a why there needs to be more primary producers than consumers

  b why there are more micro-organisms than fish

4 Explain the possible impacts on the marine ecosystem of:

  a increased melting of Arctic ice

  b increased over-fishing

5 **Exam-style question** Explain the possible effects of siltation on the marine food web shown in the diagram. (6 marks)

+ In this section you will learn how climate change impacts directly and indirectly on oceans.

Climate change, in the shape of global warming, is the most serious threat to the oceans. If ocean temperatures rise by only 3°C, oceans could suffer irreversible destruction by the end of this century.

Direct impacts

Direct impacts of climate change relate to increases in temperature. These will affect oceans as much as the land. Many ocean ecosystems, such as coral, are vulnerable to changes in ocean temperatures.

Climate change can also lead to extreme weather, such as storms and floods. These damage ocean ecosystems by increasing pollution and siltation.

As temperatures on land rise, glaciers and frozen land will melt. This will increase the amount of freshwater reaching the oceans – making the water less salty and less dense. This could affect the ocean currents that distribute heat. The result is that ocean temperatures could actually decrease in some places, and increase in others.

Indirect impacts

These relate to rises in sea level. Warmer water temperatures cause the oceans to expand (thermal expansion). Melting glaciers and ice sheets also add to the volume of water in the oceans. In the 20th century, the oceans rose by an average of 15 cm. By the end of the 21st century, this rise could be between 20 cm and 1 metre. Rising sea-levels could affect coastal mangrove swamps and salt marshes. As sea levels rise, coastal erosion could increase and erode these ecosystems. Some tidal areas will be permanently flooded, destroying the ecosystems which exist there today.

On your planet

+ Oceans act as a **carbon sink** by storing carbon dioxide. However, increasing windiness is stirring up the oceans and carbon dioxide is being released back into the atmosphere at a greater rate than it is being absorbed.

Coral bleaching

Corals are an animal, called a polyp, which lives in a shell which it secretes (together forming a coral reef). They feed on algae within the ocean, and their waste encourages other algae to survive within the coral shell. It's those algae which give a coral reef its colour and produce oxygen on which the polyps rely. But it all depends on a temperature balance – the sea must be cooler than 30°C. If ocean temperatures rise by 1-2°C, the algae cannot survive, and so to the coral polyps die. The loss of colour means they turn white, known as coral bleaching (see photos on page 110).

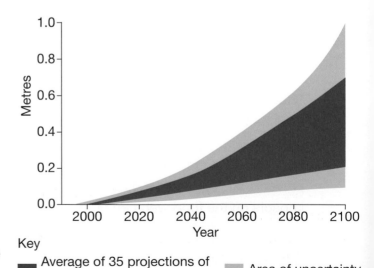

Key

■ Average of 35 projections of greenhouse gas emissions

■ Area of uncertainty

▲ The range of predictions for sea-level rise (1990-2100).

Species migration

Warming oceans and seas are causing marine species to migrate. Fish, shellfish, plankton and marine mammals move to colder waters if their home waters become too warm. The North Sea has warmed by about 1°C since 1960. This may not sound much, but the area of sea with an average temperature of 10°C has moved north by 22km every year since the 1960s (see maps). Warmer oceans are already having an impact:

- Cold-water species, like cod in the North Sea, will move north. These will be replaced by warm water species, like anchovies and sardines.
- Diseases, like New England's lobster shell disease, seem to become more common as oceans are warming.
- Jellyfish plagues have been linked to nutrient pollution, but also to warming waters — as jellyfish seem to be able to adapt faster than other species.
- Dangerous species, like the warm water, venomous, 'Portuguese man o' war' jellyfish are now often found in the North Atlantic.
- Alien species, like the Chinese mitten crab, can spread more rapidly as water temperatures rise.

Major changes in the distribution of fish and marine species will affect whole food webs. This is because the balance of predators and prey will change. Some species will find that their prey has disappeared, while others are threatened by new predators.

A useful overall indicator of the health of our oceans is how much fishermen catch. The United Nations estimates 86 million tonnes of marine fish were caught in 2000, compared to 77 million tonnes by 2010. Perhaps this falling catch is telling us something?

> **+ Alien species** are plants and animals which are foreign — as they have been accidentally or deliberately introduced to a new area.

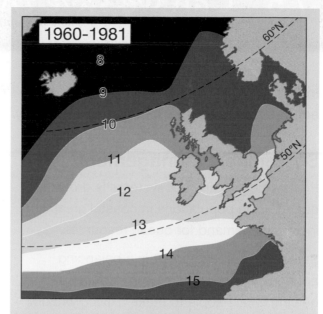

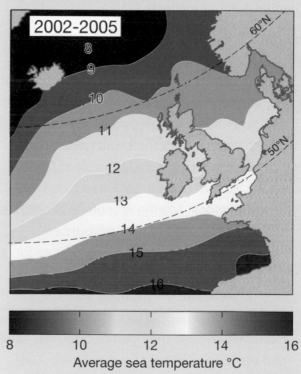

Average sea temperature °C

▲ Average temperatures in the North Atlantic and North Sea in 1960-1981, compared to 2002-2005.

your questions

1 Explain why sea levels are rising globally, and why ocean temperatures are increasing.
2 Describe the impacts of climate change on the oceans.
3 Exam-style question Using named examples, examine the impact of warming oceans on marine species and food webs. (8 marks)

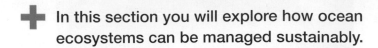

In this section you will explore how ocean ecosystems can be managed sustainably.

Sustainable management

Pressure on marine ecosystems is growing, due to rising populations in coastal areas and increased demand for ocean resources.

Sustainable management is a balancing act between ecosystem conservation, and helping local people to make a living without overharvesting resources.

For ocean ecosystems, this involves:

- Using fishing equipment which doesn't damage delicate habitats, such as coral reefs.
- Harvesting marine resources – fish, shellfish and other marine organisms – at a rate which won't destroy them for future generations. For example, if net size is too small, many young fish are caught, which reduces the breeding stock. Marine resources should be sustainable – unlike **finite resources**.
- Allowing poor people to use resources for subsistence activities. This often conflicts with government or business interests.
- Involving local people (community-based management), so they can decide how fishing and other uses should be managed. Just creating protected areas, which prevent locals from carrying out their livelihoods, is bound to lead to conflicts.

> **+** Resources which will one day run out are called **finite resources**. Examples are oil and coal.

> **+ Community-based management** involves local people in the management of natural resources.

St. Lucia: A sustainable success story?

St Lucia in the Caribbean pioneered the idea of **community-based management** of ecosystems. In 1986, 19 areas were declared Marine Reserve Areas (MRAs). These included coral reefs, turtle breeding grounds and mangroves. Fishing Priority Areas were also created (FPAs) (see the map opposite). However, the boundaries were never fully defined (always difficult in the ocean), so conflicts arose.

Why was protection needed?

- St Lucia is a volcanic island with most of its population concentrated along narrow coastal plains. Land-based damage of the ocean is therefore likely. The population is rising at nearly 2% a year. Densities average 300 per km^2.
- The continental shelf is narrow, which leads to overfishing. Most fishermen are subsistence fishermen who don't have boats for deep sea fishing. Their methods – such as placing pots on coral reefs, or chasing fish into nets by throwing rocks into the water – can be very damaging to the reef.
- The tourist industry has developed very rapidly, leading to problems of waste disposal and pollution. Tourism earnings contribute nearly half of St Lucia's annual earnings.
- More tourists have led to more snorkelling, diving and yachting. Soufrière is a popular destination for reef-based tourism.
- 20% of people live below the poverty line. Many have no jobs. They harvest mangroves for charcoal, hunt wildlife, and catch fish. This puts pressure on local resources.
- Forests are cut down for banana plantations causing heavy siltation, especially during storms.

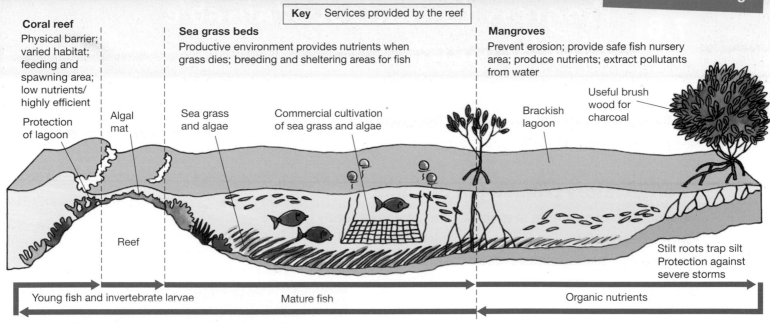

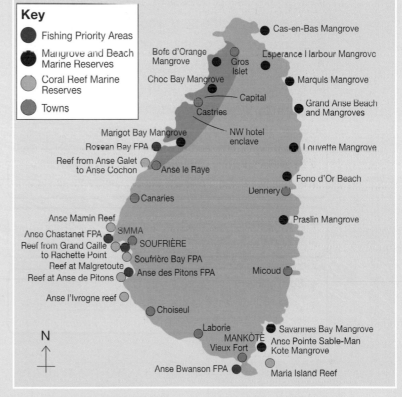

Key   Services provided by the reef

Coral reef
Physical barrier; varied habitat; feeding and spawning area; low nutrients/highly efficient

Protection of lagoon

Algal mat

Reef

Sea grass beds
Productive environment provides nutrients when grass dies; breeding and sheltering areas for fish

Sea grass and algae

Commercial cultivation of sea grass and algae

Mangroves
Prevent erosion; provide safe fish nursery area; produce nutrients; extract pollutants from water

Useful brush wood for charcoal

Brackish lagoon

Stilt roots trap silt
Protection against severe storms

Young fish and invertebrate larvae   Mature fish   Organic nutrients

▲ How the marine ecosystems of St Lucia (coral reefs, mangroves and sea grass beds) provide important services.

Managment conflicts in Soufrière and Mankòtè

In St Lucia, the coral reefs of Soufrière and the mangrove forests of Mankòtè were suffering damage. For both areas, any management plan had to allow for community participation.

In Soufrière, the conflicts were between fishermen, divers, snorkellers and yacht owners. They all had the potential to damage the reef.

In Mankòtè, there was a lot of unregulated hunting and fishing in the 1980s. It had become a site for rubbish disposal and a target for mosquito eradication – this involved spraying with insecticide. The Mankòtè mangrove is St Lucia's largest remaining mangrove forest, so it has ecological importance. It also provides social and economic services for local communities – subsistence and commercial fishing, fuelwood/charcoal, honey and salt.

Key
● Fishing Priority Areas
◕ Mangrove and Beach Marine Reserves
● Coral Reef Marine Reserves
● Towns

Cas-en-Bas Mangrove
Bois d'Orange Mangrove
Gros Islet
Esperance Harbour Mangrove
Choc Bay Mangrove
Capital
Marquis Mangrove
Castries
NW hotel enclave
Grand Anse Beach and Mangroves
Marigot Bay Mangrove
Roseau Bay FPA
Louvette Mangrove
Reef from Anse Galet to Anse Cochon
Anse le Raye
Fond d'Or Beach
Dennery
Canaries
Praslin Mangrove
Anse Mamin Reef
Anse Chastanet FPA   SMMA
SOUFRIÈRE
Reef from Grand Caille to Rachette Point
Soufrière Bay FPA
Reef at Malgretoute
Anse des Pitons FPA
Micoud
Reef at Anse de Pitons
Anse l'Ivrogne reef
Choiseul
N
Laborie
Savannes Bay Mangrove
MANKÒTÈ
Anse Pointe Sable-Man Kòte Mangrove
Vieux Fort
Anse Bwanson FPA
Maria Island Reef

▲ The protected areas of St Lucia.

your questions

1 Explain what is meant by sustainable management of the oceans resources.
2 Describe the conflicts that existed in St Lucia over the uses of the sea and its resources.
3 Referring to the diagram at the top of the page and other information, explain why healthy marine resources are important to the people of St Lucia.
4 **Exam-style question** Using examples, explain how ecosystem conservation can cause conflicts with local communities. (6 marks)

+ In this section you will learn about the stages of participatory planning.

Participatory planning in St Lucia

Participatory planning is essential if all **stakeholders** are to benefit. Both Soufrière and Mankòtè have been successful, to an extent, because they have supported the incomes of local people and involved them in managing resources in the marine reserves.

At Soufrière, fishermen have been provided with modern boats and a refrigerated ice house to improve fish processing. In Mankòtè, land has been set aside as a woodlot for fuel and charcoal (to take pressure off the mangroves). An agricultural plot for growing vegetables provides an extra source of food. In both cases, **ecotourism** has been encouraged as an extra source of employment and income.

> **+** In **participatory planning**, the whole community are involved in the development of a scheme.

> **+** A **stakeholder** is anyone who has an interest in ensuring the success of a project.

> **+** **Ecotourism** is a type of sustainable development. It aims to reduce the negative impacts that tourism can have on the natural environment.

▲ Soufrière Town, St Lucia. Today, Soufrière is more dependent upon tourism than agriculture.

▼ The key stages for successful participatory planning.

Assessment	**Making management decisions**	**Capacity building**	**Finalising institutional arrangements**
• identifying all stakeholders • establishing their rights and needs • looking for options	• zoning • resource use • means of finance	• training stakeholders in new skills carried out by an independent organisation	• management agreements • legal frameworks carried out by an independent organisation

Soufrière Marine Management Area (SMMA)

In 1992, the St Lucia Department of Fisheries (in the Caribbean) brought the following people together:

- the local town council
- local hotel owners
- water-taxi owners
- dive businesses
- fisherman
- marine managers.

They went out in boats and cruised along the coast discussing how coastal zoning could work. The SMMA developed from this trip.

Overall, the SMMA has been successful as a model of sustainability. But there have been problems. It's very difficult to get stakeholders with so many different interests to agree. For example, establishing a marine conservation area, means part of it has to be a no-go area for fishermen. The fishing community became angry seeing divers in the conservation area. They didn't believe that a conservation area would help to conserve future fish stocks.

Local people had to be trained and educated to manage the scheme. The rangers who police the area had to be equiped, and this costs money. Fees from divers and yacht owners now make the scheme self-funding, but some problems remain:

- The area has become so popular that the marine environment is threatened by mass tourism.
- Rapid development in Soufrière encourages siltation and pollution.

The good news is:

- The numbers, sizes and diversity of fish species have increased
- Many stakeholders are now involved in marine conservation.

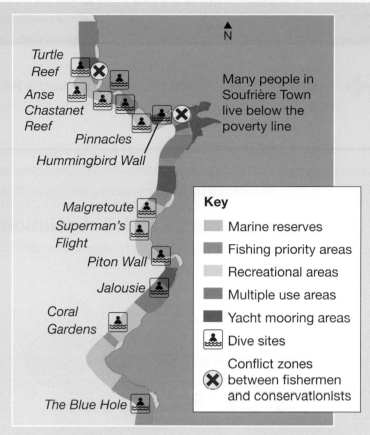

Many people in Soufrière Town live below the poverty line

Key
- Marine reserves
- Fishing priority areas
- Recreational areas
- Multiple use areas
- Yacht mooring areas
- Dive sites
- Conflict zones between fishermen and conservationists

▲ Zoning in the SMMA.

▲ Fishermen at Soufrière hauling in their catch.

your questions

1 In pairs, decide what have been the SMMA's three biggest successes, and why.
2 Why is involving local people essential for schemes like the SMMA to work?
3 **Exam-style question** Using examples, explain how marine areas can be managed sustainably. (6 marks)

➕ In this section you will learn about Marine Protected Areas and also look at management of fish stocks in the North Sea.

Sustainable management needs to take place at local, national, international and global scales. At a global scale, countries can get together to provide legal frameworks for regional and local schemes.

Marine reserves – a global solution?

> ➕ A **Marine Protected Area** (MPA) is an area of sea or coast where marine life is protected from damage or disturbance. In 2008, the UK had only one-fiftieth of 1% of our seas as MPA's.

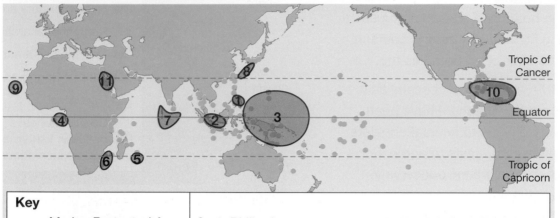

◀ The distribution of Marine Protected Areas

Key

Symbol	Description
•	Marine Protected Area
⬭	Marine hotspots
×	Not protected
?	Partially protected
✓	Good protection

? **1** Philippines
? **2** Sundaland
? **3** Wallacea
? **4** Gulf of Guinea
? **5** Mascanene
✓ **6** Eastern South Africa
× **7** North Indian Ocean
? **8** Southern Japan/Taiwan
× **9** Cape Verde
× **10** Western Caribbean (tourism)
? **11** Red Sea (oil spills)

As you can see from the map, there is no truly global network of sites. Indeed, 35% of countries have no **Marine Protection Areas** at all. The map also only shows the distribution of MPAs – with no indication of their size. Of the 500 marine reserves, most are only 1 km² in size. There are few reserves above 1000 km² – and big is beautiful when it comes to marine conservation. Kiribati in the South Pacific has just created the world's largest reserve – the size of California. In addition, the map tells us nothing about the quality of the reserves. Many are very poorly managed. Another problem is that many of the **marine hotspots** – the areas of highest biodiversity under the greatest threat – aren't currently protected.

How to set up a successful reef reserve

- Make and enforce laws to limit damage to the reefs.
- Clean up pollution from sewage outfalls.
- Provide incentives to stop people polluting.
- Find other jobs for unemployed fishermen.
- Help local fishermen to sell their goods.
- Educate people about reef management.
- Limit numbers of boats and divers using the reef.
- Charge people for using the reef. Use the money to manage the reserve.
- Use the results of scientific research to conserve species.

Managing fish stocks in the North Sea

This regional-scale case study illustrates the need for international cooperation. The 'EU Common Fisheries Policy' has tried to bring back fish stocks from catastrophically low levels.

Every year, the EU reviews its fisheries policy. The fishermen want to fish as large a quota as possible, but marine scientists argue that only 'no fishing' marine reserves will save species like the cod. These are very expensive to set up, because you have to compensate the fishermen for loss of earnings – you have to pay them not to fish! The fishermen also argue that the calculations about fish stocks aren't correct.

A **whole ecosystem approach**, not a single species approach, might save the North Sea. This would include:

- ensuring that the mesh of nets won't catch undersized young fish
- limiting the number of hours and days that fishing boats can operate each year
- quota management – each year a limit is placed on the number of tonnes of fish from various species which can be caught, based on a 'state of stock' survey
- discard management – fewer unwanted fish are discarded (by catch)
- setting up marine reserves which protect all species – some of which could be temporary to protect spawning and nursery grounds
- further research into how fishing affects the whole ecosystem of the North Sea, e.g. a lack of sand eels results in fewer puffins.

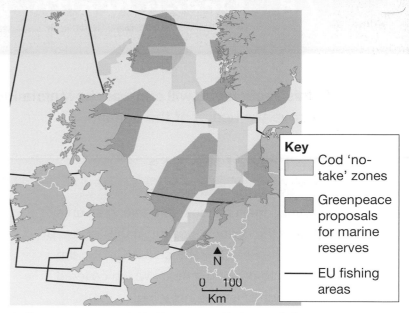

Key

☐ Cod 'no-take' zones

■ Greenpeace proposals for marine reserves

— EU fishing areas

▲ European waters are divided into sectors. Each year a limit (quota) is set on the number of each species that may be caught in each sector.

Global warming: The impact upon cod stocks

It's not just a case of too many cod being caught by efficient factory trawlers. Global warming has also played a part. The cold water species of plankton, which bloomed in the spring at just the right time for baby cod to eat, has been replaced by warm water species. These bloom in the late summer, when baby cod need larger food. This partly explains the drop in numbers of cod surviving to adulthood since 1980. Also, as the North Sea warms, cod will migrate to Arctic areas.

What do you think?
+ Has our cod had its chips?' It's up to us as consumers, too. Should we stop eating cod and change to less threatened species which are guaranteed to be sustainably produced?

your questions

1 Why is fishing a difficult industry to manage?
2 Draw a large spider diagram. Label on it as many problems as you can think of facing the North Sea.

3 In a different colour, write as many solutions as you can think of to solve these problems.
4 **Exam-style question** Using named examples, explain how marine areas are managed at a regional scale. (8 marks)

✚ In this section you will consider how global actions could help to improve the health of the oceans.

The Law of the Sea

The **Law of the Sea** was developed to prevent certain nations from taking an unfair share of the ocean's wealth. At the Third Law of the Sea Conference (UNCLOS), the aim was to develop a treaty which covered a wide range of issues including:

- fisheries
- navigation
- continental shelves
- the deep sea
- scientific research
- pollution of the marine environment.

The treaty was finally ratified in 1994. 40% of the ocean was placed under the law of the adjacent coastal states. These were defined in four zones of increasing sizes.

1 Territorial seas, with a 12-mile limit of sovereign rights
2 Contiguous zones, with a 24-mile limit – controlled for specific purposes.
3 Exclusive Economic Zones, with a 200-mile limit. Rights are guaranteed over economic activity, scientific research and environmental conservation.
4 Continental shelves, where nations may explore and exploit without infringing the legal status of the water and the air above.

The traditional 'freedom of the seas' remains for the other 60% of open oceans. Within this, the deep seabed area is designated 'the common heritage of mankind', and is controlled by the International Seabed Authority.

UNCLOS addresses the main sources of ocean pollution:

- land-based/coastal activities
- continental shelf drilling for gas and oil
- seabed mining
- ocean dumping
- pollution from ships.

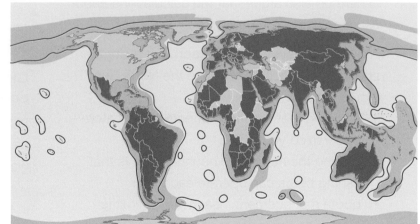

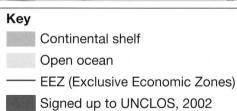

Key
- Continental shelf
- Open ocean
- —— EEZ (Exclusive Economic Zones)
- Signed up to UNCLOS, 2002

▲ The impact of UNCLOS – how the Law of the Sea controls oceans.

New challenges are putting even more pressure on the oceans.
- New technologies allow deep sea exploration.
- Increasingly complex pollutants are reaching the oceans from advanced industrial processes.
- And, of course, global warming has introduced new pressures – such as opening up the Arctic to potential increases in resource exploitation.

Marpol

The **International Convention for the Prevention of Pollution From Ships** (Marpol for short), is a global agreement that came into force in 1983. In 2012 152 countries had signed up to Marpol representing 99% of world shipping. Marpol is a set of standards for the safety and design of ships, as well as a set of rules to stop ships dumping waste at sea. Marpol has been quite successful:

- Large oil spills from ships have fallen from 25 per year in the 1970s to 2.5 per year today.
- Every year in the 1970s about 400,000 tons of oil was spilled in the oceans from ships compared to 50,000 tons today.
- Dumping rubbish and sewage from ships has been reduced too, but this is hard to monitor.

The state of the oceans

Ocean ecosystems are not protected as well as those on land. In 2010:

- Marine Protected Areas covered 1% of the ocean, but the UN target is 10%.
- About 6% of territorial seas and 3% of Exclusive Economic Zones were protected.
- Only 0.01% of oceans are 'no-take zones' for fishermen, but many scientists think 20-30% of oceans should be protected in this way.

Setting up, managing, policing, and monitoring marine protected areas is difficult and costly. But without more protection, the future looks bleak for our oceans and marine life.

▲ There are many similarities between conservation of oceans and land-based conservation, but the challenges are even greater.

On your planet
+ The 'Great Pacific Garbage Patch' is a vast area of floating rubbish in the ocean — possibly 100 million tons of the stuff.

your questions

1 In groups of 2-3, draw a labelled spider diagram to show reasons why the world's oceans are threatened. Feed back in class and decide the 3 or 4 biggest problems that all the groups identified.

2 Take two of these problems. For each make a copy of the table on the right. In it, explain how far the limits shown would help to solve the problem.

3 **Exam-style question** Using named examples, explain how global actions can help maintain ocean health. (8 marks)

Problem: (e.g. pollution)	Whether these will help to solve the problem and why
12 mile exclusion zones	
24 mile exclusion zones	
200 mile exclusion zones	
Continental shelf exclusion zones	

✚ In this section you will learn what polar climates are like and how plants and animals survive there.

North and South

The polar regions are the:

- **Arctic** – north of the Arctic circle (at **latitude** 66°33'N)
- **Antarctic** – south of the Antarctic circle (66°33'S).

Both are regions of intense cold, with few people. These high latitude regions have:

- cold temperatures – due to low levels of solar radiation from the sun (see diagram below)
- the **polar night** – when the sun does not rise at all in mid-winter
- the **midnight sun** – when the sun does not set all in mid-summer.

> ✚ **Latitude** means how far north or south from the equator a place is – measured in degrees. 'High latitude' refers to places more than 60° north or south, in other words close to the poles.

> *On your planet*
> ✚ Many polar areas are as dry as the hottest deserts. Antarctica's dry valleys are the driest places on earth; they receive no precipitation at all each year.

During the northern hemisphere winter:

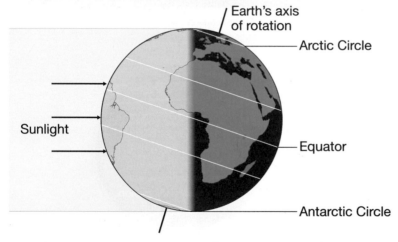

Earth's axis of rotation
Arctic Circle
Equator
Antarctic Circle
Sunlight

Sunlight (solar radiation) is spread over a large area at the poles – compared to at the equator – because of the angle of the earth's surface. During the northern hemisphere winter, the Arctic is tilted away from the sun. Even as the earth rotates, this area remains in darkness (the polar night). At the same time, the Antarctic is always in sunlight (the midnight sun) even when the earth rotates 180°.

▼ The Arctic and Antarctic may have similar climates, but in other ways they are quite different.

	Arctic	Antarctic
Geography	Has some ice-sheet covered areas (such as Greenland), surrounded by barren tundra lands. Partly enclosed by the Arctic Ocean.	Covered in vast ice sheets with almost no ice-free land and few plants. Surrounded by the wild Southern Ocean.
Countries	Eight countries have some lands within the Arctic Circle; Canada, Denmark, Finland, Iceland, Norway, Russia, Sweden, and the USA.	Governed by the Antarctic Treaty since 1961. Countries claiming lands in Antarctica cooperate to manage the region.
People	Home to many indigenous communities, like the Inuit and Sami, who have inhabited the region for thousands of years.	Has no permanent population – and never has had – only scientists who live there temporarily.

Polar climates

Polar locations can have continental or maritime climates. Svalbard is a group of islands owned by Norway in the Barents Sea. It has a mild climate by Arctic standards (see table) because it is surrounded by ocean. Inuvik, in northern Canada, has much lower winter temperatures, but warmer summers – a more continental climate.

> **+** The **tundra** biome has no trees. It is basically a polar grassland with dwarf shrubs, mosses, lichens and flowers.

Month		J	F	M	A	M	J	J	A	S	O	N	D
Climate data for Svalbard, Norway	Average temperature (°C)	-13	-14	-13	-10	-2	3	6	6	1	-5	-7	-10
	Precipitation (mm)	22	28	29	16	13	18	24	30	25	19	22	25
Climate data for Inuvik, Canada	Average temperature (°C)	-29	-29	-24	-14	-1	11	14	11	3	-8	-22	-26
	Precipitation (mm)	16	11	11	13	19	22	34	44	24	30	18	17

Polar regions have low precipitation – less than 50mm a year. It often falls as snow in the long winter due to the sub-zero temperatures. Precipitation is caused by air rising, but as cold, polar air is dense, it sinks towards the ground. This creates high pressure – and therefore a lack of precipitation. Many polar places have a summer of only 4 months. For the other 8 months the average temperature is below 0°C.

▼ Plant adaptations to polar extremes.

Arctic flora and fauna

The Arctic is an extremely fragile environment – it would take little to threaten it.

Flora – The Artic has boreal (coniferous) forests, and tundra vegetation. Polar species are vulnerable to global warming because:

• Warmer temperatures alter flowering times, migrations and hibernation.
• Animals rely on floating sea ice to move about, but this is melting.
• Arctic tundra could be squeezed out as coniferous forests move north.

Fauna – Arctic animal adaptations include:

• Migrating into the region in spring, but leaving before winter.
• Thick, white fur – for camouflage in snow.
• Dens dug into snow or caves – digging into frozen permafrost is impossible.
• Thick blubber to act as insulation.

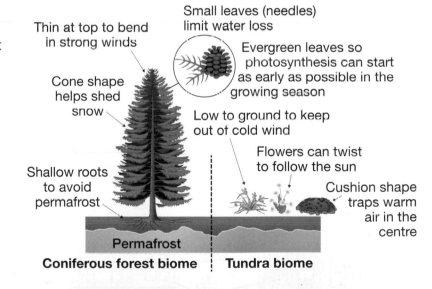

Thin at top to bend in strong winds

Small leaves (needles) limit water loss

Evergreen leaves so photosynthesis can start as early as possible in the growing season

Cone shape helps shed snow

Low to ground to keep out of cold wind

Shallow roots to avoid permafrost

Flowers can twist to follow the sun

Cushion shape traps warm air in the centre

Permafrost

Coniferous forest biome | **Tundra biome**

your questions

1 Use the data in the table to draw a climate graph for Svalbard and Inuvik; label the similarities and differences between the two places.

2 **Exam-style question** Explain how polar flora and fauna have adapted to the extreme climate. (4 marks)

✚ In this section you will learn about the unique lifestyle of polar people.

Adapting to extremes

The Arctic has a permanent population, spread across several countries. There are two main types of people:

- **Indigenous groups**, who still live traditionally in some places although many have moved into towns.
- **Immigrants**, who have moved into the region to work in mining, oil and gas drilling and other industries.

Arctic people face a number of difficulties because they live in such an extreme place:

- Extreme cold for months on end.
- Darkness and isolation in the winter.
- **Permafrost**, which is a constant hazard.

> ✚ **Permafrost** is permanently frozen ground. In polar areas, the permafrost can be 10s of metres thick. Permafrost areas are often waterlogged because water cannot drain through the frozen ground.
>
> ✚ **Active layer** is a thin layer of ground, close to the surface, which freezes in the winter, but melts in the summer.

A house in the Arctic which has been built on top of the permafrost. It has started to lean as the permafrost has been melting.

▼ Special building techniques are needed to prevent permafrost melting.

Built on a gravel pad, so heat from the house does not melt the permafrost

Water, sewage and power are in a utilidor box above ground; repair access is easier – permafrost is not melted

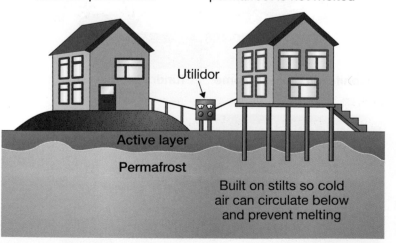

Utilidor

Active layer

Permafrost

Built on stilts so cold air can circulate below and prevent melting

Polar people have to adapt to the problems in a number of ways.

Building styles

Triple glazing and very thick insulation is needed to keep heat in, and cold out. Roofs need to be sloping to shed heavy snow (see diagram). Traditionally houses were either temporary igloos or mobile animal skins.

Clothing

Traditional Inuit clothing is made from animal skins sewn together with sinew; today modern hi-tech insulated clothing has replaced this. Gloves, hats, and multiple layers are essential to protect against frost-bite and hypothermia.

Transport

Driving is easier in winter, when the ground is frozen solid. In summer it is boggy and unstable. Skis, traditional dog-sleds and modern snowmobiles are often used. 'Ice roads' cross frozen lakes, rivers and even the sea, allowing heavy goods to be transported on trucks in the winter.

Energy use

Oil and gas are the main energy sources, because renewable energy such as solar and hydro-electric power (HEP) is unsuited to polar areas. Energy demand for heating and transport is high.

Farming and food

Arable farming is not possible in polar areas, but reindeer herding is traditional among the Sami people of Arctic Europe. The traditional Inuit diet (40% protein and 50% fat) included whale, seal, polar bear and caribou, as well as berries and seaweed. Most food has to be imported these days, and is expensive.

▲ A reindeer herder in the Russian tundra.

Unique cultures

Many indigenous Arctic people have become 'westernised'. However, it is common for them to keep old traditions alive. There are good reasons for valuing their unique culture and way of life:

- The **Inuit** have immense knowledge of Arctic animals and the environment, therefore, they often know before anyone else if the Arctic environment is changing for the worse.
- The **Nenets** people of northern Russia have an animist religion - meaning that rocks, trees and animals, as well as people, have souls and spirits. This respect for the natural environment is something the rest of the world could learn from.
- The **Aleut** people of Alaska and Russia are skilled in using every part of a seal or whale – even the bones are used for making needles or weapons.

Many western people who migrate to the Arctic to work suffer from Seasonal Affective Disorder (SAD) and depression because the cold, dark, long winters are so hard to cope with.

your questions

1 Explain why permafrost makes building in the polar environments a challenge.
2 Why should we value the culture of indigenous people in the Arctic?
3 **Exam-style question** Using named examples, explain how people have adapted to life in extreme polar environments. (6 marks)

➕ In this section you will learn what desert climates are like in Australia.

Next stop Cook!

Twice a week, the Indian-Pacific train trundles through the Nullarbor Plain on its way across Australia from Perth to Sydney – and twice back again. The Nullarbor covers an area the size of the UK! And it's just as extreme as Antarctica in terms of challenging people's survival skills. Summer temperatures are above 40 °C, rain is rare, and soils are thin and infertile.

The train is full of tourists – the journey takes 3 days, so it's hardly a commuter route! But for some outback residents, it's a lifeline for on-line shopping and supplies. When the train stops at Cook to refuel, it brings supplies for the two people living there … and others who've driven over 100 km to this, their nearest settlement.

▼ Cook, in South Australia. Once it was a town, with a secondary boarding school for 'local' farm children (up to 500 km away), a swimming pool and shops. Isolation and a harsh environment forced people away.

The 'back of Bourke'

Bourke is a small town in the outback of New South Wales. Australia's outback – as its semi-deserts are known – is huge! Fly to Sydney from Asia and you're over it for 4-5 hours – making it as big as the Atlantic! The outback is one of the world's most barren and least populated places. Mostly it's desert or semi-desert, with scattered cattle farms (or 'stations'). Mining towns are linked by a few tarmac-covered roads and dirt tracks. Australians rarely travel there, referring to its remoteness as the 'back of Bourke'.

After Antarctica, Australia is the world's driest continent; more than one-third is desert. There is some rainfall, but most evaporates quickly in the heat. The deserts vary in appearance and vegetation, so they are given different names, as shown below.

▼ This shows the extent of Australia's deserts and semi-deserts. The desert isn't true desert like the Sahara, but it's extremely barren.

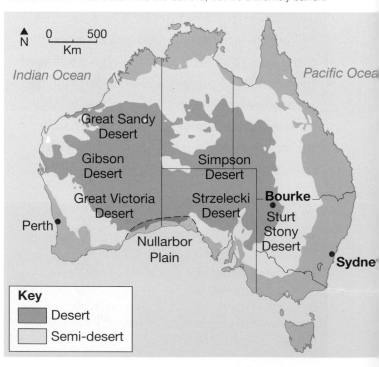

Key
Desert
Semi-desert

Why is most of Australia desert?

Normally, rain-bearing winds blow across the Pacific towards Australia. The mountains that border the coast – the Great Dividing Range – cause this air to rise and cool rapidly. This leads to condensation, and then rain (see right).

As the air descends from the mountains, it is drier – and a 'rain shadow' is created. This results in low rainfall in western areas. The further west the winds blow, the drier they are – so the driest areas are in Western Australia.

Hot arid areas have desert or semi-desert climates. In a true desert, rainfall is less than 250 mm a year and temperatures are often over 40°C. Most of the world's hot deserts occur in bands 15-30° north and south of the equator. Semi-arid (or semi-desert) regions are found around the fringes of deserts. These areas get about 500-700mm of precipitation per year.
Most of this falls in a short rainy season, and the rest of the year is dry.

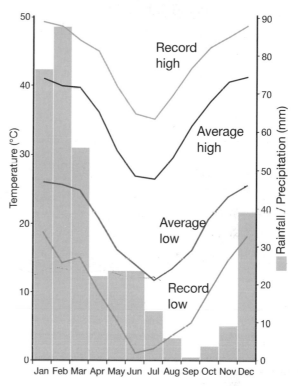

▲ This graph shows rainfall, and average/record temperatures for Marble Bar in north-western Australia.

> ## On your planet
> ✚ Australia's highest temperature was recorded at Oodnadatta in South Australia – 50.7°C on 2 January 1960!

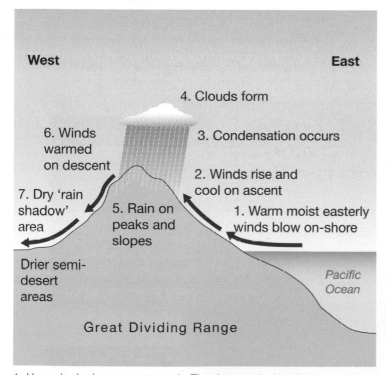

▲ How rain shadow areas are made. They turn much of central and western Australia into desert.

your questions

1 In pairs, discuss and write down the benefits and problems of living in places like Cook for: **a** young children, **b** families, **c** the elderly.

2 Place these phrases in order, to show why Australia's deserts get so little rain:
clouds form, condensation, cooler temperatures, deserts form, little moisture left, on-shore winds from the Pacific, rain falls, winds descend, winds forced to rise.

3 **Exam-style question** Describe the climate of a hot arid area you have studied. (4 marks)

+ **In this section you will learn how plants (flora) and animals (fauna) survive Australia's desert climate.**

Australia's **biodiversity** (its range of plants and animals) is unique. It has over 1 million species – many of which are found nowhere else, because of Australia's geographical isolation. The unique plants and animals have to be able to cope with the arid (dry) climates of Australia's deserts.

Desert flora

The three main ways in which desert plants have adapted to arid climates are:

- succulence
- drought tolerance
- drought avoidance.

▼ Australia's desert is actually semi-desert, as shown here. Plants have to be capable of resisting extremes of temperature and drought. Often they survive near streams.

Succulence

Australia has over 400 **succulent** species. They store water in fleshy leaves, stems or roots. Desert rains are infrequent, light, short-lived and evaporate quickly, so water must be captured and stored.

- To survive, succulents can very quickly absorb large amounts of water through extensive, shallow root systems. They can store this water for long periods.
- Their stems and leaves also have waxy cuticles (surface layers) which make them almost waterproof when their stomata close.
- Their metabolism slows down during drought and their stomata remain closed, so that water loss almost stops. Growth, therefore, stops during drought.
- However, their water stores make them attractive to thirsty animals, so most have spines or are toxic; some are camouflaged.

Australian eucalyptus trees, like those in the background, have very deep roots, and also grow close to river courses that hold water for long periods. Their leaves have few stomata, to minimise water loss.

Most desert plants resume growth within a few hours of rain.

Drought tolerance

Drought tolerance means having mechanisms that help to survive drought.

- During drought, plants of this type shed leaves to prevent water loss through transpiration. They become dormant. Others, like eucalypts, remain evergreen but have waxy leaves with few stomata, to minimise water loss.
- These plants have extensive, deep roots which penetrate soil and rock to get at underground water.
- They photosynthesise with low leaf moisture levels, which would be fatal to most plants.

Drought avoidance

Most drought avoiders are annuals – they survive one season, have a rapid life cycle, and die after seeding.

- Their seeds last for years, and germinate only when soil moisture is high.
- Some germinate during autumn, after rain and before winter cold sets in. The seedlings survive winter frost and flower in spring.

Desert fauna

Australia's deserts are not the world's driest, but rain is unpredictable. Several years can pass between showers. Many animals settle near small watercourses – known as billabongs. Desert animals have had to evolve to survive (see right).

your questions

1 Define *succulence, drought tolerance* and *drought avoidance.*
2 Draw a spider diagram to show the different ways in which different animals, birds and plants survive drought.
3 **Exam-style question** Using examples from hot arid areas, explain how flora and fauna have adapted to the extreme climate. (6 marks)

The Bilby

A small marsupial (mammal with a pouch), bilbies once lived throughout Australia's deserts. Having been hunted by domestic cats, dogs and wild foxes (all imported to Australia), few survive.

How it survives
- It is nocturnal, sheltering from the daytime heat to avoid dehydration.
- It burrows for moister, cooler conditions.
- It has low moisture needs, obtaining enough from its food, such as bulbs, fungi and insects.

The Perentie

This is a giant lizard. They can grow up to 2.5 metres in length, and weigh up to 15 kg. It's one of the desert's top predators.

How it survives
- To escape the desert heat, it digs burrows or hides in deep rock crevices. It emerges from these to hunt.
- It hibernates from May to August to avoid cold.
- Like the bilby, it has low moisture needs.

The Red Kangaroo

The Red Kangaroo, the world's largest marsupial, is one of many kangaroo species.

How it survives
- It survives by hopping (a fast, energy-efficient form of travel) to find food in the sparsely vegetated desert.
- It feeds at dawn and dusk when the air is cooler, and sleeps during the heat of the day.
- Dew is an important part of its water intake.
- Rain triggers a hormonal response in females, so that breeding only occurs during rains.

✚ **In this section you will learn how people cope with living in extreme hot arid conditions.**

If you're a white-skinned European, there's one rule for the midday Australian sun – stay out of it. Early European explorers walked across Australia's interior and few survived. They failed to use the lessons of Australia's own aboriginal peoples – to live sustainably with the environment.

However, white settlers do live in Australia's most extreme environments. The problem is making a living. The soils are poor, with little organic material (which retains moisture) and few nutrients. Poor soils mean that plants have low nutrient content. And many plants resist being eaten by animals because they have thorns or toxic sap. If water is available, there is just about enough grass to feed cattle or sheep – but only in very low numbers. So outback farms are huge – some the size of Wales.

Managing supplies of water

With little rain, farmers have two water sources:
- Most farms have dams and reservoirs to store water which cattle and sheep can drink.
- Farms also use **boreholes** to tap into underground **artesian water**. Rain soaks into the desert soils and **percolates** (trickles) down into the bedrock. Over many years, water gradually collects. If you drill a borehole, it comes up – either under natural pressure or by using **windpumps** (see right). This water can be used for domestic supply or for animals.

The system of obtaining and storing water by boreholes and reservoirs is fragile. Many people question now whether water and the landscape are being used sustainably.
- Recent droughts have put pressure on the landscape. Animals still graze, but the grass starts to die in the arid conditions. As the roots die, nothing binds the soil together – so it's eroded by windstorms.
- Underground water is also being over-used, beyond the amount of annual rainfall which recharges it. Water tables are falling each year.

◀ A windpump in the outback – these draw water from underground boreholes, which lead to aquifers of artesian water (see below). In front of the windpump is a reservoir for storing water for sheep.

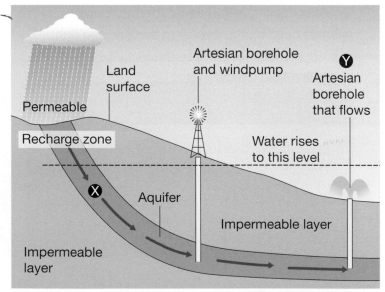

▶ Artesian water – how it's formed and extracted.

Adapting to a hot arid climate

Building styles

Traditional desert houses have very thick walls to keep summer heat out, and winter warmth in. Flat roofs make a good outdoor sleeping area – as pitched roofs are not needed in low rainfall areas. Buildings in Coober Pedy, in South Australia, (see right) are underground to keep cool. Verandas are a good way of keeping the baking sun away from walls and windows.

Energy use

Keeping cool in hot arid areas is essential. Modern buildings have air conditioning, but this uses a lot of energy. Luckily solar panels (photovoltaics) work very well in hot, sunny climates. Desert areas often have strong winds, making wind turbines a good option too.

Transport

Crossing deserts can be a struggle, even for modern vehicles. Traditionally camels were used because they can go days without water. People travelled in groups with camels in a 'desert caravan'. Modern vehicles are usually 4x4 as getting stuck in sand is common. Light aircraft and even trains, such as the Tran-Australian Railway that crosses the Nullabor Plain, are more reliable than cars.

Clothing

Clothing in hot arid areas has traditionally been head to foot, loose fitting and light coloured. This prevents sunburn, reduces water loss by sweating and reflects the sun's heat. Head coverings – like the Tuareg, tagelmust and Arab, keffiyeh – keep the head cool, and keep sand and dust out.

your questions

1 What makes it so difficult to survive in the Australian outback?

2 Copy the diagram showing artesian water. Explain:
   a how rainwater gets underground at point X
   b why water comes up without any need for a pump at point Y.

3 In pairs, list reasons why farming in the outback would be difficult: **a** economically, **b** socially, and **c** environmentally.

4 **Exam-style question** Using examples from hot arid areas, explain how people have adapted to extreme environments. (6 marks)

✚ In this section you will learn about Aboriginal culture in Australia and the value of desert cultures.

Each afternoon, an aboriginal man takes his seat on the ground in Circular Quay in Sydney. It's the place where you can catch ferries for tours around Sydney's harbour. He plays a didgeridoo, a traditional instrument – but he plays to an electronic dance rhythm mixed with sounds of the outback, such as kookaburras. His creative mix earns huge applause and he sells copies of his CD. He probably wonders how many customers have been to the outback, or met aboriginal people.

He is one of the few aboriginals who white Australians are likely to meet. Aboriginal people remain hidden, both within certain parts of cities and also in remote outback camps. White Australians are more likely to watch TV programmes about aboriginal life instead.

Healthy eating, aboriginal style

The traditional aboriginal diet varies, depending on the area of Australia they are in. Australia has a growing 'native food' industry, based on traditional aboriginal knowledge of what's edible in the outback. Several plants have multiple uses as food, medicine, utensils, tools, musical instruments or weapons.

The aboriginal population:

- has only two-thirds of the life expectancy of white Australians – 52 years instead of 78
- has Australia's worst drug and alcohol abuse; and homelessness is also a problem
- sees its traditional lifestyle disappearing – there is a risk that aboriginal customs, knowledge and beliefs will disappear unless action is taken to preserve them.

Fruits
- Bush tomatoes taste of tomato and caramel, and are used for chutney or as a seasoning for meat.
- Desert limes have a strong citrus flavour and are used in jams and sauces.
- Quandongs (pictured) are also called 'native peaches'. They are bright red berries, high in Vitamin C, which taste a bit like apricots.
- Bush bananas are vines with long fruit that tastes like green peas and avocado.

Seeds
- Wattle (acacia) seeds from wattle trees, like the one on the right, are used in biscuits, in drinks or as dressings.
- Sandalwood nuts are eaten.

Grubs
- Witchetty grubs are the larvae of moths and beetles, which are eaten raw or cooked. They taste like scrambled eggs and peanut butter with a crispy 'chicken skin' coating. (see left)

Meat
- Traditional wild animals, such as kangaroo, crocodile or emu.

Aboriginal beliefs and lifestyle

Aboriginal beliefs focus on the land. They see it as sacred and something to be protected. We think of aboriginal people as being nomadic, but in fact their tribal groups followed strict paths according to the seasons.

Traditionally, aboriginal people survived by **hunting and gathering** – finding edible plants and animals.

- They used fire to drive out animals for hunting, to clear wood and allow grass to grow. As a result, fire-tolerant plants (eucalyptus trees) came to dominate the landscape. These re-grow quickly after fire. The seeds of some species actually need fire to burst and germinate.
- Aboriginal crafts were based on hunting (boomerangs) or music and tribal celebrations (didgeridoos).
- Their customs and stories were spoken, never written. It's only now that some stories are being written down.
- As younger generations of Aboriginals are moving away to live in towns and cities, stories about care of the land are being lost.

Valuable insights

Traditional cultures in desert and semi-desert regions are very valuable. Many are mobile and nomadic, such as the Bedouins and Tuareg, in the Sahara (Africa), and Australia's traditional-Aboriginal people.

- They live with few possessions – because of their mobile lifestyle – and waste nothing.
- They have learnt to conserve water, firewood and other scarce desert resources.
- Traditional desert houses have superb natural insulation which helps them to stay cool during the day and warm at night, without the need for heating or air-conditioning.
- Desert peoples have deep knowledge of the medical properties of plants, which could be useful to everyone.

Like people who live in polar areas, they too, are likely to notice the impact of global warming and other environmental problems before the rest of us.

On your planet

+ Aboriginal Australians generally don't believe in owning land – they see themselves as caretakers.

your questions

1 List as many ways as you can to show how well aboriginal people have adapted to Australia's extreme environment.

2 In pairs make two lists **i** ways in which the aboriginal lifestyle is unique, **ii** ways in which understanding this lifestyle is valuable to us as a human race.

3 Research 'aboriginal food' on the internet. Create a pamphlet or poster about how aboriginal people use their local resources for food and medicine.

4 **Exam-style question** Using examples from polar and hot arid areas, explain why traditional cultures are unique and valuable. (8 marks)

✚ In this section you will learn about the threats facing traditional cultures and natural systems, in extreme environments.

Tourism

Extreme environments – both desert and polar area – are surprisingly popular with tourists.

- The Australian desert outback site of Uluru (also known as Ayres Rock) is sacred to the aboriginal, Anangu, people – a popular tourist destination.
- In the Arctic tourists visit to watch whales and view glaciers, often from cruise ships.

However, extreme environments are often fragile – only able to cope with small numbers of tourists:

- Aboriginal cave paintings at Uluru could easily be damaged by tourists.
- Desert and polar plants can be trampled, and wildlife breeding disrupted.
- Arctic coastal settlements can be swamped by visitors from large cruise ships.
- Traditional cultures have often had only limited contact with outsiders before tourists arrive.

There is a danger of **cultural dilution** as the Aboriginal and Inuit people 'put on a show' for tourists. They could produce souvenirs to suit tourists' tastes, or dress up in traditional costumes. This might be a good way to earn money, but over time their beliefs and values could be lost.

▼ Several kilometres from the rock itself, the view from the car parks at sunrise or sunset is startling. Coach tours offer tourists the 'champagne at sunset experience'.

On your planet

✚ Australia's oldest aboriginal settlements are 20 000 years old – they were already 15 000 years old when the Egyptian pyramids were built!

> ✚ **Cultural dilution** means that traditional beliefs, dress, art, music and family relationships are degraded by contact with outsiders, such as tourists.

Year	Visitors to Ularu
1961	5000
1984	100 000
1987	180 000
2000	380 000
2005	400 000

▲ Visitors to Uluru: 60% of tourists come from overseas.

Tourism has been managed at Uluru. Today visitors are asked not to climb the rock, because it is against Anangu spiritual beliefs. The Aboriginal Cultural Centre educates visitors about aboriginal history and walks. Bush tucker guides are led by aboriginal peoples themselves. Over 30 aboriginal people are employed within Uluru-Kata Tjuta National Park.

Resources and pollution

Extreme environments are often exploited for their natural resources.

- The Super Pit gold mine in Kalgoorlie, Australia, is so large it can be seen from space.
- Iron and manganese ore are mined in the Pilbara region of Australia. This is increasing the demand for water and housing for miners to live in.
- Oil and gas are already extracted from the Arctic – especially Alaska and Siberia.
- The Mary River iron ore mine, on Baffin Island, Canada, is being expanded to mine 18-30 million tons per year.

Mining and drilling provide many jobs, but there are risks to the environment:

- Mine waste, and spoil heaps, can scar the landscape.
- Mining uses a lot of water, a scare resource in arid places, and the water ends up polluted.
- The Trans-Alaskan oil pipeline in the Arctic has leaked 4 times since opening in 1997.
- In 1989 the Exxon Valdez oil tanker spilled 11 million gallons of oil into Prince William Sound in Alaska, which killed 250,000 seabirds and 3,000 sea otters.

Leaving the land

Extreme environments have limited economic opportunities. Although jobs can be found in mining and tourism, mining areas experience 'boom and bust', depending upon the price of oil, gold and other resources. In 2011-2012 the number of visitors to Uluru dropped by 19% - not good news for tourism jobs. Droughts in Australia have reduced farming jobs, causing **out-migration**.

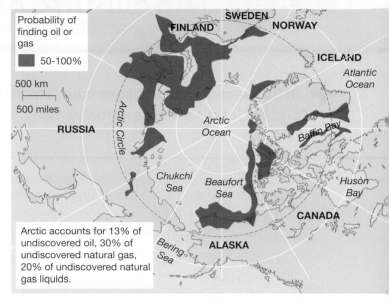

▲ Oil and gas fields in the Arctic. Estimates suggest 90 billion barrels of oil could be in the Arctic.

Arctic accounts for 13% of undiscovered oil, 30% of undiscovered natural gas, 20% of undiscovered natural gas liquids.

Despite differences between hot arid, and polar, regions, there are common trends:

- Mining and oil drilling areas have growing populations, whereas farming areas and the settlements of indigenous people, have declining populations
- In both the Arctic, and the Australian outback, there are few large towns and cities. Young people and indigenous people tend to be leaving, attracted to the cities by jobs and education opportunities.

+ **Out-migration** is when people leave an area and move to live somewhere else. Young people often leave to look for a job in cities.

your questions

1 Draw a line graph to show the increase in tourist numbers at Uluru since 1961. Why do you think the numbers have increased so much?
2 Explain why the traditional cultures of people in extreme environments are interesting for tourists.
3 Explain why mining and oil drilling are a threat to the environment in extreme areas like the Australian outback and the Arctic.
4 What types of people are mostly likely to migrate out of extreme environments and why do they leave?
5 **Exam-style question** Using examples, examine the threats facing culture and the environment in extreme environments. (8 marks)

In this section you will learn how climate change threatens polar and hot, arid extreme environments.

Australia's deserts

Global temperatures are increasing. Temperatures in the outback are predicted to be 1.4-5.8°C higher by 2100. This will make the outback a more difficult place to live:

- Droughts will become more frequent, reducing water supply for people and farmers.
- Evaporation will increase, especially in the north and east of Australia, reducing water supply even more.
- Bushfires will become more common in the drier and hotter climate.

The Australian Bureau of Meteorology predicts that rainfall will decrease in south and southwest Australia, expanding the desert 100-200 km further south.

There is some evidence that parts of Australia are already getting drier (see graph). Australians call the period between 2001 and 2010 'the Big Dry'. This was a long period of drought that cost the government £3 billion and stretched water supply to the limit. Could this become more common if climate changes in the future?

On your planet

+ The area of Arctic summer sea ice lost between 1979 and 2012 is about the size of the European Union.

+ **Land degradation** means a drop in the quality of land, especially the soil. Degraded land is not as good for growing crops and grazing animals.

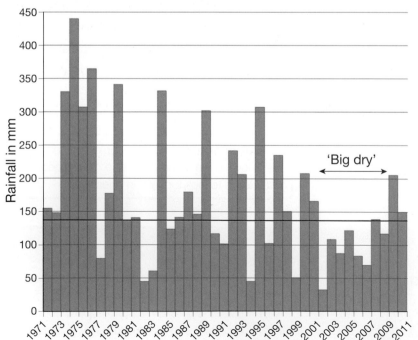

▲ Rainfall trend 1971-2011 for Marree Station in South Australia. Average rainfall is 142 mm per year. The lowest recorded was 32 mm in 2002.

Impacts on the land and local economy

Over 60% of Australia's farmland suffers from **land degradation**. **Soil erosion** – caused by the removal of too much vegetation – and **overgrazing**, by sheep and cattle, is a major cause of land degradation. During a drought the erosion gets worse. This is because there are too many farm animals grazing the parched land, so even more grass is eaten. During the 'Big Dry' many farmers abandoned their land because farming was impossible. This led to more out-migration in search of jobs elsewhere.

Warming Arctic

In the Arctic, global warming has already taken hold. The Arctic Ocean is covered with floating sea ice. This expands in winter, to cover a huge area, and then melts back in the summer. In September 2012 the sea ice shrank to the smallest area ever recorded (see graph). Other impacts of the changing climate upon the Arctic include:

- A rise in temperatures by 1-2°C since the 1960s. Elsewhere on earth they have only risen by 0.5-1°C.
- Large areas of permafrost have melted making the ground unstable and waterlogged.
- Winters have become less cold, with mild spells happening when it would usually be well below freezing.
- In summer 2012, 97% of the surface of the Greenland ice sheet was melting, greater than at any time in the last 30 years.

Climate change in the Arctic is already changing people's lifestyles.

These changes are likely to continue, having both positive and negative impacts (see table).

Impacts on ecosystems

As the sea ice retreats, polar bears will find it harder to hunt, and may starve. As a result, seal numbers could rise, which in turn could mean more fish are eaten. The whole food web of the Arctic might alter. Other changes are already happening:

- Beetle infestations are increasing in coniferous forests, as warmer winters allow beetle larvae to survive.
- Caribou grazing lands are shrinking as forests spread over the tundra; caribou herds are an important food source.
- The timings of many animal migrations have changed by up to 15 days for some species in the last 20 years.

Widespread and permanent changes to permafrost, the timing of seasons, sea ice cover and ecosystems will make life very hard for traditional Inuit and Sami communities.

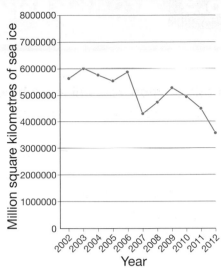

▲ Arctic summer sea ice cover 2002-2012.

Impacts on the local economy

Indigenous people	Traditional ice fishing and hunting will decline as sea ice retreats.
	Moving about will be harder as permafrost melts and ground remains unfrozen and waterlogged in winter.
	As migration times change, and breeding areas shift, people will have difficulty hunting.
Oil, gas and mineral companies	Oil exploration will become easier, increasing the risk of spills and other pollution.
	As ice retreats on land, new areas are exposed that can be mined, bringing more jobs.
Shipping companies	The Arctic Ocean will be open to shipping, cutting journey times between Europe and Asia.
	This will increase the risk of oil and waste being dumped from ships.
Tourism	The tourist season will be longer, so opportunities to make money increases.
	The already fragile environment could be swamped by visitors
Fishing	As seas in the Arctic warm, cod could be replaced by shrimp as the main species so fishing practices will need to change.

your questions

1 Describe the rainfall trend for Marree Station since 1971 and comment on how this might have affected people in the area.

2 Explain why the reduction in Arctic sea ice is both good and bad news for different people in the Arctic.

3 **Exam-style question** Using named examples, examine how climate change threatens people who live in extreme environments. (8 marks)

In this section you will learn how people can take local actions to help them adapt to climate change.

Climate change in Africa

Africa is already 0.5 °C warmer than a century ago, and temperatures have risen even more inland. Droughts are common. Areas that were already arid are now drier. Africa is also less able to deal with climate change than other areas of the world. Of its 54 countries, two-thirds are among the world's poorest 50 countries. These are least able to cope with change.

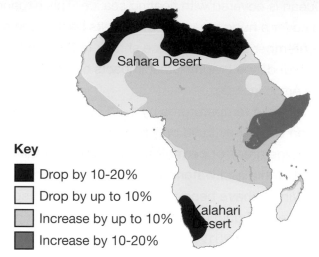

Key

■ Drop by 10-20%
□ Drop by up to 10%
▨ Increase by up to 10%
▨ Increase by 10-20%

▲ How rainfall in Africa is predicted to change by 2030. Rainy seasons are likely to be more unreliable. There will be less rainfall in some areas, and more in others.

Desertification in the Sahel

The Sahel has been badly affected by climate change. It lies on the southern edge of the Sahara Desert, in a belt extending 3000 km east to west and 700 km north to south. It's a transition zone between the dry Sahara to the north, and wetter grasslands further south. It depends on the monsoon rainy season for its rainfall, but this varies hugely from year to year (see below).

In semi-arid areas like the Sahel, land can be degraded by desertification (turning into a desert). As the population has grown, farmers have grazed too many animals on the grasslands. This exposes the soil to erosion during the rainy season. Any forest and scrub is cut down for fuel wood. In order to get water for crops, boreholes and wells have been dug to tap groundwater. This water evaporates quickly in the heat leaving salt deposits on the soil (salinisation) and poisoning it. The barren dry soil is vulnerable to wind erosion. In many areas the deserts are spreading over what was once farmland.

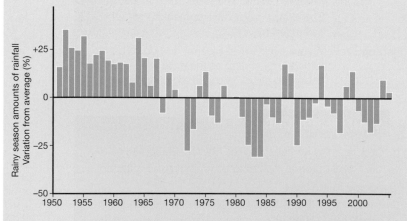

▲ How rainfall varied in the Sahel, 1950-2006. The '0' line is the average for the period, and shows whether rainfall fell above or below that average. Anything above the line shows wet years, and anything below shows dry.

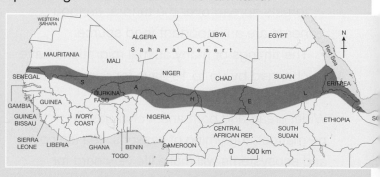

▲ The Sahel region is vulnerable to desertification.

Rainfall variation is a major cause of poverty in the Sahel. For people living there, climate change threatens their survival.

- If the rains don't arrive, the grass dies – exposing the soil to removal by wind erosion.
- When the rains do arrive, heavy rains erode and wash away the soil.

Diguettes in Siguin Voussé

Siguin Voussé is a small village in Burkina Faso in the Sahel. It was badly affected by drought. Deforestation and over-use of the land left villagers unable to grow enough food to feed themselves. Trees and grass were cleared for farming, so that whenever it rained, the precious topsoil was washed away.

In 1979, Oxfam started a project called *Projet Agro-Forestier*. It aimed to:

- prevent further soil erosion
- preserve as much rainfall as possible.

Local farmers were encouraged to build **diguettes**, a type of **intermediate technology**, to form barriers to erosion. A diguette is a line of stones, laid along the contours of gently sloping farmland. It slows down rainwater and gives it a chance to soak into the hard ground. The diguettes also trap soil, which builds up behind the stones. Soil erosion is therefore reduced.

The diguettes were a success. Almost everyone in the village had improved crop yields. And families now feed themselves. Over 400 villages in Burkina Faso have now built diguettes. In one test done by farmers, soil depth in an area *without* diguettes decreased by 15 cm; while in a field *with* diguettes, it increased by 18 cm.

▲ Building diguettes in Burkina Faso.

Awa Bundani, from the village, explains why diguettes are important to her community.

'Last year the rains were good. But in some years they stop, and crops die. The diguettes have made a huge difference. Before, the compost and soil were washed away. And when the rain was poor, the soil would dry out quickly. We knew it was a problem, but we didn't know what to do about it. Since we built the diguettes, the land produces more. We would have had only one bag of groundnuts, where now we get two.'

On your planet

+ Diguettes are an example of intermediate technology – little know-how is needed, the materials exist locally, and labour is free. It's a cheap solution to a problem.

your questions

1 Use an atlas to identify: **a** five countries in Africa which will become wetter with climate change, and **b** five countries that will become drier.
2 What problems will each group of countries face? Make a list and explain your ideas.
3 **a** Write a 300-word speech to persuade a village elder in Burkina Faso that diguettes are the way to go.
   b Draw a poster to show how diguettes work.
4 **Exam-style question** Using a named example, explain how intermediate technology can help people adapt to the changing climate. (4 marks)

+ In this section you will examine how local and global actions can protect extreme environments from the threat of climate change.

Climate change in extreme environments

Climate change is a reality in hot arid and polar extreme environments. The Sahel is the region most vulnerable to climate change because people are poorest here. In the polar Arctic, the effects of global warning are more obvious than anywhere else. This is because the Arctic is warming faster than anywhere else on earth.

Adapting to climate change – the local way

People living in polar and hot, arid regions only produce about 5% of CO_2 emissions, yet they could be the worst affected by global warming. In many cases, the only option for them is to try and adapt to the changes which climate change brings.

In hot, arid regions, such as the Sahel, farmers are using intermediate technology (e.g. the construction of diguettes - sees page 140-141) and **conservation farming** methods to help trap moisture in the soil, and minimise soil erosion and drought. Conservation farming includes **multi-cropping** — an alternative to single crop farming (see table) — where farmers plant several different crops on the same area of land (see diagram). Multi-cropping reduces the risk of crop failure and has helped to increase crop yields by 10 times.

Single crop farming	Conservation farming
Plough all the land.	Plough only where you plant crops.
Moisture easily evaporates from all of the land.	Moisture evaporates only from parts that have been dug.
One crop planted.	Several crops planted, mixed together.
Harvesting done all at once.	Harvesting done over the whole year.
If the crop fails all income is lost – food aid is needed to feed those who might starve.	If one crop fails there are others – so no one starves.

▲ How conservation farming differs from single crop farming.

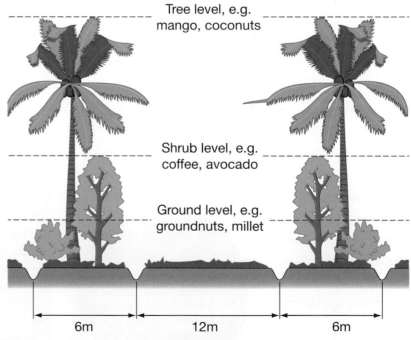

Tree level, e.g. mango, coconuts

Shrub level, e.g. coffee, avocado

Ground level, e.g. groundnuts, millet

6m   12m   6m

▲ How conservation farming brings many crops together.

Global agreement

One global action could be an agreement to slow down climate change by cutting greenhouse gas emissions (carbon dioxide, methane and nitrous oxide). This was attempted in 1997 and was called the Kyoto Summit (see textbox).

Kyoto has not been very successful. Even if all the countries that signed up stuck to their targets, it has been estimated that global temperatures would only be reduced by 0.1°C.

In 2009 countries met again, in Copenhagen, to come up with a new agreement. They agreed that,

- climate change was a major challenge
- action should be taken to limit global warming to no more than 2°C

However, no binding targets for reducing greenhouse gas emissions were agreed, so some countries have since set their own targets. One positive aspect of the Copenhagen Summit was that big developing countries like China, India and Brazil were involved whereas they had not been involved in the Kyoto Summit in 1997.

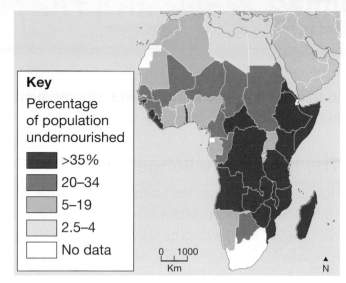

Key

Percentage of population undernourished

- >35%
- 20–34
- 5–19
- 2.5–4
- No data

0　1000 Km

▲ 'Hunger hotspots' in Africa. These countries are also the poorest and the ones most vulnerable to climate change.

Cutting emissions – the Kyoto Summit

In 1997, the Kyoto Summit was held to cut greenhouse gas emissions by 5.2% by 2012. By 2011, 191 countries had signed up to a protocol. These countries fall into four groups:

Group 1: Signed, and now meeting Kyoto targets – UK, France, Germany, Sweden, Poland.

Group 2: Signed, but not meeting Kyoto targets – Denmark, Spain, Portugal, Italy, Ireland, Japan, Norway.

Group 3: Signed, but weren't set targets – China, India, and other LICs (low income countries). HICs (high income countries) produce most emissions, and LICs ought to be given time to develop targets.

Group 4: Didn't sign – USA and Australia. Following a change of government in 2008, Australia has now signed. In December 2011, Canada renounced the protocol, having previously signed it.

your questions

1 Explain how conservation farming might help farmers in the Sahel adapt to a changing climate.

2 In about 100 words, explain why Africa and the Arctic are vulnerable to climate change.

3 Draw up a table of the achievements and failings of the Kyoto and Copenhagen Summits.

4 **Exam-style question** Using named examples, explain why local and global actions against climate change are needed in extreme environments. (8 marks)

Unit 2 People and the planet

✚ In this section you'll learn about the themes in Unit 2, and how they link together.

The human challenge!

How many people can the Earth support? In the late 18th century, Thomas Malthus believed that the world was already way beyond its ability to support the human population. He thought that mass starvation would result, because he believed that the food supply could never keep up with population growth.

Britain's population in 1801 was just over 5 million people, but rising fast. The world's population at the time – though no one is certain about this – was probably about 800 million. Today, Britain's population is about 63 million, and the world's population has passed 7 billion.

But although there are famines somewhere in the world from time to time, these are usually caused by war. In the world as a whole, there is enough food for everyone.

Malthus wasn't alone in thinking what he did – and many agree with him now! The four themes in this unit are all about the things that he was concerned about over 200 years ago.

- Many people think that the world's **population** cannot survive as things are. But we have got this far – can't we go on?
- People consume **resources** – from food and water, to oil and metals. The world's oil won't last forever. But aren't we able to invent other ways of providing what we need?
- The world has become a **globalized**, interconnected place, where almost every country in the world relies on goods and services from another country.
- Economic **development** differs for people in different parts of the world. Many countries rely on farming, others rely on industries, and some on services. Some countries get great benefits from their wealth, while others struggle to make a living.

On your planet
✚ By the year 2050, will there be 9 billion people or 15 billion people in the world? Even the experts can't agree!

▲ Children in Rwanda – how many people will there be in the world when these children are 64?

▼ How much longer will the world's oil last?

▲ As countries industrialise, many people move to the cities, like Shanghai, to work in the factories there.

▲ China has developed rapidly, but are all of its people gaining from better incomes?

Theme	What's it about?	What challenges are there?
Population dynamics	• How the world's population is changing. • How some countries control their population.	• How big will the world's population get? • Should immigration be encouraged?
Consuming resources	• Resources around the world. • Growing use of resources. • Sustainable use of resources.	• How can inequalities between rich and poor be reduced? • How well the world can cope with increasing use of its resources?
Globalisation	• How the economy of the world varies in different places. • Changes that are taking place to the economies of different countries.	• Which countries will be the winners and losers in the global economy? • How do countries and international companies depend on trade?
Development dilemmas	• Economic development in different parts of the world. • How the differences in development between rich and poor can be reduced.	• How can we narrow the gap between the world's richest and poorest countries?

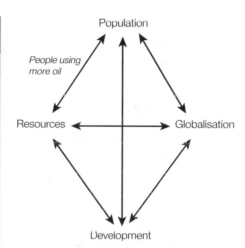

▲ Diagram showing links between the four themes in Unit 2.

your questions

1 List any resources that you've consumed already today.

2 Write down three ways in which you have put pressure on the earth's resources already in your lifetime.

3 Draw a large version of the diagram above showing the links between different parts of Unit 2. It will need to be a whole page. Label ways you can think of to show how the themes are linked. One has been done for you.

4 Which do you think might be the bigger problem – if the world's population increases as it is now, or if we keep using up its resources? Explain your reasons.

+ In this section you'll learn that the world's population was growing very quickly, but is now beginning to slow down.

Billions of people!

In 2013 there were just over 7.1 billion (that's 7,100 million, or 7,100,000,000) people living on planet Earth. According to the United Nations, the seven billionth person was born in October 2011 — and the eighth billionth will probably arrive sometime in 2027. In your lifetime there will have already been another one billion people added to the world's population. These people all require food, water and shelter. They all have ambitions similar to yours and they all want access to education, opportunities and the latest technology.

A book, called 'The Population Bomb', written by a leading **demographer**, said that by 8000 BC it had taken the human race a million years to double in number. Yet, in 1968, (the year the book was published) the population was 3.5 billion and the doubling time had fallen to 35 years. In other words, the bigger the population, the faster it grew. This is called **exponential** growth.

Year	World population
1500	500 million (half a billion)
1804	1 billion
1927	2 billion
1960	3 billion
1974	4 billion
1987	5 billion
1999	6 billion
2011	7 billion

Many people became worried that the world's population was growing too fast. People talked about a fear that the world could become **overpopulated**. There were estimates that the number of people could reach 10 billion, 20 billion, or perhaps 50 billion by the year 2050! How could all of these people be fed, be housed, or find work?

World population reaches seven billion

The United Nations celebrated the '7 billion' milestone on 31st October 2011. The UN Secretary General, Ban Ki-moon, called out to world leaders to meet the challenges that a growing population faces such as ensuring that there is enough food and clean water for all. Mr Ban noted that the world's population passed 6 billion in 1998 (only 13 years earlier) and could grow to 9 billion by 2043. The celebration was purely symbolic because, with three babies being born every second, it would have been impossible to know exactly when the seven billionth baby was born.

+ A **demographer** is a person who studies population growth.

+ An area which has too many people for the resources available is said to be **overpopulated**.

Something's changing

The fears of an overpopulated planet seem to be disappearing. In the 1960s nearly every country on the planet had an expanding population, but from 1971 the number of babies born in developed countries dropped. Many of these countries — such as Sweden and Italy, are now seeing their populations falling slightly. Other countries — such as China and India, introduced controls to limit their populations. The number of babies being born in developing countries overall is therefore beginning to level off.

Year	World population
2027	8 billion
2043	9 billion
2185	10 billion

▲ No country in the European Union is producing enough babies to stop their population declining

Population growth uncertain

Each month, the number of babies born equals the population of Portugal, whilst the number of people who die is equivalent to the population of Norway. The difference between the two is roughly the population of Libya — about 6.4 million people. That is the number by which the human race grows every 31 days.

The United Nations estimates that the world's population will grow to about 9 billion by the middle of the century, then stabilise, and slowly begin to decline.

This is because people are having fewer babies than expected, especially in developing countries. Indian women are now having fewer babies than American women were in the 1950s.

▲ Adapted from newspaper articles, October 2011

On your planet
+ Poorer countries now experience nearly all of the world's population growth.

What do you think?
+ Why are people living in the European Union choosing to have fewer children?

your questions

1 Draw a line graph to show world population growth 1500-2185.
   a Use data for 1500-2011 as a solid line.
   b Then use the estimates for 2027-2185 and draw a dotted line for these. Link 2011 to 2027 with a dotted line also.
   c What does your graph show you about global population growth?
2 Look at the table showing world population 1500-2011.
   a Starting with 1804, calculate how long it takes to add 1 billion people.
   b What do you notice about how long it takes each time?
   c What word describes this growth?
3 Now look at the estimates for global population 2027-2185.
   a Starting with 2027 calculate how long it takes to add 1 billion people.
   b What do you notice?
4 Exam-style question Study the line graph you have drawn for Question 1. Describe the different rates of population growth between 1500 and 2185. (2 marks)

✛ In this section you'll learn how to compare population change between countries.

It's all about babies

The population of a country is constantly changing. In some countries the population will be growing, in others it may stay level, or may even decline. The change of population in a country is the difference between the number of babies being born and the number of people who die. This difference is called the **natural increase**. If this is a positive number, there are more births than deaths (**population increase**). If it is a negative number, there are more deaths than births (**population decline**). If birth and death rates are almost equal, the country will have a **population balance**.

Population balance: Denmark (High-income country)
Total population: **5 515 575**

Birth rate: 10.29 per 1000
Death rate: 10.19 per 1000
Natural increase: **0.1 per 1000 (0.01%)**
This increase is so small that it means that Denmark's population effectively remained level in 2010.

Population increase: Senegal (Low-income country)
Total population: **12 323 252**

Birth rate: 36.73 per 1000
Death rate: 9.26 per 1000
Natural increase: **27.47 per 1000 (2.75%)**
For every 1000 people living in Senegal, 27 more people were added to the population in 2010.

Population decline: Japan (High-income country)
Total population: **127,579,145**

Birth rate: 8.54 per 1000
Death rate: 9.05 per 1000
Natural increase: **-0.51 per 1000 (-0.05%)**
For every 2000 people living in Japan, 1 person was lost from the population in 2010.

✛ **Birth rate** is the number of babies born alive for every 1000 people in one year.

✛ **Death rate** is the number of people who die for every 1000 people in one year.

✛ **Natural increase** is the number of people added to, or lost from the population for every 1000 people in one year.

Note: It is easier to compare countries by measuring birth rate, death rate and natural increase per 1000 of the population, because the total population of each country is different.

Why so different?

Birth rates and death rates vary greatly between countries. Some of the many factors that can affect differences between birth rates and death rates are the:

- level of **development** of a country
- **religious views** of people in a country
- **policies** of the Government.

For example, birth rates can be affected by the availability of, and attitudes towards, contraception – while death rates can be affected by the level of access to health care and medicines.

How to work out natural increase

Natural increase = birth rate – death rate

United States of America, 2010
Total Population: 308 282 053
Birth rate: 13.68 per 1000
Death rate: 8.38 per 1000

The natural increase for the USA in 2010 is:
13.68 — 8.38 = 5.3 per 1000 = **0.53%**
(Note that the natural increase is often given as a percentage)

Country	Total population 2010	Birth rate (per 1000)	Death rate (per 1 000)
Brazil	201 103 330	17.79	6.36
Cambodia	14 453 680	25.4	8.07
China	1 330 141 295	12.29	7.03
Egypt	80 471 869	24.63	4.82
Haiti	9 648 924	24.4	8.21
Iraq	29 671 605	28.81	4.82
New Zealand	4 252 277	13.68	7.15
Senegal	12 323 252	36.73	9.26
UK	62 348 477	12.29	9.33
USA	308 282 053	13.68	8.38

On your planet
+ There are on average 4 births and nearly 2 deaths somewhere in the world every second!

◀ Population data for selected countries, 2010.

your questions

1 Write down what each of the following terms means:
   a natural population increase
   b natural population balance
   c natural population decline
2 Use the population data table for 2010 to:
   a work out the natural increase for each of the 10 countries
   b list the countries in rank order by their natural increase
   c describe any patterns that you may notice in the rank order

3 Copy and complete the following table:

Country	Birth rate (per 1000)	Death rate (per 1000)	Natural increase (per 1000)
Austria	8.7	9.9	
Bolivia		7.4	15.0
Canada	10.3		2.7
Thailand	13.6	7.2	
Zambia	40.5		19.2

4 **Exam-style question** Explain how natural increase is affected by both, birth rates and death rates.
(4 marks)

In this section you'll learn how birth rates and death rates influence how the total population of a country can change over time.

The Demographic Transition Model

The Demographic Transition Model (DTM) is based upon the work carried out by an American demographer, Warren S. Thompson in 1929. The model is a simplified representation of population change over time on which any country in the world could theoretically be placed. It supposes that a country could be placed in any one of five stages of population change depending on its birth and death rates.

STAGE 1 – High stationary	
Death rate: HIGH	**Birth rate:** HIGH
Reasons: Diseases, famines, conflict, a lack of clean water, poor medical care.	**Reasons:** Lack of birth control, low age of marriage for women, children may work which adds to the family income.
Rate of natural increase: LOW — the total population stays low as the high death rate cancels out the high birth rate.	
UK position: The UK was in Stage 1 before 1760	
Examples: Few places are in Stage 1 today — perhaps a few remote tribes in the tropical rainforest.	

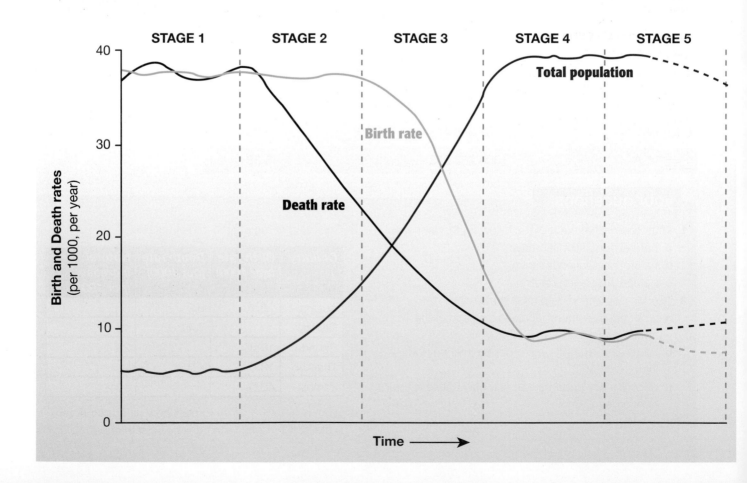

STAGE 2 – Early expanding

Death rate: FALLING

Reasons: Improved medicine, cleaner water, better food, improved sanitation.

Birth rate: HIGH

Reasons: Lack of birth control, low age of marriage for women, children may work which adds to the family income.

Rate of natural increase: HIGH — the total population increases rapidly as less people are dying, but the birth rate is still high.

UK position: The UK was in Stage 2 between 1760 and about 1900, during the Industrial Revolution.

Examples: Today, some LICs (low-income countries) are in this stage e.g. Bangladesh and Niger.

STAGE 3 – Late expanding

Death rate: FALLING

Reasons: Continued improvement in medical care and sanitation, improving food and diet.

Birth rate: FALLING

Reasons: Women are staying in education and getting married later — so delaying having children. Birth control is available so people choose to have fewer children. Children are less likely to work — going to school instead — so do not contribute to the family income.

Rate of natural increase: HIGH — The population continues to grow as the birth rates are still higher than the death rates, but it is beginning to slow down as birth rates are falling too.

UK position: The UK was in Stage 3 between about 1900 and 1950

Examples: Today, many MICs (middle-income countries) are in this stage e.g. India, Brazil, Mexico.

STAGE 4 – Low stationary

Death rate: LOW

Reasons: High standards of medical care, healthy lifestyles, excellent sanitation services.

Birth rate: LOW

Reasons: High cost of bringing up children, women choosing to have careers before marriage, good access to birth control.

Rate of natural increase: VERY LOW / STABLE — with low birth and death rates, there is little, or no, natural increase in the population.

UK position: The UK has been in Stage 4 since about 1950.

Examples: Many HICs (high-income countries) are in this stage, e.g. USA, France, Denmark.

STAGE 5 – Declining

Death rate: RISING

Reasons: A greater proportion of the population is now elderly so the death rate is rising slightly.

Birth rate: VERY LOW

Reasons: Remains low due to lifestyle changes — more people are choosing to have fewer children later in life.

Rate of natural increase: NEGATIVE — the total population begins to decline slowly as there is a higher death rate than birth rate.

UK position: The UK could enter Stage 5 soon.

Examples: Some HICs have already reached Stage 5, e.g. Japan, Russia, Germany. This stage was not on the model when it was first devised — it has been added to show recent developments in population change.

your questions

1 Describe what happens to each of the following throughout the stages of the Demographic Transition Model:
  a The birth rate
  b The death rate
  c The total population
2 Explain how the birth rate and death rate influence the changing total population at each stage of the model.
3 What stage of the Demographic Transition Model do you think each of the following countries are in? Give reasons for your answers.

Country	Birth Rate	Death Rate	Natural Increase
China	12	7	5
Italy	9	10	-1
Mexico	20	5	15
Sierra Leone	39	15	24
Sweden	12	10	2
United Kingdom	13	9	4

4 **Exam-style question** Using examples, describe how natural increase in population can change as a country's level of development increases. (6 marks)

In this section you'll learn that population change is occurring at different rates, in different countries.

What's happening where?

Although the population of the world is expected to continue to grow until about 2100, this growth will not be even.

- **Population increase** is mainly happening in Africa, the Middle East, and parts of South America and South Asia.
- **Population balance** is mainly found in North America and Europe.
- **Population decline** is happening in Russia and parts of Central and Eastern Europe.

It is important to understand that:

- higher levels of population increase are occurring in developing (**low-income** and **middle-income**) countries
- lower levels of population increase, population balance, or even population decline, mainly occur in developed (**high-income**) countries. See the table.

The **World Bank** groups countries by their Gross National Income (GNI) per person. This is the average amount of money earned by a country in one year, divided by all the people who live there.

In 2011, the groups were:

High-income countries (HICs): $12 476 or more
Middle-income countries (MICs): $1 001 - $12 475
Low-income countries (LICs): $1 000 or less

	Average GNI per person	Average % natural increase of population (1990–2005)	Average estimated % natural increase of population (2005–2020)
Low-income countries	567	2.0	1.7
Middle-income countries	4 857	1.1	1.2
High-income countries	41 144	0.7	0.4

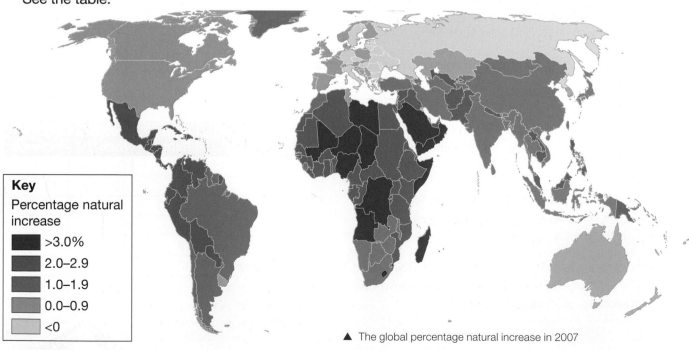

Key
Percentage natural increase

- >3.0%
- 2.0–2.9
- 1.0–1.9
- 0.0–0.9
- <0

▲ The global percentage natural increase in 2007

Let's compare two countries

Russia (upper middle-income country)	Yemen (lower middle-income country)
Total population (2012): 142.5 million	Total population (2012): 24.7 million
GNI per person: $10 400	GNI per person: $1 070
Birth rate: 12.3	Birth rate: 32.6
Death rate: 14.1	Death rate: 6.8
Natural increase: -0.2	Natural increase: 2.6

- In 1950, the total Russian population was 103 million. In Yemen there were just 4.3 million. This meant that there were 24 Russians for every Yemeni.
- In 2012, Russia's population had grown to 142.5 million — and Yemen's population to 24.7 million.
- By 2057, Russia's population is expected to fall to about 104 million, while Yemen's population is expected to rise to 105 million. Between now and 2057, the population of Yemen will equal that of Russia – and then overtake it!

Russia's population will decline because of:
- falling life expectancy for men (60 years) caused by industrial disease and alcoholism
- **outward migration** of young men and women
- a low **fertility rate** of 1.2 children per woman.

Yemen's population will grow quickly because of:
- early age of marriage: 48% of women are married by the age of 18
- low literacy rates among women: as most girls marry early, they rarely complete secondary school
- a high fertility rate of 6.7 children per woman and increasing life expectancy due to improved child vaccinations.

Country	GNI per person $US (2012)	% Natural population increase (2012 est.)
Brazil	10 720	0.86
Cambodia	830	1.69
Chile	12 280	0.88
D. R. Congo	190	2.58
Greece	25 030	0.06
Mongolia	2 320	1.47
Nicaragua	1 170	1.07
Norway	88 890	0.33
Portugal	21 250	0.18
Senegal	1070	2.53
Sierra Leone	340	2.28
UK	37 780	0.55

+ **Fertility rate** is the average number of children born to a woman in her lifetime.

your questions

1 Look at the table above. Draw a scatter graph for the 12 countries, showing their GNI and percentage rates of natural increase:
   a Plot the GNI per person on the X axis.
   b Plot the % natural increase on the Y axis.
   c Draw a line of best fit.
   d Using the criteria for high, middle and low-income countries in the text box opposite, divide your graph into 3 sections: high-income, middle-income, and low-income countries.
   e Describe the pattern that your graph shows.
   f Try to explain what the pattern means.
2 Draw a table to compare the reasons for Russia's falling population, with the Yemen's rising population.
3 Why do you think low literacy rates among women can cause a high fertility rate?
4 **Exam-style question** Referring to examples, explain the factors that can lead to *either*, a natural population increase *or*, a natural population decrease. (4 marks)

➕ In this section you'll learn that Japan has a population which is both declining, and ageing.

The silvering of the yen

Japan has the oldest population in the world:

- Over-65s make up 23.9% of the population.
- There are 30.5 million pensioners.
- The average age is 45.4 years — the highest in the world.
- The birth rate remains below **replacement level**.
- Under-15s make up just 13.5% of the population.

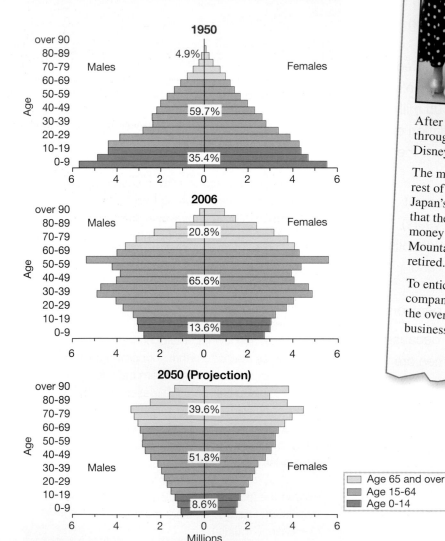

▲ The changing structure of population in Japan.

Japan runs out of children for Disneyland

After 25 years of successfully luring children through the gates of the Magic Kingdom, Tokyo Disneyland has decided to chase the 'silver yen'.

The move comes because Disney, along with the rest of Japan, is running out of children. With Japan's birth rate in decline, Disney has accepted that the largest group of customers with the money and the time to spend a day on Splash Mountain or Pooh's Hunny Hunt is mostly retired.

To entice these people on to the rides, the company is offering a cut-price season ticket for the over-60s and has made them a target for new business.

▲ Adapted from newspaper articles, February 2008

➕ **Replacement level** is the average number of children required to be born, per woman, to ensure that the population remains stable (it is 2.1).

Why is Japan's population structure changing?

There are two main reasons:

- People in Japan are living longer. The average life expectancy is 81 for men, and 87 for women. This is due to a healthy diet (low in fat and salt) and a good quality of life. Japan is one of the richest countries in the world, with a good health and welfare system. Japan spends 9.3% of its Gross Domestic Product (GDP) on healthcare — and there are 206 doctors for every 100 000 people.

- The birth rate in Japan has been declining since 1975, as the graph shows. This is partly due to the rise in the average age at which women have their first child. The average age rose from 25.6 years in 1970 to 29.2 in 2006. Throughout this period, the number of Japanese couples getting married has fallen, and the age at which people get married has risen (see the table).

What does this mean for Japan?

An increasing elderly population, together with a shrinking younger population, means that there will be a number of challenges in the future:

- An increase in the cost of pensions. More elderly people, living longer, will require pensions for longer. With the falling birth rate, there will be fewer workers in the economy, so higher taxes will be needed to fund those pensions. The Government has already raised the pension age from 60 to 65.

- A rising number of elderly people living in nursing homes. Since 2000, everyone over 40 has had to contribute £20 a month to pay for care for the elderly.

- An increase in the cost of healthcare, as more elderly people require medical treatment. The Government has already raised patient contributions for medical expenses from 10% to 20%.

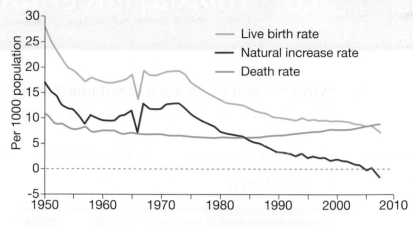

▲ The natural increase of population in Japan.

Year	Men	Women
1950	25.9	23.0
1960	27.2	24.4
1970	26.9	24.2
1980	27.8	25.2
1990	28.4	25.9
2000	28.8	27.0
2010	30.5	28.8

▲ The average age of first marriage in Japan.

▶ A Japanese bride and groom in their traditional wedding clothes.

your questions

1. Look at the three population pyramids for Japan.
   a Describe the shape of each pyramid.
   b Explain the changes between 1950 and 2050.
2. Look at the graph above showing the natural increase in population. Describe the changes that it shows.
3. Explain why the population of Japan is:
   a getting older b decreasing
4. Exam-style question a Describe what is meant by an 'ageing population' (2 marks)
   b Using a named example, outline two problems faced by countries with an ageing population. (4 marks)

Population change in Mexico

✚ In this section you'll learn about Mexico's increasing population and the country's high percentage of young people.

▼ Adapted from: www.imediaconnection.com, 2007.

What's happening to Mexico's population?

Mexico (middle-income country)
Total population: 115 million
GNI per person: US$ 9240
Birth rate: 18.9
Death rate: 4.9
Natural increase: 1.4%

- Mexico has a large, youthful population. Under-15s make up 28% of the population. Almost 7% of the population are over 65 (see the population pyramids below).

- The population grew from just 20 million in 1940, to 70 million by 1980 and is fast approaching 115 million.

- The fertility rate in 1970 was as high as 7.1, but has now fallen to 2.27 — still above replacement level.

- The average age in Mexico is just 27.

- 47% of Mexico's population are now entering childbearing age.

Mexico online: Targeting the young population

Mexico has the second largest online market in Latin America (behind Brazil), with more than 20 million Internet users, and it ranks 13th out of the top 20 Internet-user countries. The online population in Mexico grew from 2.7% in 2000 to 19.2% in 2006.

The main reason for this massive growth is that 53% of Mexico's population is under 24 years old, and Internet users in Latin America tend to be younger than the population as a whole. The growing number of young Mexicans are reading books, newspapers and journals, surfing the Internet and using multimedia education in order to improve their chances of success in the over-crowded job market.

Mexico is proving to be one of the key markets for online advertisers to break into, due to the growing population of young people.

▼ The changing structure of population in Mexico.

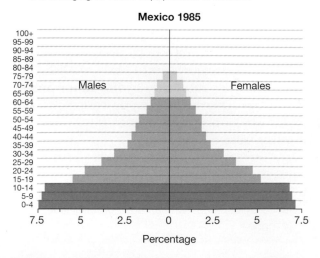

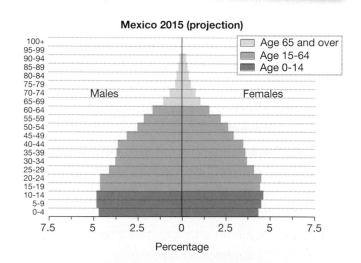

Why is Mexico's population changing?

There are several reasons why Mexico's population structure is changing:

- A low death rate – just 4.78 deaths per 1000. Not only are more babies being added to the population, but people are living longer as well! Since 1990 vaccination programmes by the United Nations have cut infant mortality. Children used to make up a high percentage of deaths. Now that Mexico has a healthier, youthful population, this has resulted in a very low death rate. In future decades, the death rate will rise again as this generation ages.
- Although the birth rate is falling, there is still a large percentage of young people. Even if these young people have fewer children than their parents did, the population of Mexico will still continue to increase. Today's children are tomorrow's parents.
- It is expected to take at least 35 years before the population structure of Mexico loses its triangular shape, and the population starts to decline.

What does this mean for Mexico?

A growing population, coupled with an increasing percentage of young people, means that there are a number of challenges and opportunities facing Mexico:

- A large youthful population requires an increase in school places.
- Large numbers of young people are unable to find work, so some migrate to the USA in order to find employment.
- There is a growing manufacturing industry. The Mexican economy is expected to grow to overtake the UK's and become the seventh largest economy in the world by 2050.
- Although Mexico is a strongly Catholic country, abortion has been legalised in Mexico City in an attempt to reduce the number of abandoned children.

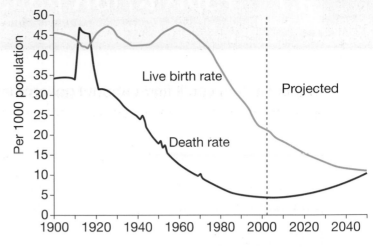

▲ Birth and death rates in Mexico.

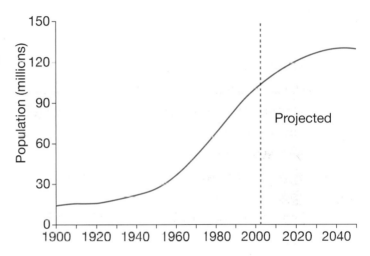

▲ Population growth in Mexico, 1900–2050.

your questions

1 Explain why the population of Mexico is increasing.
2 Internet companies and advertisers have already noticed the potential of an increasing number of young people in Mexico. Give other examples of the opportunities that a young population may provide for a country.
3 **Exam-style question** Look at the population pyramids for Mexico.
   a Describe the shape of each pyramid. (2 marks)
   b Explain the changes between 1980 and 2015. (2 marks)
4 **Exam-style question** Look at the graph showing the birth and death rates in Mexico.
   a Describe the changes that the graph shows. (2 marks)
   b How would the natural increase in population have changed over this period? (2 marks)

In this section you'll find out how governments try to influence population growth in their countries.

Population control

Many countries around the world have introduced **population policies** to encourage people to have either more or fewer children.

ANTI-NATALIST POLICY:
India
(Middle-income country)
Total Population: **1 205 million**
Population continues to grow

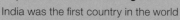

India was the first country in the world to introduce a population policy in 1951, aiming to reduce the fertility rate to 2.1. The government began by offering contraceptive advice and sterilisation. Throughout the 1970s forced sterilisation was used on men. This policy has now ended and the government now uses incentives to encourage couples to have only two children. However, the fertility rate still remains high at 2.58.

PRO-NATALIST POLICY:
Sweden
(High-income country)
Total Population: **9 million**
Population is ageing and has been decreasing

Sweden's fertility rate was at a peak of 2.1 in 1989 but fell throughout the 1990's to 1.5 in 1999. Since then the government have introduced a range of benefits to encourage couples to have more children:

Paternal Leave: Available for 13 months at 80% of earned income

Speed Premium: An extra payment if there are less than 30 months between children

Cash Child Benefit: Currently €900 per child per year

Sick Child Care: 120 paid days per child per year

All day child care and all day schools: Enabling parents to work full-time.

By 2012, Sweden's fertility rate had increased slightly to 1.7.

PRO-NATALIST POLICY:
Singapore
(High-income country)
Total Population: **5 million**

GNI per person: US$ 49 044

Birth Rate: 7.72 per 1000

Death Rate: 3.41 per 1000

Natural Increase: 4.31 per 1000

+ Population Policies: Measures taken by a government to influence population size, growth, distribution, or composition.

+ Pro-natalist Policies: These are policies that are used to encourage people to have **more** children by offering incentives, such as financial payments.

+ Anti-natalist Policies: These are policies that are used to encourage people to have **fewer** children by offering incentives, such as free state education for only the first child in a family.

IMMIGRATION POLICY:
Russia
(Middle-income country)
Total population: **143 million**
Population is decreasing

Russia introduced its population policy in January 2007. The aim is to stabilise the population and ensure that there are enough workers to maintain economic growth. The government has used two strategies:

- To offer a cash incentive, to encourage Russians who have moved abroad to come back and live in Russia.

- To encourage migrant workers to come and work in high skilled professions in Russia. 6.5 million immigrants were welcomed in 2007.

IMMIGRATION POLICY:
United Kingdom
(High-income country)
Total Population: **63 million**
Population rising more rapidly since 2001

The population policy of the UK Government has, until recently, been that of 'no intervention' with fertility rates or migration. However, since 2010, the policy is to reduce immigration from non-EU countries. The government's view on population growth is officially one of acceptance, though many believe that migration of younger EU migrants helps to balance an ageing population and high government spending on state pensions.

Pro-natalist population policy: Singapore

Singapore's government is committed to ensuring a strong economy and a good quality of life for its 5 million residents. However, at current birth rates, the number of Singaporeans, aged 65 and over, will triple to 900 000 by 2030. This will leave a much smaller working population, that will be both ageing, and in decline. To try and address this imbalance, Singapore has introduced a pro-natalist population policy.

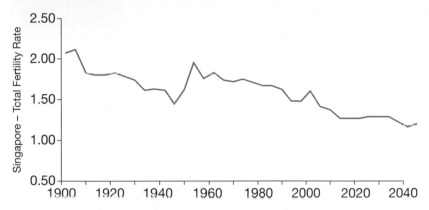

▲ Singapore's total fertility rate has been declining

What is the policy?

The Marriage & Parenthood Package was launched in 2008. It encourages Singaporean couples to have at least two children through a range of measures including:

- Helping single people find their life partner by providing social interaction opportunities.
- Providing housing schemes to help couples set up home.
- Offering a baby bonus scheme (see diagram below).
- Offering maternity leave of up to 16 weeks.
- Providing paid child care leave – for working parents.
- Offering state subsidised child care.

your questions

1 In pairs, discuss why some countries need to use pro-natalist population policies, whilst others need to use anti-natalist population policies.

2 What other population control policies might countries use, and why?

3 Look at the graph above which shows Singapore's declining fertility rate from 1900-2040.
   a Describe the changes that the graph shows.
   b How might the 'Baby Bonus' scheme influence the future fertility rate in Singapore.
   c What problems might Singapore face in the future if not enough children are born?

4 Think about how population policies affect the lives of people living in a country where the policy operates. How do you think people might react to these policies?

5 Exam-style question Using named examples, explain why some countries choose to increase their population, but others choose to reduce it. (6 marks)

A middle-income Singaporean household with two children

can enjoy:

$20,000 in Baby Bonus cash and co-savings

$53,000 in infant care and child care subsidies

$10,000* in tax savings

4 months of paid maternity leave per child

6 days of paid child care leave per year per parent

the equivalent of about:

$142,000 until both children turn 7

*This excludes about $18,000 in additional tax savings under the Marriage & Parenthood Package which is typically utilised beyond the child's first 7 years.

▲ Singapore's Marriage and Parenthood Package

➕ In this section you'll learn about China's 'one-child' population policy

Anti-natalist policy: China

The best-known population policy is China's one-child policy. China's population grew exponentially throughout the 1950s and 1960s. The birth rate reached 5.8 per 1000, which was unsustainable given China's available natural resources of food, water and energy. In 1979, the government introduced rules to limit population growth. Couples who only have one child receive financial rewards and welfare benefits. Heavy fines are imposed on those who have more than one child.

What is the one-child policy?

The rules of the policy are complex:

- The one-child policy only applies to Han Chinese (90% of the population). Ethnic minorities are exempt.
- Men have to be over the age of 22, and women over the age of 20, in order to be granted permission to have a child.
- Parents who follow the rule receive a certificate and extra money when they retire.
- Two babies are permitted in rural areas, or if the first child is female.
- Breaking the rule results in a heavy fine.
- State officials, who have more than one child, automatically lose their job.

ANTI-NATALIST POLICY:
China
(Middle-income country)
Total Population: **1,343 million**
GNI per person: US$ 4 940
Birth Rate: 12.31 per 1000
Death Rate: 7.17 per 1000
Natural Increase: 5.14 per 1000

The history of China's population policy

1949 – 1958: The new government of the People's Republic of China encouraged growth of the population to increase the workforce. In 1957, Chairman Mao Zedong stated that he wanted the Chinese population to stay at 600 million for many years.

1959 – 1963: The famine of these years saw the government implement a birth control campaign focused on supplying contraceptives – and a mass media campaign promoting late marriage and low birth rates.

1964 – 1977: The national Family Planning Office was set up with the aim of lowering the fertility rate. Officials encouraged a low birth rate in major cities, by monitoring residents and enforcing policies. Rural areas remained largely untouched.

1978 onwards: The one-child policy was introduced.

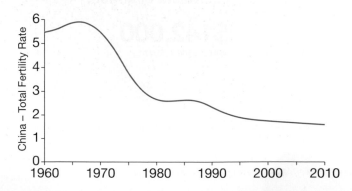

◀ China's total fertility rate has declined rapidly since the 1970s.

Impacts of the Policy

The policy has been successful in preventing 300 million births. However, it has also had some negative effects, both economic and social.

- There is now a serious imbalance of men to women (see right). Couples have often used illegal methods to ensure that their one child is a boy, as boys are traditionally more able to care for their parents when they are elderly. Unwanted baby girls have often been abandoned or left in orphanages.

- Some people now believe that China's rapidly growing economy will not have enough skilled and well-educated workers to keep it going. The number of young people starting work between the ages of 20 and 24 will drop by half from 2010 to 2020 – another reason why the policy may need to be relaxed.

- Babies born under the one-child policy will have two parents and four grandparents to look after as they reach old age. This is known as the '4-2-1' problem.

- The one-child policy is often presented as a cure-all for rapid population growth. However, even when the policy was first introduced in China, the Chinese fertility rate had already fallen to 2.7. It is potentially wrong to say that this fall was due to this policy, when the bulk of the fall came before it was in place.

In future, the Chinese Government may be forced to relax the one-child policy to address the problems it has created.

On your planet
+ Even with the one-child policy, around 50 000 babies are born in China every day.

China faces population imbalance crisis

China will be short of 30 million brides within 15 years, according to a report from the State Population and Family Planning Commission in China. About one in every 10 men aged between 20 and 45 (equivalent to almost the population of Canada) will be unable to find a wife. By 2020, the report says, the number of men of marriageable age will outnumber women by 30 million.

▲ Adapted from newspaper articles, January 2007

▲ One of China's problems — there are now fewer young people to support an elderly population.

your questions

1 Describe the changes in China's fertility rate from 1960-2010.

2 As people in China become wealthier, they are expected to live longer. Why should this be?

3 What problems will the 4-2-1 problem create for people living in China?

4 **Exam-style question** Explain how some countries use population policies to limit their population growth.
(6 marks)

✚ In this section you'll learn how the UK tries to control its number of immigrants.

Britain is changing!

Immigration is never far from the headlines. The results of the 2011 census showed that for the first time:

- White Britons had become a minority in London.
- Britain has become increasingly diverse, with the proportion of ethnic minorities, in England and Wales, increasing to more than 10%.
- 13% of the 56.1 million people living in England and Wales were born abroad — the top two countries being India and Poland.

These figures are the results of an immigration policy which has brought nearly four million immigrants to England and Wales in ten years.

Tony Blair: Immigration has been good for Britain

Up to 90,000 students 'in Britain illegally

A major minority: New census stats reveal changing face of Britain

▶ Adapted from newspaper headlines, December 2012.

Britain shames itself by detaining immigrants indefinitely

Impacts of immigration

Immigration can be both, positive or negative, depending on your point of view.

Some people argue that immigration can be positive because:

- It promotes tolerance between different groups of people and connects different places around the world.
- It can provide an economic boost to a country. It is estimated that Polish people living in the UK have contributed three times more money to the UK economy, than they have cost.
- It ensures that the economy has a skilled and well qualified workforce.
- It contributes to the cultural diversity of the country, bringing such as benefits as Indian restaurants and Chinese herbal medicine.
- Migrants are often young — they often marry and have children, which counters the problem of an ageing population — leading to a more balanced population pyramid.

Others argue that it can be negative because:

- It puts a strain on welfare and healthcare systems.
- It can mean that local people can find it difficult to compete for jobs in an overcrowded job market.
- Immigrants can often remain isolated and not mix with the wider community — often leading to negative stereotypes.

▲ A sign of the times? Sikh soldier Jatenderpal Singh Bhullar becomes the first guardsman to parade outside Buckingham Palace wearing a turban instead of the famous bearskin.

The UK's migration policy 1997-2010

Different governments since the 1990s felt that the UK economy needed a boost from migrants to fill jobs that needed doing, especially in London. Until 2010, the UK operated an increasingly **open-door policy** because of the expansion of the European Union. EU membership allowed all people within the Union to move freely between each country. When new countries from Eastern Europe joined the Union, such as Poland, Britain allowed people from these countries to come to the UK in order to fill the shortage of workers. However, those from non-EU countries have never had this freedom.

The UK's migration policy since 2010

The UK has now adopted a more **quota** based approach, with the introduction of **skills tests**. This is because some people argue that Britain has accepted too many immigrants since 1997. The government has now set a target of bringing the number of immigrants to below 100 000 each year – quite some feat seeing that the number in 2011 was 216 000. However, the government cannot stop Britons and Europeans from coming and going, due to the European Union laws on free movement. The focus has therefore been to reduce the number of migrant workers and students – the very people that some argue that Britain really needs!

+ An **Open-door** approach is when a country encourages migrants to move in, often for economic reasons.

+ **Quota** based migration is where a country sets limits on the number of legal migrants that it is prepared to allow in each year.

+ **Skills tests** are tests which are used to assess the qualifications and skills that potential immigrants have — to see if they will match the jobs and skills that are needed in a country.

The UK points-based, skills test scheme

The scheme awards points for skills that you have. The more skills you have which are in demand, the more points you are awarded. There are five 'tiers' in the scheme:

- Tier one: **Highly skilled**
  — including entrepreneurs, top scientists and business people — they have the best opportunity to stay permanently in the UK. The system regards them as having the potential to earn the most money.
- Tier two: **Skilled with job offer** — people with qualifications or important work-related experience. Employers are required to sponsor an individual
- Tier three: **Low skilled** — jobs in hospitality, food processing and agriculture. This tier is not open to people outside the European Union.
- Tier four: **Students** — this is for fee-paying students from overseas who are issued a student visa from their place of study.
- Tier five: **Temporary workers** – this is mainly for professional sports people or musicians who want to work in the UK for a specific limited period of time.

your questions

1 What is meant by an 'open-door approach' to immigration?
2 Produce a table showing the positive and negative impacts of migration.

3 Explain why you think the British government has now set limits on the number of immigrants welcome to the UK.
4 **Exam-style question** Describe **two** potential advantages and **two** potential disadvantages of having a large immigrant population? (4 marks).

✚ In this section you'll learn how the USA tries to control its number of immigrants.

An immigration policy in action

The USA has had some form of immigration policy ever since its founding in 1783. Since 2006, more legal immigrants have settled in the USA than in any other country in the world. There are now some 37 million foreign-born US residents: 12.4% of the total population. However, at least a third of these are thought to have entered the country illegally (see photo).

USA	
(High-income country)	
Total Population: **316 million**	
GNI per person: US$ 48 620	
Birth Rate: 13.7 per 1000	
Death Rate: 8.4 per 1000	
Natural Increase: 5.3 per 1000	

'America has constantly drawn strength and spirit from wave after wave of immigrants...They have proved to be the most restless, the most adventurous, the most innovative, the most industrious of people.'

Former US President Bill Clinton

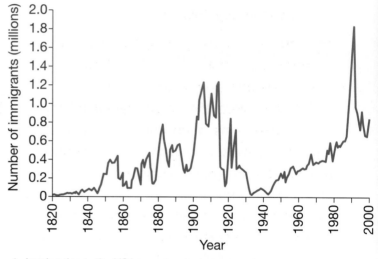

▲ Immigration to the USA.

Key

Immigration rate relative to total population of sending country, 2001–2005 (rates per thousand)

- ■ 10.00–53.89
- ■ 3.00–9.99
- ■ 1.00–2.99
- ■ 0.30–0.99
- ■ 0–0.29

▲ Source countries of legal migrants to the USA, 2001–2005.

Migrating to the USA

In 1990, the US Government passed the Immigration Act. The aim of this was to help businesses attract skilled foreign workers. An annual limit of 675 000 permanent immigrant visas (permits to enter the USA) was set. It was also agreed that 125 000 refugees would be admitted each year.

Anyone who wishes to migrate to the USA has to obtain a visa. There are two types of visa:

- an *immigrant* visa for people who intend to live and work permanently in the USA
- a *non-immigrant* visa for people who live in other countries and wish to stay temporarily in the USA, e.g. tourists or students.

The immigration policy of the USA aims to:

- Admit workers with specific skills for jobs where workers are in short supply.
- Admit immigrants who already have family members living in the USA.
- Provide refuge for people who face racial, political or religious persecution in their country of origin.
- Increase ethnic diversity by admitting people from countries with previously low rates of immigration to the USA.

However, the long land border with Mexico is difficult to police, and there are many people who try to enter the country illegally (see opposite). There are an estimated 12 million illegal immigrants living in the USA, of which 57% are from Mexico.

Advantages and disadvantages of immigration

In a survey in 2004, 55% of Americans said that legal immigration should remain at current levels or be increased. 41% said it should be decreased.

Some of the advantages are that:

- immigrants add more than $30 billion to the US economy
- foreign-born workers fill unskilled, low-wage jobs
- 40% of US PhD scientists were born abroad
- immigrants are more likely to start up businesses and provide further employment – immigrants start up 40% more new businesses than non-immigrant Americans
- over their working lifetime, immigrants contribute nearly $80 000 more tax revenue than US-born Americans.

However, many believe that immigration can cause tensions within the host country. Some of these include:

- wages are forced down by migrants' willingness to work for less
- health and welfare systems are strained by the demands of a larger population
- immigrants often do not integrate into broader American society and culture, becoming isolated and misunderstood by the majority of Americans.

your questions

1 List reasons why the USA would want to control the number of immigrants entering the country each year.

2 Take each of the four aims of US immigration policy and explain why you think these might be suitable for the USA.

3 Look at the graph showing immigration to the USA.
   a When were the peak periods for immigration?
   b What is the current trend for immigration?

4 Look at the map showing the source countries of legal migrants to the USA, 2001-2005. Describe where the highest number of migrants came from.

5 Exam-style question Using examples, explain how some countries try to increase immigration, whilst other countries try to reduce it. (6 Marks).

10.1 Different types of resources

In this section you'll learn how resources can be classified into different categories.

+ A **resource** is a naturally occurring substance (e.g. water, minerals) which can be used in its own right, or made into something else.

How can we classify resources?

Resources are all the natural things we need to live and work. The different types are:

- **Physical** – Natural materials found at or below the Earth's surface, such as soil and rock. Many rocks are used as energy sources (e.g. coal) or contain minerals (e.g. iron) and can become essential for human activity and making products.
- **Energy** – Resources used specifically for heat. Fossil fuels (so-called because they are formed from fossilised plant or animal matter over millions of years) – such as coal, oil or natural gas – can be burned, either for heat, or to create steam to drive generators to produce electricity. Nuclear energy uses heat from a nuclear reaction in uranium or plutonium.
- **Mineral** – Materials normally quarried or mined from the ground in raw form (e.g. iron ore) and then heated and purified to become materials that are used (e.g. steel is manufactured into cars).
- **Biological** – Resources with a biological origin that have developed through growth and development (e.g. trees or other natural vegetation), or resources grown for human use (e.g. crops).

Resources can also be divided into two main categories, which refer to their ability to last over time:

- **Renewable** – These will never run out and can be used over and over again, e.g. wind and solar power. They are infinite resources. Some of these resources are also categorised as **sustainable**, e.g. bio-fuels and hydrogen-powered cells (see Section 10.10). They meet the needs of people now, without preventing future generations from meeting their own needs.
- **Non-renewable** – These are being used up and cannot be replaced, e.g. coal and oil. They are sometimes known as finite resources.

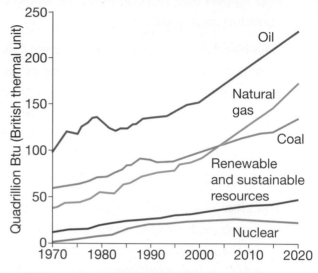

▲ Global energy use by fuel type, 1970 – 2020 (projected).

Renewable energy: wind power in the USA

Wind turbines convert the power of the wind into electricity. There are now more than 13 000 large wind turbines in California, and hundreds of homes and farms across the state are also using smaller wind turbines. In 2012, wind energy provided 5% of California's total energy requirements.

Sustainable energy: biogas in India

Biogas plants convert organic matter, such as wood chips and animal dung. This ferments, releasing methane gas. This is collected in a tank and can be burnt to provide electricity or gas for cooking. There are over 2.5 million biogas plants across India, providing 57% of the country's energy (also see Sections 12.7 and 16.8).

Non-renewable energy: natural gas in Europe

Natural gas is used for electricity production, heating and cooking. In Britain, gas is collected from underneath the North Sea (see photo). However, much of this gas has now been used up. In 2006, the UK imported 50% of its gas supplies. This is expected to increase to 80% by 2020. Like many other countries in Europe, Britain is now dependent on gas supplied from Eastern Europe and Russia. Large pipelines carry the gas across the continent to where it is needed. In 2004, the Government produced a report identifying that global gas supplies will fall after 2030. Other energy sources will need to be found after this date.

your questions

1 Make a copy of the table below and classify the following resources into the correct columns: solar power, oil, apples, wave power, vegetable oil, iron ore, limestone, grass, milk, sand, wind power, water, chicken, coal, natural gas, biogas.

Type	Renewable	Sustainable	Non-renewable
Physical			
Energy			
Mineral			
Biological			

2 Look at the three examples of energy resources in the coloured boxes. Research one more type of resource, describe where it's found and explain whether it is:
   a renewable, sustainable or non-renewable
   b physical, energy, mineral, or biological.

3 Exam-style question Look at the graph opposite.
   a Describe what is happening to global energy use. (3 marks)
   b Outline two pieces of evidence to show that global energy use is not sustainable. (4 marks)

✚ In this section you'll investigate the impacts that the world's growing population might have on resources.

More people = more demands

In 2007 and 2008, there were street protests and riots in West Africa, and outbreaks of violence in Mexico, Morocco, Yemen, Haiti, Egypt, India, Indonesia and other places (see the map). The reason? **Food insecurity**. The cause? Rising standards of living and global population growth mean that there is greater demand for food. At the same time, there is a world shortage of wheat, rising milk and rice prices, and drought in parts of Africa. The result is that billions of people struggle to afford basic foodstuffs. With over 7 billion people on the planet, there is a delicate balance between the resources that these people need and the resources that are available.

+ **Food security** is the ability to obtain sufficient food on a day-to-day basis. People are considered to be 'food secure' when they do not live in fear of hunger.

+ **Food insecurity** is when it is difficult to obtain sufficient food. This can range from hunger through to full-scale famine.

Food crisis threatens security

Ban Ki-Moon, the UN General Secretary, issued a gloomy warning yesterday. He said that the deepening global food crisis, which has triggered riots and threatened hunger in dozens of countries, could have serious implications for international security, economic growth and social progress.

▲ Adapted from a newspaper extract, April 2008.

▼ Global price protests and their main causes.

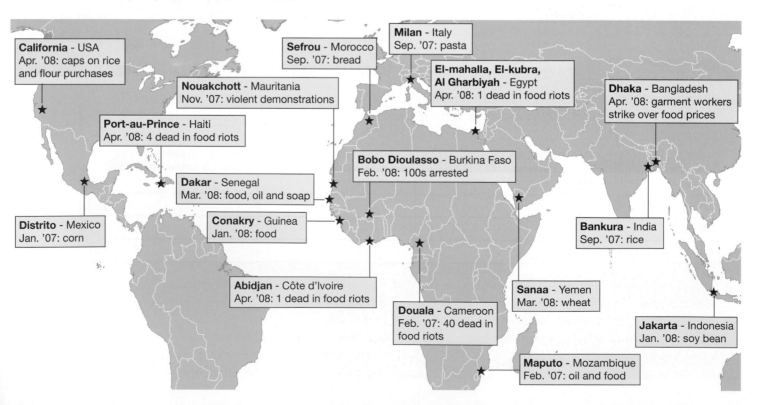

California - USA
Apr. '08: caps on rice and flour purchases

Nouakchott - Mauritania
Nov. '07: violent demonstrations

Port-au-Prince - Haiti
Apr. '08: 4 dead in food riots

Distrito - Mexico
Jan. '07: corn

Dakar - Senegal
Mar. '08: food, oil and soap

Conakry - Guinea
Jan. '08: food

Abidjan - Côte d'Ivoire
Apr. '08: 1 dead in food riots

Sefrou - Morocco
Sep. '07: bread

Milan - Italy
Sep. '07: pasta

El-mahalla, El-kubra, Al Gharbiyah - Egypt
Apr. '08: 1 dead in food riots

Bobo Dioulasso - Burkina Faso
Feb. '08: 100s arrested

Douala - Cameroon
Feb. '07: 40 dead in food riots

Sanaa - Yemen
Mar. '08: wheat

Maputo - Mozambique
Feb. '07: oil and food

Dhaka - Bangladesh
Apr. '08: garment workers strike over food prices

Bankura - India
Sep. '07: rice

Jakarta - Indonesia
Jan. '08: soy bean

The population of the world will continue to grow. The United Nations predicts that it will peak at around 10 billion by 2185. That's another 3 billion people in the world, compared to now. Those people are likely to face challenges of:

- more expensive food
- more expensive fuel – more people will mean greater demand for oil, etc.
- climate change – more people will mean a greater release of CO_2 into the atmosphere
- water shortages – already many people in the world lack access to safe water
- more migration – many people will be born in some of the world's poorest countries, and they will want to move to where they can achieve a better quality of life
- political instability and war – with more people competing for fewer resources, individuals and governments may resort to desperate means.

▲ Police quelling food riots in Haiti in April 2008.

China's growth could spark political tensions

China's booming economy is expected to consume more than half of the world's key resources within a decade.

The rapid development of China's economy means that it is likely to consume the majority of the world's supply of all the major metals and minerals, potentially leading to clashes with other countries over controlling access to resources. China already accounts for 47% of all iron ore consumption, 32% of aluminium and 25% of copper.

▲ Adapted from a newspaper article, January 2008.

Will the world cope?

Some people believe that there are two possible outcomes:

- A future where there are not enough resources for the global population. This will lead to mass starvation, and ultimately a fall in population. This is called the Malthusian view (see page 170).
- A future in which people successfully use technology in order to provide resources for the growing population. This is called the Böserupian view (see page 171).

your questions

1 Ban Ki-Moon, the UN General Secretary, stated that he was concerned that food shortages could have severe global impacts. How far do you agree with him? Use evidence from the information in this section.

2 As you have seen (in Section 9.8), China's population is still growing. What impact could this have on its future use of resources?

3 What challenges does a bigger world population create?

4 Exam-style question Look at the world map opposite.
   a Describe the locations of the countries where most of the protests took place. (3 marks)
   b Outline two possible reasons why these countries might be most affected by resource shortages. (4 marks)

✚ In this section you'll learn that there are different views about the relationship between population and resources.

Malthus – 'We're all doomed!'

Thomas Malthus (1766-1834) was born into a wealthy family. He was educated privately and studied at the University of Cambridge. He later married and had three children and became Professor of Political Economy at the East India Company's college in Hertfordshire. His students affectionately referred to him as 'Pop' or 'Population' Malthus.

In 1798, Malthus wrote an influential essay about population. He believed that population grew **exponentially** (doubling at each stage – 1:2:4:8:16, etc.), but that food production grew **arithmetically** (adding one unit at each stage – 1:2:3:4:5, etc.). This meant that population would eventually outstrip food supply (see the graph). At this point, the population would decrease through starvation. Malthus called this a 'natural check' on population growth. Other 'natural checks' were war, disease and morality.

Where does morality come into it?

Malthus had deep religious beliefs that affected his work. He believed that people had a moral duty to keep the population low. This could be achieved by restraint from marriage and sexual relations. He called this 'delaying the gratification of passion from a sense of duty'.

According to Malthus, whenever population outstripped resource supply, 'natural checks' would come into play. The population would be reduced to a more manageable level and would then continue to grow again until the next 'natural check'.

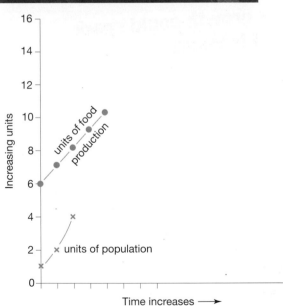

In 2000, the United Nations said that 12.5% of the world's population were starving (down from 17% in 1990). Despite this fall, 40% of the world's population were still living on just US$2 a day – or less.

'I think I may make two points: first, that food is necessary for the existence of man. Secondly, that passion between the sexes is necessary and will always remain. Assuming this then, I say that the power of population is greater than the power in the earth to produce food for man'.

Thomas Malthus, 'An essay on the principle of population', 1798.

The Honourable East India Company

The East India Company was founded in 1600 by Elizabeth I, and traded cotton, silk, tea, spices and opium that it sourced from Asia. For many years, the company effectively governed India. This came to an end in 1858, when the British government took direct control of India after the Indian Mutiny.

Malthus's views often influenced the policies of the East India Company. When famines struck in India, the British authorities would do nothing to relieve them. Famine was seen as a 'natural check' on population growth. This became known as the 'Malthusian' policy of the East India Company.

Böserup – 'Necessity is the mother of invention'

Ester Böserup (1910-1999) was a Danish economist who worked for the United Nations. In 1965 she published *The Conditions of Agricultural Growth*. This book opposed the ideas of Malthus. Böserup did not like the idea of 'natural checks'. She argued that food production does not limit or control population growth. Instead, she said that population growth controls farming methods. She believed that people would try not to give in to disease or famine. Instead, they would invent solutions to the problem. She used the term 'agricultural intensification' to explain how farmers can grow more food from the same piece of land using better farming techniques and chemical fertilisers. The graph below shows Böserup's view.

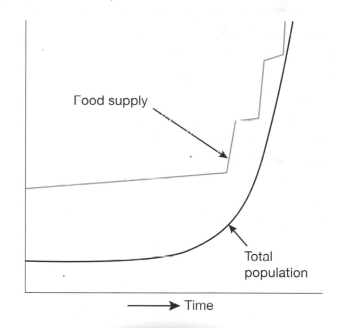

your questions

1 Look at the graph that shows Malthus's prediction.
   a Copy the graph on page 170. Extend the lines. Remember that population doubles at each stage. Food production only gets bigger by one unit at each stage.
   b Shade in the area where there would be too many people and not enough food. Where this happens, people would die of starvation. Label this area on your graph.
   c Do you think that this is the situation in the UK today? Why?
   d Is it true of other parts of the world?
2 Look at the graph that shows Böserup's view.
   a Describe what the graph shows.
   b Explain why there is no point on the graph where there are too many people and not enough food.
3 **Exam-style question** Explain why some people believe that the world's resources will run out soon, while others think that will not happen. (6 marks)

What do you think?

+ Although the global population is now 14 times larger than it was at the time of Malthus – at over 7 billion – the food supply has grown even faster. So was Böserup right?

Patterns of resource supply and consumption

✚ In this section you'll learn that resource supply and consumption are not evenly spread.

The trade in metals

It's January 2013, and slow economic growth dogs the economies of Europe and the USA. One of the main problems is high metal prices – they've reached record highs since 2008. The mining companies are enjoying record profits, but something must be driving prices upwards. Which countries use these metals, and where do they all come from?

The two maps on the right have been drawn to show which countries export and import metals. Metals include nickel, zinc, copper and aluminium. They also include metal items such as tools and cutlery. Of all the money spent on trade worldwide, 3.8% is spent on metal exports. On each map, the size of each country has been adjusted to show how much metal it either exports or imports.

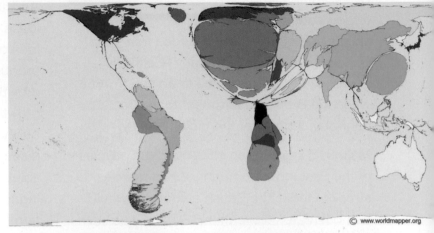

▲ Countries that **export** metal and metal products (top) and countries that **import** metal and metal products (bottom)

A new enemy for 4x4 owners

Owners of 4x4s now have a new enemy – thieves who slip under their car and saw off its catalytic converter. The police believe that thousands of the exhaust parts have been stolen from 4x4s in the past six months by criminal gangs who want the precious metals inside. The prices of these metals have soared over the past two years.

The thieves, who sell the catalytic converters on to metal traders for between £100 and £200, are the first stage of an international trade in scrap metal that takes the stolen car parts to India, China and Eastern Europe to be recycled and sold on.

▲ Adapted from a newspaper article, August 2008

Does every cloud have a silver lining?

In June 2008, the *New Scientist* magazine reported in detail about why the price of metals on the world market was increasing. Metals are being used up (see right). The report claimed that if metals continue to be used at the current rate, there will only be 45 years' supply of gold left, 46 of zinc, 29 of silver and 59 of uranium. Worse still, if the rest of the world were to use metals at the rate of the USA, things could get even worse – there would only be 36 years' supply of gold, 34 of zinc, 10 of silver and 19 of uranium!

Stocks of silver have declined worldwide to about 300 million ounces. That's a seventh of what they were 10 years ago. Recycling of silver is difficult. Some can be recycled, but most of it is lost forever once it has been used. The result is that the price of silver continues to rise (see right).

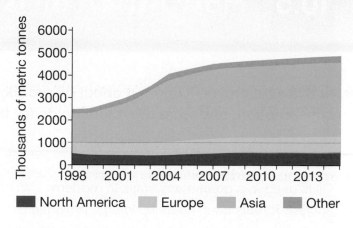

▲ Worldwide metal consumption by continent 1998–2015 (estimated)

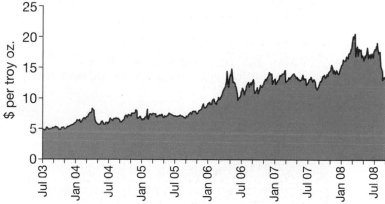

▲ The price of silver on the world market, July 2003 to September 2008.

▲ Silver has the highest thermal conductivity of all metals. It is used in computer keyboards, mobile phones, washing machines, batteries, TVs, and cameras.

your questions

1 Look at the maps (opposite) showing metal exports and imports.
   a Describe the distribution of countries that export metals.
   b Explain why the distribution of countries importing metals is different from that of countries exporting metals.

2 Look at the graph (top of page) showing worldwide metal consumption by continent.
   a Which continent uses the largest quantity of metals?
   b Using the export map (opposite), explain where most of these metals come from.

3 Explain why the price of silver continues to rise on the world market.

4 **Exam-style question** In the future, it may be more difficult for countries to obtain the metals needed to make everyday products. Explain:
   a how countries that export metals may benefit from this. (3 marks)
   b the problems faced by countries which import metals. (3 marks)

In this section you'll find out about oil production and consumption, and what the future holds for this 'black gold'.

Black gold!

Oil is used in a great many ways in modern society. It fuels cars, heats buildings, provides electricity, and makes the plastics that we use in everything from milk containers to computers. Oil is a **finite resource** – in other words, it will run out someday. There is only so much oil under the ground, and, once it is all used up, it will be gone forever. The question is, how much oil is left?

Oil consumption rose from less than a million barrels a day in 1900, to 87 million barrels a day by 2011. The International Energy Agency predicts that demand will rise to 116 million barrels a day by 2030. Unfortunately, the oil industry believes that it is impossible to produce more than 100 million barrels a day. This is the problem that the world faces — oil is needed for energy and transport, but there is not going to be enough of it to go round!

The map and table on the next page show which countries have the largest known oil reserves.

> + **Black gold** is another name for oil, because it is such a valuable commodity.
>
> + A **finite resource** is one that is limited or restricted.

On your planet

+ Transport accounts for over half of all global oil consumption. 40% of the recent increase has come from the road-freight (trucks) industry.

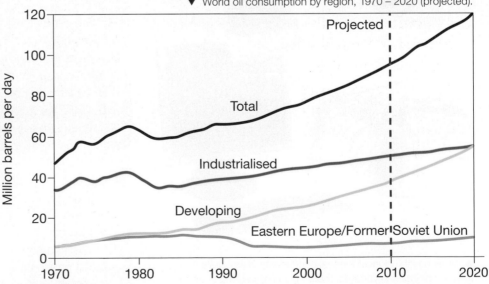

▼ World oil consumption by region, 1970 – 2020 (projected).

Y-axis: Million barrels per day (0 – 120)
X-axis: 1970, 1980, 1990, 2000, 2010, 2020

Projected

Total

Industrialised

Developing

Eastern Europe/Former Soviet Union

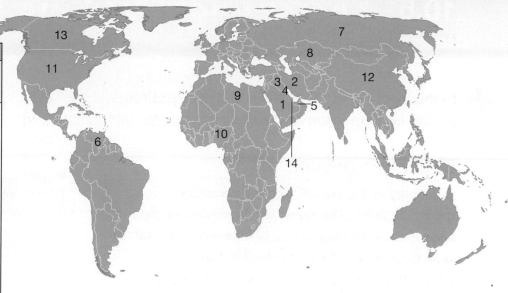

World reserves of oil		
	Billions of barrels	Percentage of world reserves
1 Saudi Arabia	262.73	22.3
2 Iran	132.46	11.2
3 Iraq	115.00	9.7
4 Kuwait	99.00	8.4
5 UAE	97.80	8.3
6 Venezuela	77.22	6.5
7 Russia	72.27	6.1
8 Kazakhstan	39.62	3.4
9 Libya	39.12	3.3
10 Nigeria	35.25	3.0
11 USA	21.37	1.8
12 China	17.07	1.4
13 Canada	16.80	1.4
14 Qatar	15.20	1.3

+ **Peak oil** is the point at which oil production reaches its maximum level and then declines.

Drilled out?

It is difficult to tell how much oil remains under the ground, or how much demand there may be in the future. However, what is certain, is that once **peak oil** is reached, it will become much more difficult and expensive to extract what is left.

Oil pessimists believe that the world has already reached – or is close to reaching – peak oil. They say high oil prices are evidence of this. Oil optimists believe that this point is still decades away, because there is so much oil yet to be discovered – including huge reserves of tar that can be refined in Canada. The chief economist at BP predicts: 'People will run out of demand before they run out of oil.'

A post-oil future

Oil pessimists argue that we are at a turning point in oil production. From now on, they argue, there will be less oil to extract. This could lead to recession and possibly war, as countries that import oil try to get access to oil reserves. In the 1970s, an earlier time of high oil prices, the USA considered military action to seize Middle-Eastern oil fields. The USA decided against such action then, but the future could be different. Many people argued that the 2003, US-led invasion of Iraq was directly linked to oil.

your questions

1 Look at the graph that shows world oil consumption.
   a Give reasons why world oil consumption has continued to rise.
   b Suggest reasons why industrialised countries now account for a smaller proportion of world oil consumption.
2 Look at the map and table showing who has the largest oil reserves. Describe the distribution of these countries.

3 In the future, it may be more difficult for countries to import oil. Explain the problems that countries may face as follows:
   a Countries that rely on oil for energy and transport.
   b Countries that were major exporters of oil.
4 Exam-style question Using examples, describe how the pattern of consumption of one resource you have studied is changing. (4 marks)

✚ In this section you'll find out about pressures on the consumption and supply of energy.

Economic growth

In January 2013, the UK reported zero economic growth for 2012. The American, Brazilian and Japanese economies grew by just 2%. Meanwhile, in the same year, China's economy grew by 7.5% and India's by 5%! The continued strong growth of these two economies means one thing – energy consumption will keep on rising. The global demand for energy is expected to grow by over a third by 2035 (see the graph) – with daily oil demand increasing to 100 million barrels (up from 87 million in 2011).

Changing international relations

The USA and energy security

The USA was the world's largest oil consumer in 2011. However, it has to import half of its oil, so it's **energy insecure**. Even the loss of one source can create shortages and drive global oil prices up, so a continuous supply is essential. This has led to a search by both oil companies and the US government for new dependable sources of oil.

- Oil companies have explored new territories in the Arctic and Canada, where there are thought to be huge reserves.
- The USA has attempted to stabilise areas of human conflict where oil supplies could be cut off.

The invasion and occupation of Iraq between 2003 and 2011 by US and Allied forces was critical to the future of US oil supplies, because Iraq has the world's third largest oil reserves. Political tensions and disputes between Iraq and its neighbours (and the wider world) posed a threat to future US and global oil security.

▶ China needs more oil to run the thousand extra cars arriving on Beijing's streets every day. Much of that oil will need to be imported.

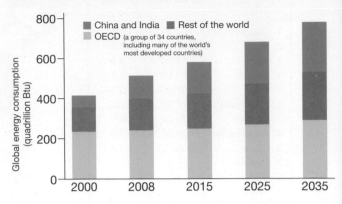

▲ The expected increase in global energy consumption by 2035.

> ✚ **Energy security** means access to reliable and affordable sources of energy. Countries with enough, or surplus, energy are said to be **energy secure**, while those without enough, are **energy insecure**.

China

China's economy doubled in size between 2005 and 2012. By 2013, it was the world's second largest energy consumer (behind the USA). The Chinese demand for energy has risen by 86% since 2000, and it now accounts for one sixth of all the energy produced globally. It needs this energy for industry, and also to meet the demands of growing Chinese car ownership. China is expected to account for 40% of the world's total energy consumption by 2030.

The USA has tried for many years to reduce its dependence on imported oil – by seeking new domestic energy supplies from previously inaccessible sources, e.g. shale gas. Gas can easily replace oil, e.g. in electricity power stations, so using gas saves oil. US domestic energy production is now rising, and imports are falling.

As the USA becomes more energy secure, the international oil market will change. It's expected that 90% of oil exports from the Middle East will go to Asia by 2035. Iraq, after decades of conflict and instability, hopes to transform itself by expanding its oil production to become an important supplier to China.

Energy tensions

Many of the world's oil and gas reserves are located in areas vulnerable to political tensions, disputes between countries – and even war (see some recent examples on the map below). As a result, global oil and gas supplies are at risk of being disrupted.

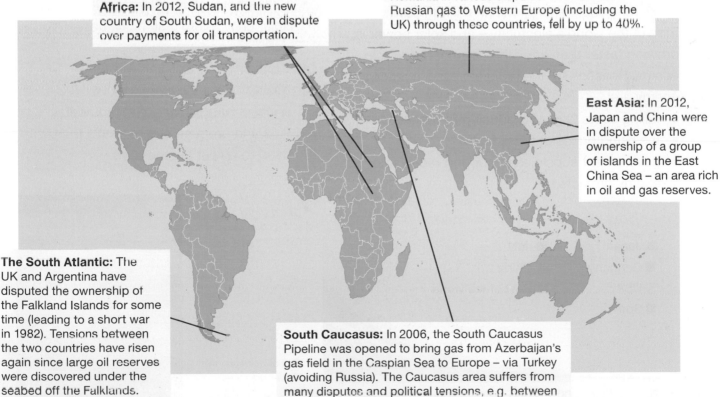

your questions

1 In pairs, brainstorm the reasons why expanding economic growth in China and India will lead to greater demands for **a** energy, and **b** oil. Think of all possible users.
2 Why does increasing economic growth in India and China have big implications for **a** car ownership in those countries, **b** their energy security, and **c** demands for oil supply globally?
3 In pairs, research 'USA shale gas supplies' and produce a 6-slide presentation (including pictures) to show **a** what shale gas is, **b** where it's found, and **c** why it causes concern for some people.
4 Choose one of the recent disputes outlined on the map. Conduct some research to find out more about the dispute, and then produce an illustrated poster (with a maximum of 250 words) to summarise and explain the risks to oil or gas supplies in that area.
5 Exam-style question Using examples, describe likely pressures on the future global supply of **either** oil **or** gas. (6 marks)

Russia: Between 2006 and 2010, Gazprom (Russia's gas company) cut winter gas supplies to Ukraine, Poland, Hungary, Bulgaria, and Romania in various disputes. The flow of Russian gas to Western Europe (including the UK) through these countries, fell by up to 40%.

Africa: In 2012, Sudan, and the new country of South Sudan, were in dispute over payments for oil transportation.

East Asia: In 2012, Japan and China were in dispute over the ownership of a group of islands in the East China Sea – an area rich in oil and gas reserves.

The South Atlantic: The UK and Argentina have disputed the ownership of the Falkland Islands for some time (leading to a short war in 1982). Tensions between the two countries have risen again since large oil reserves were discovered under the seabed off the Falklands.

South Caucasus: In 2006, the South Caucasus Pipeline was opened to bring gas from Azerbaijan's gas field in the Caspian Sea to Europe – via Turkey (avoiding Russia). The Caucasus area suffers from many disputes and political tensions, e.g. between Azerbaijan and Turkmenistan.

In this section you'll look at the consumption and supply of renewable energy – in particular, solar power.

Developing renewable energy

Fossil fuels still generate 85% of global energy supplies, as the pie chart shows. However, as demand for energy grows (see Section 10.6), and 'peak oil' is reached (see Section 10.5), renewable resources (such as solar and wind power) could help to fill the potential energy gap.

Solar power generation increased by 86% in 2011 – admittedly from a small base (see graph below). Most growth took place in Germany and Italy, but China is also developing domestic solar power (as well as building a new 200 MW solar power PV plant). The table lists the top ten countries, in terms of solar power production and consumption, for 2011.

The International Energy Agency predicts that, by 2035, renewable energy will be used to produce a third of the world's electricity – and solar power is growing faster than any other form of renewable energy. Reasons for the rapid increase in renewable energy are:
* falling costs (e.g. of producing PV panels)
* the rising prices of fossil fuels
* government subsidies to support renewable energy projects.

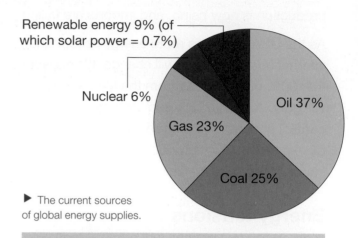

▶ The current sources of global energy supplies.

Renewable energy 9% (of which solar power = 0.7%)
Nuclear 6%
Gas 23%
Coal 25%
Oil 37%

Solar energy

Solar energy can be used to heat water, and to heat and cool living and working spaces. It can also be used to create electricity:
* Solar photovoltaic (PV) directly converts solar energy into electricity, using PV cells grouped into panels.
* Concentrating Solar Power (CSP) devices generate electricity by concentrating solar energy to heat a fluid and produce steam (that's then used to power a generator).

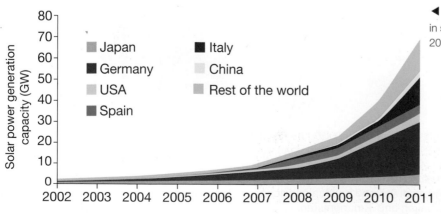

◀ The growth in solar power 2002-2011.

▶ Solar power – the top ten countries in 2011 (by amount of installed solar energy).

	MW
Germany	10 000
Spain	3500
Japan	2700
USA	1800
Italy	1300
Czech Republic	600
Belgium	450
China	400
France	350
India	200

Global supply

Some parts of the Earth receive far more solar energy than others, as the map shows. Solar energy has huge potential – especially for developing countries in tropical areas, e.g. Ghana (see right). In tropical areas, the high angle of the sun makes solar energy more intense – and the cost of developing solar power could be less than extending conventional electricity grids and power lines into rural areas.

The advantages for countries in producing their own energy from renewable sources include:

- Increased energy security – with less dependence on imported energy.
- Protection against changes in international relations.
- 'Future-proofing' their energy supplies (in a world with declining fossil fuels), by making use of energy sources that will not run out.
- Producing the energy needed for industrial growth.

Africa's largest solar power plant to be built in Ghana

Ghana has the fastest-growing economy in sub-Saharan Africa – with GDP growing at over 14% in 2011. But it needs energy to drive its economy. So, it's planning to use the sun's energy to produce electricity.

Blue Energy is the company behind the $400 million project. It says that the 155 MW solar photovoltaic plant will be operational by the end of 2015. This power plant will be the fourth biggest of its kind in the world, and will help Ghana to meet its target of increasing its renewable energy use by 2020, from the current 1%, to 10%.

▲ Adapted from a newspaper article, December 2012.

On your planet

+ Just 0.0005% of the Earth's surface would need to be covered with solar panels in order to power the entire planet.

▼ The average amounts of solar energy reaching the Earth, 1990-2004. The highest levels of solar energy are shown in red, the lowest levels in purple and pink.

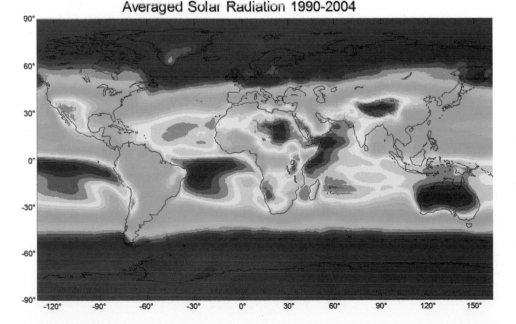

Yearly Mean of Daily Irradiance in W/m²

Averaged Solar Radiation 1990-2004

your questions

1 In which countries is solar power generation increasing? Suggest two reasons why.

2 Complete a table to show the advantages and disadvantages of countries obtaining their own energy from solar sources.

3 In pairs, draw a spider diagram to outline the possible reasons for and against offering government subsidies for solar panels to **a** industry, **b** individual homes.

4 Explain **a** why energy demand is rising in Ghana, and **b** how the development of solar energy offers advantages over oil for developing countries like Ghana.

5 Explain how far you think solar energy has the potential to become the world's number one fuel, replacing oil.

6 **Exam-style question** Describe the pattern of the sources of global energy supplies shown in the pie chart opposite. (3 marks)

✚ In this section you'll weigh up the theories of Malthus and Böserup, by considering global food demand and supply.

Lots of oil, but no food

Abu Dhabi is the largest of the seven emirates that make up the United Arab Emirates (UAE). It is a major producer of oil, producing over 2.7 million barrels a day. However, it is a desert country. Abu Dhabi has very little rainfall or land that is suitable for growing crops. The emirate relies on the profits made from selling oil to import food resources for its people. There was a global food shortage in 2008, coupled with price rises in wheat and grain. This led the Abu Dhabi Fund for Development (ADFD) to search for a sustainable solution to its food security requirements.

The answer has been to develop 30 000 hectares of farmland in the country of Sudan. The government of Sudan has agreed to lease the land free of charge, in exchange for technological expertise in improving farming techniques. Sudan lies close to the Equator, but has a plentiful water supply from the River Nile. It also has large areas of land that are suitable for agricultural development. Crops, such as wheat, potatoes, beans and alfalfa, are grown and exported to Abu Dhabi. This solution suits Sudan, which needed new ways of earning money after it lost its oil reserves when South Sudan became independent.

The ADFD is worried about global warming and continued population growth. The time may come when, even if you have money, buying resources such as food may not be easy. Abu Dhabi is committed to developing guaranteed food supplies, with similar projects in Uzbekistan and Senegal.

▲ Adapted from a newspaper article, July 2008

Can the planet feed us?

The world's population reached 7 billion in 2011 – more than doubling from 3 billion in less than 40 years. Even so, our food supply has kept pace with this growth. Not only has it kept pace, but more people are consuming more calories than 40 years ago (see the graph below):

- world cereal production has doubled since 1970
- meat production has tripled since 1961
- the number of fish caught grew more than six times between 1950 and 1997

Malthus (in Section 10.3) predicted that population would grow faster than the production of food. This would provide a 'natural check' on population growth. In recent times, this has not been the case. Böserup's view, that 'Necessity is the mother of invention', would seem more accurate. But the huge growth in food output was only possible by giving nature a 'helping hand' through the use of fertilisers, irrigation systems and pesticides.

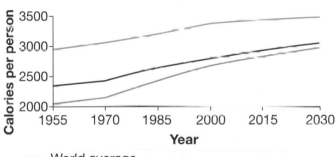

- World average
- Developing countries — Developed countries

▲ Calories consumed per person per day.

▲ Genetically modified corn has been developed to be resistant to pests.

Can the rate of food production continue to grow?

The world's population is expected to peak at around 10 billion people by 2185. Most of the world's best farmland is already being used, or even overused. Pesticides are beginning to have less effect than in the past, as pests develop resistance. Perhaps the future will depend on bio-technology? Genetically modified (GM) crops are being developed to be drought- or pest-resistant. The concern is whether such foods will be safe for people to eat, or have other unforeseen consequences.

your questions

1 Read the extract opposite about Abu Dhabi.
  a Give reasons why Abu Dhabi is aiming to lease land from other countries.
  b Do you think that this is a sustainable solution for Abu Dhabi? Give reasons for your answer.
2 Using the UAE to illustrate your answer, explain how some countries can be rich in one resource, yet poor in another.
3 a Using the Internet, research i what is meant by GM foods, ii the arguments for and against their use.
  b Do you think that GM crops are the answer to a sustainable food supply?
4 Exam-style question Look at the graph showing calorie consumption levels per day.
  a Describe the trends shown in the graph. (3 marks)
  b Outline two reasons why the number of calories consumed has continued to increase, despite the world's population continuing to grow. (4 marks)

+ In this section you'll learn about some ways in which governments can manage resource consumption.

How much longer can we keep this up?

In 2007, WWF-UK analysed the average **ecological footprint** of people living in the UK's 60 largest cities. They calculated the resources needed for them to travel, to produce the food they ate, and manage all their waste. As a result, WWF-UK worked out that if everyone in the world lived the same way as the average Briton, it would take three planets to support the world's population! Worse, if everyone lived in the same way as the average American (the world's richest country), they would need 9.6 planets to support them! The graph shows how the global trend in ecological footprint is rising.

Clearly this can't go on forever. As standards of living continue to rise in Asia, demands on the planet will increase. Consumption has to be managed in order to prolong the life of finite resources, such as oil or metal ores. National and local governments can help to reduce resource consumption – it's a matter of changing behaviour, which involves **education**, **conservation**, and **recycling**.

Education

Every UK local authority has ways of encouraging schools to recycle. For example, Ealing in west London offers:

- 240 litre wheelie bins for paper recycling, which are emptied weekly for free
- advice on re-using material, e.g. using cardboard boxes as recycling containers beside photocopiers
- 240 litre wheelie bins for recycling drink or food cans
- free compost bins for fruit and vegetable peelings, and a garden-waste collection service.

+ An **ecological footprint** is a calculation that estimates the area needed to supply resources to an individual, or a group of people, to maintain their lifestyle. If people live within the ability of the Earth to supply resources, their footprint is 1. The higher the figure, the more space is needed – and the greater the inability of the Earth to supply all they need.

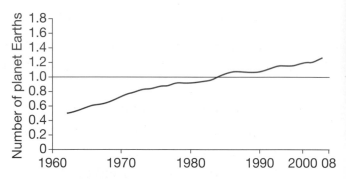

▲ Humanity's Ecological Footprint. This graph compares the resources people use and the Earth's ability to provide them. In 2008, it took about 1.25 planets to supply all the resources used.

▲ Wheelie recycling bins in a school. How far do you think these teach future citizens to recycle as a habit?

Conservation

Reducing domestic energy consumption (i.e. within our homes) is one way to reduce resource use. Improving the insulation in houses reduces energy consumption as more heat is trapped inside the house. The UK government offers two 'Green Deals' to help households to do this:

- 100% grants – available to low-income households, or those who live in an area with high fuel poverty (where the cost of the energy used accounts for more than 10% of a household's income).
- Loans – repaid in instalments that are added to the household's electricity bills over several years. These loans pay for home improvements to cut energy consumption, e.g. secondary glazing.

Until 2010, the UK government also offered large subsidies for installing private and commercial wind turbines, and solar panels:

- Householders could install PV solar panels and be paid for every kwh of energy that was added to the National Grid (plus receiving low-cost or free energy). The subsidies were reduced in 2010, but still exist.
- Energy companies could receive grants towards building wind turbines.

Both subsidies were criticised, because they didn't encourage people to reduce their energy consumption (although they are renewable sources). Also, onshore wind turbines are often criticised for their impact on the countryside, and are unlikely – without large numbers – to bring about a reduced consumption of fossil fuels.

▼ Onshore wind farms are controversial, because some people claim that they spoil views of the countryside. However, every kwh of energy they produce means less fossil fuel is needed.

Recycling

UK households are now recycling more. In the ten years between 2001 and 2011, the percentage of household waste recycled in England rose from 11% to 40%. However, this still means that the average person in England disposes of 263kg of waste every year (half of which goes to landfill).

- The UK produced 7 million tonnes of food waste in 2009, which – if used as bio-fuel – could produce the same amount of electricity as a new power station. That would save oil.
- Recycling plastic water bottles (made from crude oil) would also save oil. To make six one-litre plastic bottles requires 1kg of oil. In the USA, the amount of oil used to produce plastic for bottled water is enough to run a million cars for a year!

Therefore, governments give local authorities recycling targets. For example, the Welsh Assembly has set Welsh local authorities the target of increasing recycling from 52% of waste in 2012-13, to 70% by 2024-25. Because central government provides over 80% of local authority funds, there could be financial penalties if the recycling targets are not met.

your questions

1 a In pairs, research 'ecological footprint' and look for a website that will help you to calculate your own footprint. Compile the results as a class.

  b Write a paragraph to describe **i** how your footprint was calculated, **ii** which parts contributed most to your overall score, and **iii** how your footprint could be reduced.

2 Draw up a table to compare the strengths and weaknesses of the government policies on education, conservation, and recycling outlined in this section.

3 In pairs, design a plan to **a** increase further the percentage of waste that is recycled, **b** reduce food waste, **c** reduce the usage of plastic bottles.

4 **Exam-style question** Using examples, explain how governments try to manage resource consumption. (6 marks)

+ In this section you'll consider whether technology and renewable resources could help to 'fix' future resource use.

Can we 'fix' our way out of trouble?

One of the largest contributors to global CO_2 emissions is oil – and one of the biggest consumers of oil is the car. Supplies of oil can only last so long (see Section 10.5), so governments across the world have tried to reduce car use – but with little success. The only way to manage car usage at present seems to be to 'fix' the technology to reduce oil consumption.

Since 2000, car manufacturers have achieved large reductions in fuel consumption by:

- Improving engine technology. Most manufacturers now produce engines with lower fuel consumption (see the table), which also results in lower CO_2 emissions. To encourage this, the UK government has removed Vehicle Licence Duty on vehicles with CO_2 emissions of less than 100 gm/km.
- Developing cars which use electric power as well as diesel or petrol – known as 'hybrids'. These use electric power on slow, 'stop-start' urban trips, and diesel or petrol on longer journeys (where electric power would run out). Most completely electric cars have only a very limited range before they need recharging.

▲ The 2013 VW Blue Motion Golf, which has an average fuel efficiency of 88 mpg and CO_2 emissions low enough to qualify for free Vehicle Licence Duty!

The hydrogen economy

Hydrogen is the most abundant element in the universe, but it's usually combined with others, e.g. carbon (oil, natural gas) or oxygen (water). It can be separated using heat, and (once separated) it provides an alternative option to oil for powering vehicles – and with no harmful emissions! Maybe the whole economy could use hydrogen instead of oil (hence the name 'the hydrogen economy'). ▶

	Car	CO₂ emission (gm/km)	Fuel consumption (miles/gallon)	Cost new (lowest-cost model) 2013
1	Vauxhall Ampera (electric)	27	235.4	£29 000
2	Toyota Prius Plug-in (hybrid)	49	134.5	£22 000
3	Fisker Karma (hybrid)	53	62.4	£95 000
4	Toyota Yaris (hybrid)	79	80.7	£15 000
5	Renault Clio 1.5 dCi 90 ECO (diesel)	83	88.3	£12 500

◀ The Top 5 most fuel-efficient cars (according to a Microsoft MSN analysis).

Hydrogen's best uses are as fuel for heating or transport, where it's used in a fuel cell that converts hydrogen into electricity. Fuel cells are more efficient than traditional vehicle engines – so a little hydrogen goes a long way. However, because hydrogen is not found as a pure element, energy has to be used to convert water, biomass, or natural gas into hydrogen in the first place. This means that it can have negative impacts, depending on the energy used to separate it.

Although hydrogen itself is CO_2-free, the energy used to separate most of it for use is natural gas – which does produce CO_2 emissions! Taking this into account, a hydrogen-fuel-cell car averages emissions of about 80 grams of CO_2 per kilometre (about the same as an average hybrid or electric vehicle charged using electricity from the grid). However, the emission levels could be zero if the hydrogen is separated using renewable electricity sources, such as nuclear or solar.

Despite the drawbacks resulting from the separation process, there are great advantages to using hydrogen:

- It's cheap to produce, and has a variety of sources.
- It has many uses, e.g. power generation, transport, and low-carbon heating.
- It's safe and already used in the chemical and other industries (e.g. in the production of methanol and margarine).
- Its only by-product is water.
- Refuelling vehicles is quick, and a full tank of hydrogen gives a range of several hundred kilometres.

Filling stations of the future!

In the USA, new hydrogen cars (such as Ford's 'Edge') produce one emission – water pure enough to drink! In 2005, to encourage consumers to buy these cars, California developed the world's first hydrogen highway (the CaH2Net). By 2013, there were 36 hydrogen filling stations in the state (mostly in Los Angeles). This infrastructure might encourage Californians to adopt hydrogen-fuel-cell vehicles as they become more affordable.

▲ A hydrogen filling station in California. It looks just like a normal filling station, but the fuel's different!

your questions

1 **a** Explain why 'hydrogen fuel' is a sustainable fuel, compared to fossil fuels.

  b Describe some examples of ways in which an increased use of hydrogen could help to create a 'hydrogen economy'.

2 Explain how a 'hydrogen economy' could reduce **a** personal, and **b** national ecological footprints (use Section 10.9 to help you).

3 Outline the advantages and disadvantages for **a** governments and **b** individuals, of encouraging lower vehicle emissions (including hydrogen cars).

4 In pairs, brainstorm a vision of the future about the possible effects of a hydrogen economy on **a** how people travel, **b** what cities could be like, **c** jobs that might be gained and lost compared to the present. Present your ideas to the class.

5 **Exam-style question** Using examples, explain how possible it is that a hydrogen economy could resolve resource shortages. (6 marks)

➕ In this section you'll contrast different employment and working conditions in Vietnam and Malawi.

Welcome to Vietnam!

It's July and the monsoon season in Ho Chi Minh City, Vietnam. That means a wet motorbike ride to work for Dang Thu Hoan, one of the city's many textile workers (see the photo). She works in a huge clothing factory there.

Ho Chi Minh City is booming – its population of 6 million in 2004 rose to nearly 8 million by 2012 (nearly the size of London). In the monsoon season, traffic gets clogged up in this busy city as the roads flood.

Vietnam is typical of many Asian countries, with growing cities full of people who have left traditional **rural** lives. Manufacturing jobs are increasing, so people are moving to **urban** areas where these jobs are located. Dang Thu Hoan's factory pays her US$8 a day for a nine-hour shift – double Vietnam's minimum wage. She works a six-day week, but it's easier than for many of her family in the countryside. In some rural areas there is now a shortage of young people, with only older family members left to farm. As industry expands in the cities, fewer people work as farmers.

▲ Dang Thu Hoan in a clothing factory in Ho Chi Minh City.

▲ A modern textile factory in Ho Chi Minh City, Vietnam.

Why is Vietnam booming?

Vietnam plays its part in an increasingly **globalised** world – one in which large companies in the world's High Income Countries make products in countries where wages are low. In Dang Thu Hoan's factory ¬- one of Vietnam's best for pay and conditions - a wage of US$8 a day means that labour is much cheaper than in the USA. Companies making clothes (e.g. Gap) and sportswear (e.g. Nike), **out-source** their production to Vietnam, contracting Vietnamese factories to manufacture products for them. In California, the minimum wage is US$8 per hour – nine times higher than in Vietnam.

Few companies are as good as the one that Dang Thu Hoan works for. Many Vietnamese workers work long hours, in sweatshops, for far lower wages. They produce goods that are **exported** (sold overseas) which increases Vietnam's **Gross Domestic Product** (GDP) – or its income.

Welcome to Malawi!

Meanwhile, in Malawi (eastern Africa), Liena sets out to work on her smallholding. Like half of Malawi's farmers, Liena has about 1 hectare of land, which she cultivates to provide almost all of her family's food. She is a **subsistence farmer** – producing just about enough to feed her family, plus some tobacco and groundnuts to sell.

Nearly 85% of Malawi's population lives in the countryside. Liena uses hand tools and relies heavily on family labour. If she had more produce to sell, she'd have more income – and with more income she could afford fertiliser or better tools. Until this happens, her income is unlikely to increase.

▲ Traditional farming methods in Malawi

Industrialisation

Since 1990, Vietnam has seen rapid industrial growth, known as industrialisation – an economic and social process:

- Economically, it involves using money (or capital) to set up factories, turning raw materials (or primary products) into manufactured goods (or secondary products). This adds value to raw materials. A piece of furniture, for example, is worth more than a block of wood.
- Socially, it relies on workers who are prepared to move from rural areas (the countryside) to urban areas (or cities). Most secondary jobs are in urban factories. Because manufactured goods are worth more than raw materials, urban wages are often higher than rural.

Many factory goods are made for sale overseas (export), which adds to a country's earnings (its Gross Domestic Product).

	Vietnam	Malawi
GDP per person (in US$)	3400	900
Where GDP comes from (%)	Agriculture: 22.0 Industry: 40.3 Services: 37.7	Agriculture: 30.3 Industry: 16.7 Services: 53.0
Percentage of people by occupation	Agriculture: 48.0 Industry: 22.4 Services: 29.6	Agriculture: 90.0 Industry and Services: 10.0
Value of exports (in US$)	95.32 billion	0.91 billion
Export goods	clothing, shoes, electronics, wooden products, machinery, rice	tobacco (53% of the total), tea, sugar, cotton, coffee, peanuts, wood products, clothing

▲ Comparing the economies of Vietnam and Malawi (2011). Countries with more industry tend to have a higher GDP.

your questions

1 Copy and complete the table about the advantages and disadvantages of Dang Thu Hoan's life and working conditions in Vietnam, compared with Liena's in Malawi.

	Dang Thu Hoan (Vietnam)	Liena (Malawi)
Advantages of her life and working conditions		
Disadvantages of her life and working conditions		

2 Explain how industrialisation has given Vietnam a higher GDP per person than Malawi.

3 **Exam-style question**

a Study the table comparing the economies of Vietnam and Malawi. Identify four pieces of evidence to show that Vietnam is a more-industrialised country than Malawi. (4 marks)

b Outline **one** benefit and **one** problem that industrialisation can bring to a country. (4 marks)

In this section you'll learn about how employment changes, and how the Clark Fisher model can explain this.

Employment and industry

Industry provides employment. Employment and industry can be classified into four main groups, or sectors:

- **Primary industry** – the extraction of raw materials from the land or sea, e.g. farming, fishing, quarrying, mining.
- **Secondary industry** – manufacturing, where raw materials are converted into a finished product, e.g. house building, car making, steel processing, food processing.
- **Tertiary industries** – or services. There is a wide range of service industries associated with both manufacturing (e.g. distribution, retailing) and people (e.g. education, nursing).
- **Quaternary industries** – provide information and expert help. They are often associated with creative or knowledge-based industries, especially IT, biosciences, media, etc.

Primary employment: 1.4%

Secondary employment: 18.2%

Tertiary and Quaternary employment: 80.4%

▲ The UK's employment structure in 2011.

Together, the balance of these is known as a country's **employment structure** (see the pie chart).

A company can employ people in different ways. Someone whose job is manufacturing cars is part of the secondary sector. But the sales people for the same company are in the tertiary sector.

In the UK, there has been a major change in the types of jobs that people have been doing over the past 40 years, as the graph shows. There has been a drop in primary and secondary employment, and an increase in the tertiary and quaternary sectors (which now account for about 80% of all UK employment). So the UK's employment structure has changed – but why?

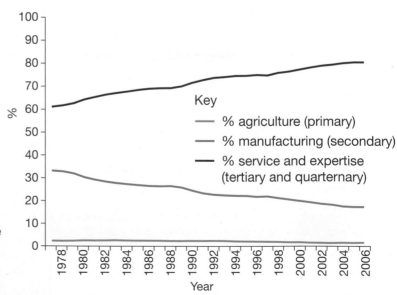

Key
— % agriculture (primary)
— % manufacturing (secondary)
— % service and expertise (tertiary and quarternary)

▲ Changing employment in the UK, 1978-2006.

The Clark Fisher model

Two economists, Clark and Fisher, produced a theory – or model – that helps to explain changes in employment structure over time. As countries develop their economies, Clark and Fisher said that they go through three stages (see the graph):

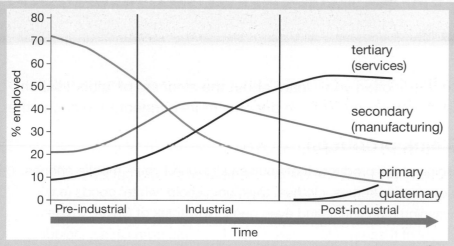

▲ The Clark Fisher model.

A **Low-income countries** are dominated by the primary sector (pre-industrial).

B **Middle-income countries** are dominated by the secondary sector (industrial). As economies develop and incomes rise, the demand for agricultural and manufactured goods increases.

C **High-income countries** are dominated by the tertiary and quaternary sectors (post-industrial). As incomes continue to rise the tertiary sector develops as people start to use more services, – leisure, banking and insurance. Finally, the quaternary sector develops as tertiary services become more specialised, such as IT, legal or medical services, the media. Many countries in this category have lost their manufacturing – to countries like Vietnam, where wages are lower (see Section 11.1).

Welcome to France!

Like many high-income countries, the French economy is changing. It has very healthy quaternary employment in cities, but it's also kept much of its manufacturing industry (Peugeot, Citroen and Renault are three of the world's largest car companies). However, unlike Vietnamese factories, most French factories rely on technology, rather than people, to produce goods like cars.

French farming is also healthy, but a lot of the work is now automated. Many French farmers rent out their land to larger farms with more machinery, or they convert their farmhouses to rent out as 'gites' for tourists. So rural primary employment has declined, while the tertiary economy is growing.

	France
GDP per person (in US$)	35 600
Where France's GDP comes from (%)	Agriculture: 1.8 Industry: 18.8 Services: 79.4
Percentage of people by occupation	Agriculture: 3.8 Industry: 24.4 Services: 71.8

your questions

1 Make a copy of the Clark Fisher model and annotate it with the current position for **a** Vietnam **b** UK **c** France. Give evidence to show why this is the case for each country.

2 Research ten job vacancies in your local area and an overseas city of your choice. Classify them into primary, secondary, tertiary, and quaternary. What do you find?

3 **Exam-style question** Study the table above. Identify three pieces of evidence to show that France is a high-income country. (3 marks)

In this section you'll learn about the meaning of 'globalisation', and how countries have become more connected to each other.

Think products!

Think of the products that you see all around you – mobile phones, computers, kettles, clothes, toys, household 'white' goods (e.g. washing machines), and even pens. Now think of China, where many of these products are made. Products from China, including those made for American brand names (e.g. Nike, Apple or Sony), are found almost everywhere. They arrive in the UK every day in large container ships (usually owned by Maersk, the world's largest shipping company).

China has become one of the largest influences in a **globalised** world, where:

- countries try to export products overseas in order to boost their GDPs
- common products are available globally, e.g. Coca-Cola, McDonald's, KFC and Starbucks
- expanding travel networks and communication technologies easily connect manufacturers with overseas buyers. Geographers call this **connectedness**.

Globalisation has meant that national borders have become less important whilst **transnational companies** (TNCs) have become more influential.

> **+ Transnational companies** (TNCs) are giant companies operating in many countries.

International trade

International trade has exploded in volume as (TNCs) invest in developing countries to manufacture goods. These are then shipped to markets in North America and Europe. Between 1970 and 2010 the global population increased 1.8 times, whereas the value of all exports grew 48 times.

▼ Sea trade flows connecting producers and consumers in 2011. The Width of the blue line is proportional to the amount of trade.

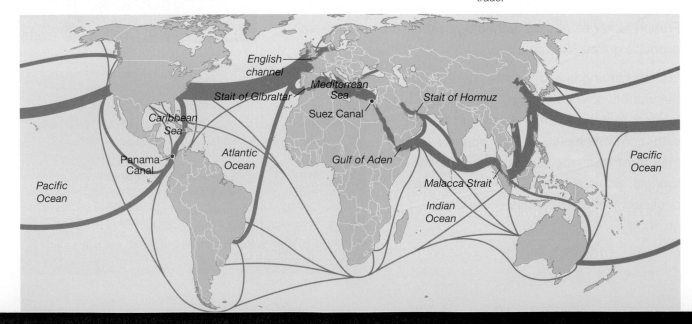

Think globalisation!

Globalisation is defined as the ways in which countries become increasingly connected to each other. This happens through:

- greater economic **inter-dependence** between countries – national borders have become less important
- an increasing volume and variety of **trade** in goods and services
- easier international money **flows** to invest in other countries
- increased spread of **technology**
- **culture**, where global media companies spread news, TV programmes, film, and music (you're now just as likely to hear Coldplay in Shanghai as in Southampton).

The rapid spread of globalisation has been helped by greater international trade, plus changes in investment:

Foreign Direct Investment (FDI)

In the 1990s, American and European TNCs invested in new factories and transport infrastructure for Chinese cities. This type of investment is known as FDI. As Chinese wages were 90% lower than the USA and Europe, TNCs were able to manufacture goods much more cheaply. The result was a huge growth in Chinese exports – by 2007, China was the world's second largest trader (overtaking Japan). By 2025, it's likely to overtake the USA.

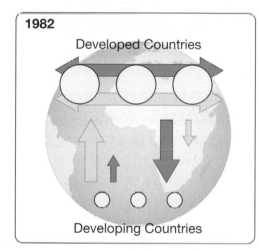

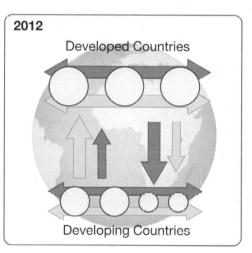

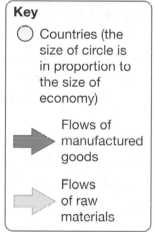

Key

Countries (the size of circle is in proportion to the size of economy)

Flows of manufactured goods

Flows of raw materials

▲ Until the 1980s, the flow of trade was either between the world's developed and developing countries, or between the developed countries themselves.

▲ Now there is a third type of trade – between the growing developing economies of countries like India and China. China now buys raw materials from Africa, and Chinese goods are exported to India and the developing economies of South America.

your questions

1 In pairs, create a mind map to show how globalisation can bring benefits to **a** China, **b** North America and Europe.

2 Explain how globalisation has led to a 48-times increase in exports, and 22-times increase in global GDP, while global population has only increased by 1.8 times.

3 Exam-style question
   a Describe the changing pattern of global trade since the 1980s, as shown in the above diagram. (3 marks)
   b Explain how globalisation has led to greater trade between countries. (4 marks)

➕ In this section you'll learn about the reasons for the rapid rate of globalisation.

Globalisation – but how?

The theory of globalisation is easy – if no country can produce everything it needs, trade with other countries becomes essential. In theory, there are so many countries and companies making televisions, for example, that prices should come down.

But, even then, how can goods produced in Asia be cheaper than those made in Europe? Aren't distances so great that transport costs would wipe out any advantages caused by cheaper labour? In fact, three trends have made globalisation easier – lower transport costs, the growth of TNCs, and state-led investment – all of which are explained below.

Lower transport costs

Transport has actually become cheaper, even if fuel has become more expensive. Three major transport changes since the 1980s have reduced the costs of transporting goods – meaning lower prices in the shops. These are, changes in shipping, containerization, and aircraft technology.

- **Shipping.** Ships transport over 90% of our goods and have become much larger. Whether large or small, a ship only needs one crew, and fuel efficiency has also reduced fuel consumption. A large ship costs only slightly more to run than a small one.

- **Containerization.** Most goods now arrive from Asia in containers, which are easier to transport to ports, to load onto ships, and to unload at the other end. Each container is bar-coded, so machines rather than people now identify its contents, and which ship it should be on. Almost all manufactured goods, smaller than a car, are now shipped in containers.

- **Aircraft.** Transport by aircraft is more expensive than by ship, so only 0.2% of our goods are transported by air. However, this makes up 15% by value. Airfreight is 70 times more valuable than goods transported by sea, e.g. electronics, medical equipment, and fruit and vegetables from California or Africa. Jet engines are also more fuel-efficient now, and airlines operate with far fewer staff.

▼ An aerial view of containers by the dockside in China's Pearl River delta.

TNC growth and merger

Many TNCs a higher income than the GDP of most countries! Of the 100 largest economies in the world, 52 are corporations and 48 are countries. TNCs play a huge part in trade – the 500 largest TNCs account for 70% of global trade, because so many now manufacture overseas where labour is cheaper. In theory, TNCs should produce goods more cheaply.

Each TNC started small and then grew through:

- **growth** in the sales of popular products. For example, Apple products have increased from the small desktop computer to a range of electronic hardware and software.
- **merger** with other companies, by consolidation or creating a conglomerate.

+ **Consolidation** involves merging with companies that make similar products. For example, in 2011 Sony bought Ericsson in order to give it a greater share of the global market for mobile phones and tablets, and to integrate it with their PlayStation products.

+ **Conglomeration** involves buying companies which make different products to create a 'conglomerate'. For example Moet & Chandon (champagne), Louis Vuitton (fashion), Bulgari (jewellery) and Pink (clothing), they're all owned by the same French company – LVMH – which specialises in buying luxury brands (see the photo).

▲ The luxury product range of LVMH

State-led investment

Along with Foreign Direct Investment (FDI – see Section 11.3), much of China's economic growth has come from **state-led investment** by the government. Unlike the UK, where banks are free to invest money as they wish, China's government keeps a tight grip over its banks. They use money from household savings and overseas trade to fund state-owned companies (controlled by the government). Large state-owned companies then borrow this money at low interest rates.

your questions

1 In pairs, research why a chosen TNC decides to manufacture overseas. Google a phrase like: 'Why Apple manufactures in China'. Produce a short PowerPoint to present to the class.

2 Research product brands from the following companies and explain how these TNCs have grown – by growth, consolidation or conglomeration: Whitbread, Apple, Ford, Virgin, BP.

3 **Exam-style question** Using examples, explain how the growth of TNCs has led to greater global trade between countries. (4 marks)

In this section you'll learn how two global institutions help to keep the global economy stable.

Greece in turmoil

In October 2012, not for the first time, thousands of protestors were on the streets of Athens in Greece (see photo). Some even threw petrol bombs at the police. Why was there such trouble in a country so popular with British tourists?

The protestors were fighting against '**austerity**' (a word meaning cuts in government spending). Between 2001 (when Greece joined the euro) and 2008, the Greek government spent much more than it raised in taxes – and got into debt. In 2012 it was trying to reduce the debt by cutting spending on, for example, pensions and healthcare, and by making many government employees redundant. With less money in their pockets to spend, people buy less and private companies start to go bankrupt as well – fuelling unemployment. By October 2012, Greek unemployment had risen to 25% – with youth unemployment at 54%

The international banks which lent money to the Greek government, now want to be repaid. Meanwhile, Greek people want jobs. It's a tough problem. If government cuts continue, unemployment will worsen and so will social unrest. Some fear that if Greece actually went bankrupt (it couldn't repay its debts), it could lead to problems in other countries in the European Union (EU), such as Spain, which also has large debts. Could the euro collapse? If that happened, the UK would suffer badly, because 70% of its trade is with the EU. It's as though globalisation has made inter-connections so great that ripples in Greece become waves across Europe.

▲ Protests in Greece over austerity in 2012

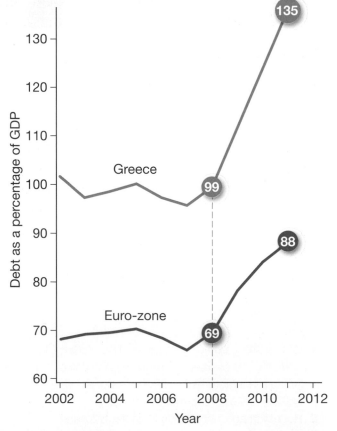

▲ Rising Greek debt

Enter the IMF!

It is in everyone's interest to prevent Greek problems from worsening, so the **International Monetary Fund** (IMF) has stepped in to help. The IMF is a US-based organisation that raises funds from the world's wealthier countries, to help countries which become economically unstable. The IMF aims to maintain stability and has four main jobs:

- **To stop existing crises** from developing and spreading. For example, the IMF is providing Greece with financial help.
- **To spot potential future crises.** The IMF's research team report on every country's economy.
- **To check investment** and how banks operate. The IMF monitors banks to try to prevent a banking crisis from happening again.
- **To reduce global poverty** by helping the world's poorest countries. The IMF claims that if these countries have more money to spend, everyone will gain through increased trade.

Enter the WTO!

The **World Trade Organisation** (WTO) aims to get countries to agree that goods traded between them will be free of duties (known as **tariffs**). These duties add to the price of goods – lower tariffs mean cheaper goods, so people will buy more. However, this can have other results too:

- Because whoever is the cheapest wins, some countries might find that their own producers lose out to cheaper overseas competition.
- Cutting tariffs does not benefit all, as the banana example on the right shows.

The banana trade

The islands of the West Indies produce bananas, mainly on small farms. Their bananas are more expensive to produce than those from large US-owned plantations in Latin America, so the EU protects them by charging an import tariff of €140 per tonne on Latin American bananas. This makes the price of bananas the same wherever they come from. Now the WTO is insisting that the banana tariff is cut. This will make Latin American bananas cheaper than West Indian bananas, so the big plantations will gain and the small farmers will suffer.

▲ Banana production in the West Indies on a small family farm.

your questions

1 In pairs, design a flow diagram to show how Greek debt could lead to problems in **a** Greece, **b** the EU, **c** the UK, **d** the rest of the world.

2 Explain how the IMF could interrupt the flow diagram you have drawn in 1, and reduce the problems.

3 Draw a table to show the advantages and disadvantages of cutting all tariffs on trade for **a** producers, and **b** consumers.

4 **Exam-style question** Using examples, explain how organisations like the IMF and WTO can help the process of globalisation. (6 marks)

➕ In this section you'll learn about the impacts of globalisation on men and women in Leeds and Bangladesh.

Leeds – a changing city

Leeds is the UK's sixth largest city. Like most northern cities in the UK, it has a strong industrial history. In 1964 (see the first table) it contained textile companies like Burton and Hepworth (now known as Next). The steel, railway and defence industries also provided skilled engineering jobs. Employment was split between the sexes – men in mining, railways and engineering; women in clothing factories, health and administrative jobs. Many jobs were skilled and came from apprenticeships served from the age of 15.

However, globalisation changed the employment structure of Leeds. The UK's clothing industry could not compete with cheaper clothes from overseas, and cheaper imports of steel and coal, led to falling demand for British steel and coal.

As a result, many factories closed during the 1980's – a process known as **deindustrialisation**.

By 2011, the employment structure of Leeds was very different to that in 1964 (see the second table). This resulted in both winners and losers in the jobs market.

- Well-qualified men and women benefitted from the growth in financial and public services.
- The decline in manufacturing reduced the jobs available for skilled men and women.
- Fewer apprenticeships available for young people often meant that only low-paid jobs were available.
- Older men and women with unwanted skills struggled to find new, well-paid jobs.

Employment Type	1964 (%)	Examples in Leeds
Primary	0.8	mining, quarrying
Secondary	45.2	clothing and footwear (37 000 jobs) engineering (32 000 jobs)
Tertiary and Quaternary	54.0	distribution companies (40 000 jobs) professional and scientific services (26 000 jobs)

Employment Type	2011 (%)	Examples in Leeds
Primary	0.02	mining
Secondary	7.0	engineering (13 250 jobs) printing, packaging and publishing (6900 jobs)
Tertiary and Quaternary	93.0	finance and business services (113 000 jobs, e.g. Yorkshire Bank, Leeds Building Society, call centres) public services (100 000 jobs, e.g. Department of Work and Pensions) call centres (e.g. BT)

▲ The BT Call Centre in Leeds

Meanwhile in Bangladesh ...

Bangladesh is developing rapidly, and is a typical example of a **Newly Industrialising Country** (NIC). Its economy grew by an average of 6% each year between 2005 and 2011 – the eighteenth fastest in the world. However, a third of the population lives on less than US$1 a day.

The main reason for this rapid economic growth is the willingness of Bangladesh's government to allow large TNCs, specialising in clothing manufacture, to take advantage of its low wages. In 2010, wages in Bangladesh were 98% lower than average wages in the EU and the USA. By 2011, about 200 big foreign companies had started operating there, the majority of which were American. The clothing industry in Dhaka alone employs 650 000. In 2011, the Bangladesh clothing industry as a whole employed 5.5 million people and earned more than US$15 billion in GDP for Bangladesh. Other foreign companies operating in Bangladesh include Tesco, Walmart, and Tommy Hilfiger.

However, this employment trend is controversial:

- Although there is a minimum wage in Bangladesh, it was only £23 a month in 2012. Poverty is common in Bangladesh, so there is no shortage of people willing to work 100 hours a week in factories for an average of £40, including overtime.

- Although working conditions have improved, there are still many **sweatshops** employing younger women – often discriminating against older women returning to work after raising children.
- Bangladesh's 2006 labour laws, means that any child can work in a factory from the age of 14 - although it this is difficult to enforce.
- Few textile jobs are skilled – about 70% of employees in clothing factories are unskilled women.
- There are now also more opportunities for men, with increasing numbers of jobs in transport and distribution and in retailing.

▲ One of many hundreds of clothing factories in Dhaka, the capital of Bangladesh.

your questions

1 Copy and complete the table below to show who has gained and lost from employment changes in Leeds since the 1960s:

	Gained	Lost
Men		
Women		

2 Now repeat the exercise for Bangladesh to show who has gained and lost from industrialisation.

3 Now summarise your work in **1** and **2** to show how far globalisation benefits **a** younger people, **b** older people, **c** women, **d** men.

4 Exam-style question
   a Using examples, describe how globalisation can impact on men and women in developed countries. (6 marks)
   b Using examples, explain why globalisation can lead to unequal impacts on men and women in developing countries. (6 marks)

In this section you'll learn how BT operates in different parts of the world, including outsourcing.

Getting to know BT

BT is one of the largest telecommunications companies in the world (operating in over 170 countries). It is British-owned and deals in telecommunications – not just telephone rentals, broadband and mobile phones – but industrial communications for factories, retailing, and subscription TV. Since 2000, BT has bought or merged with several companies in the USA, Italy, Germany, South America and South Africa.

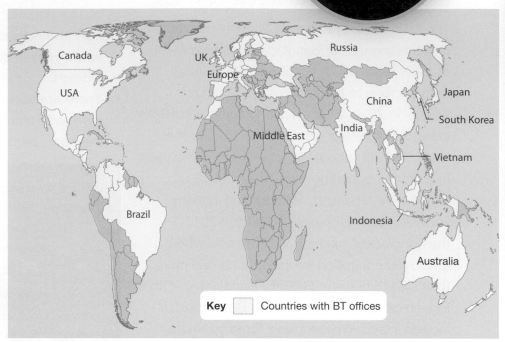

Key ☐ Countries with BT offices

▲ The countries where BT has offices.

BT and outsourcing

In the 1990s, BT – like other similar TNCs – began '**outsourcing**' (moving services to a location where wages are lower). Three types of outsourcing occurred:

- English-speaking foreign nationals took over services like call centres. Most Indian call-centre employees are graduates earning £3000 per year (just 20% of the amount BT would have to pay in the UK).
- Software development skills in India enable BT to develop, and support, its Broadband product out of Bangalore and New Delhi.
- Locations with administrative skills manage services like company accounting.

Although BT's Indian headquarters are located in New Delhi, new technology development takes place in Bangalore. Bangalore is India's 'Silicon Valley', and has experienced an IT boom. In its high-tech buildings are found every major global technology company which attract young professionals from other Indian cities.

Three factors have helped Bangalore's growth as an outsourcing centre:

- Its English-speaking university provides science and technology graduates.
- The city has attracted software development companies like Infosys – creating a 'Silicon Valley' which, in turn, attracts other companies like BT.
- It offers reduced taxes to attract companies.

▲ A software company in Bangalore's 'Technology City'– an area which has attracted major IT companies, developing a kind of separate city.

The new economy

BT is typical of many companies which are part of the 'new economy' – based on the sale of services, rather than manufactured products. Most of the world's high-income countries have moved towards the new economy to earn their GDP. Unlike companies which specialise in manufacturing, companies in the new economy are not fixed to locations where raw materials are available, or close to ports from which goods can be shipped. They are 'footloose', i.e. they can locate anywhere – as long as high-quality communication links are available such as superfast broadband and fibre-optic cable.

BT relies on being able to employ skilled, creative, well-qualified people with high levels of technical knowledge. Companies like this are more likely to locate close to university cities and research centres in science parks – anywhere in the world! They can just as easily locate in Bangalore as in Birmingham – it all depends on where the skills are and upon wage costs. A low level of skill, such as a call centre – requiring a month or so of training, is more likely to operate from a low-wage economy.

BT and the new economy ...

... is about focusing on the development of knowledge, ideas and services. The company provides highly specialised services, based on technology, research and development.

Examples: Jobs in ICT, TV production, broadband and fibre-optics technology.

Characteristics:

- It is globally based and interconnected – people are just as likely to discuss work with someone overseas as with the person sitting at the next desk.
- High quality recruitment is vital in order to supply talented specialists.
- Most creative thinking is done at global headquarters, in cities like London, with other operations at lower-wage locations around the world.
- It is usually English speaking.
- There tends to be equal male and female employment.

your questions

1 In pairs, use the internet to identify various images of Bangalore. Develop a presentation which tries to 'sell' Bangalore as a location where BT should outsource.
2 Copy and complete the following table to show who gains and loses from outsourcing services to Bangalore.

	Gained	Lost
BT		
Bangalore		

3 How much is Bangalore **a** gaining, and **b** losing from globalisation?
4 **Exam-style question** Using examples, explain how outsourcing can affect different countries in different ways. (6 marks)

➕ In this section you'll learn how Nike operates in different parts of the world, including outsourcing its manufacturing overseas.

Just do it!

Every year, the Forbes directory ranks the world's most valuable brand names. In 2012, the most valuable sportswear company by far was Nike – valued at over US$10 billion. Its instantly recognizable 'swoosh' logo and 'Just do it' slogan are its greatest assets, known throughout the world. Of its total sales in 2012 (nearly US$25 billion!), 90% came from selling products with the Nike logo. It's the only sportswear company in the world whose market share has increased since 2008.

Nike operates and sells in over 140 countries; 46 of these also manufacture goods for Nike (see the map). It is a major employer. It directly employs 44 000 people worldwide – but over 20 times this number work in factories under contract to produce Nike products. Although small amounts of Nike clothing are made in the USA and Europe, the majority of factories are located in Asia (see the table). Most employ a majority of women. In China and Vietnam (the two largest producers), over half of the workers are migrants from other parts of the country, who live in workers' hostels.

USA: Head office – Nike World Campus at Beaverton, Oregon. Product design is carried out at the Oregon head office.

China (1980s): Nike began production in China, to take advantage of cheaper labour there.

Japan (1960s): Training shoes were first made in Japan when labour was cheap in the 1960s.

Vietnam (2000s): Now that China's currency is worth more, it is cheaper to make many items in Vietnam, so production has increased there.

Thailand and Indonesia (late 1980s): South Korean companies, with whom Nike had developed a long-term relationship, moved operations south to Thailand and Indonesia in search of cheaper labour.

South Korea, Taiwan (1970s): Nike was attracted by cheap labour in South Korea and Taiwan. Instead of owning its own plants, Nike outsourced production to South Korean and Taiwanese companies.

▲ The countries where Nike has manufacturing workers.

	Total number of Nike manufacturing workers	Percentage of female workers	Average age of workers
China	272 000	71	29
Vietnam	198 000	84	25
Indonesia	126 000	81	28
Thailand	51 000	73	28
India	32 000	51	25
Sri Lanka	20 000	80	24
Brazil	18 000	63	28
Pakistan	10 500	8	29
Mexico	10 000	60	28
Honduras	10 000	73	26

▲ Nike's ten largest manufacturing countries by contracted employees. (Note that most of these people do not actually work for Nike – Nike's manufacture has been outsourced to them.)

Nike and outsourcing

Nike started in 1964 when Phillip Knight began importing running shoes to the USA from Japan, where labour was cheaper. His aim was to compete with German brands like Adidas and Puma. Nowadays, manufacturing overseas is common in the clothing and shoe industries (and in others like toys and electronics). But decision-making responsibilities such as management or design and marketing, are kept in the USA. Most Asian outsourcing countries get the less-profitable, production activities.

Many clothing and shoe factories in Asia have been investigated by organisations shocked by employees' working conditions and low wages – as little as US$2 a day. There have also been reports of neglect of health and safety, physical and sexual abuse, and persecution of workers who have tried to organise trades unions. Although Nike inspects those companies outsourced to make Nike goods, some people feel this is not enough.

> + A **sweatshop** is a factory where workers are expected to work very long hours, with low pay and poor working conditions.

Breakdown of costs of a $65 pair of Nike training shoes	US$
Production labour	2.50
Materials	9.00
Factory costs	3.25
Supplier's operating profit	1.00
Shipping costs	0.50
Cost to Nike	**16.25**
Nike costs (research and development, promotion and advertising, distribution, admin	10.00
Nike's operating profit	6.25
Cost to retailer	**32.50**
Retailer's costs (rent, labour, etc.)	22.50
Retailer's operating profit	10.00
Cost to consumer	**65.00**

Campaigning against Nike

Since 2000, many campaigns have encouraged Nike to improve conditions for workers in factories making Nike products. Unlike many Western countries, where trade unions campaign for better pay, these campaigns have been organised by consumers. A 'boycott Nike' campaign was supported by members of the US Congress, as well as pressure groups campaigning for change. Campaigns also tried to fight unfair working conditions in what were alleged to be **sweatshops**.

One of the problems for consumer groups is that many countries in which Nike's products are made are not democratic, and there are few workers' rights. However, they still feel that Nike should ensure factory conditions, working hours and wages are kept at a decent level. Nike now publishes data about supplier inspections on its website.

your questions

1 Explain why Nike outsources manufacturing overseas.
2 What are the risks for companies who outsource like this?
3 Why should countries like Vietnam want to attract TNCs like Nike? List the benefits.
4 Compare Nike and BT (see Section 11.7) in:
   a how they operate in different parts of the world,
   b their methods of outsourcing, **c** who they use as outsourced workers, **d** how they bring benefits and problems from outsourcing.

5 How far would either Nike's own attitudes, or the campaigns against Nike, affect your buying habits? Why?
6 Exam-style question
   a Using the table below the map, describe the characteristics of Nike's manufacturing workforce. (4 marks)
   b Using examples, explain how TNCs operate in different parts of the world. (6 marks)

+ In this section you'll learn about different ways of defining and measuring development.

What are development dilemmas?

'Development' means change economically (in terms of income and the economy) and socially (affecting people). It brings jobs and trade, but it can also benefit some people more than others. Who benefits most – those in urban or rural areas? Which parts of the country benefit most? These questions present problems, or **dilemmas**, for countries.

Measuring development

According to the United Nations (UN), Malawi (see map) is one of the world's 25 poorest countries. The UN uses different **development indicators** to work this out (see table opposite). Development indicators, when combined, give an overall picture of a country's **level of development**, and can be used to compare countries.

Economic development indicators include:

- **GDP** – In 2011, Malawi's annual GDP was only US$900 per person. Even though its cost of living is low, this makes Malawi extremely poor. By contrast, in 2011 the UK had a GDP per person of US$36 600.
- **Poverty line** – Half of Malawi's population lives below the poverty line.
- **Overseas debt** – In 2012, Malawi owed US$1.3 billion overseas (36% of its GDP was used for debt repayments). Its debts are being reduced by the International Monetary Fund (IMF) and the World Bank (similar to the IMF). But it still depends heavily on aid.

However, as the table shows, development can also be defined using **social** indicators. These are linked to economic indicators. e.g:

- Countries with high GDP tend to spend more on education, so literacy figures are higher.
- GDP can also influence how much a person eats each day. In 2010, a third of Malawi's population was underfed.

▲ Malawi is about two-thirds the size of England, but with less than a third of England's population.

▼ A comparison of social development indicators.

Data	Malawi	UK
Population		
Fertility rate (average number of births per woman)	5.4	1.9
% of population aged 0-14	52.0	17.3
% of population aged 15-64	45.3	65.8
% of population aged 65 and over	2.7	16.9
Dependency ratio	120.7	52
Health		
Life expectancy in years (2012)	52.3	80.2
Infant mortality per 1000 live births (2012)	79	4.6
Maternal mortality per 100 000 births (2010)	480	12
Number of doctors per 100 000 population (2008)	2	274
Education		
Average number of years in school	9	16
Literacy rate (%)	75	99

Some **environmental** indicators are also used. For example, 83% of Malawi's population had access to safe drinking water in 2010 – something that a person in the UK would take for granted. This still means that people (usually women) often have to walk long distances to obtain water. The pump in the photo provides safe water, but is some distance from the village.

Know your development indicators

+ **GDP (Gross Domestic Product)**: The total value of goods and services produced by a country in a year. Dividing GDP by the total population gives you GDP per person (sometimes called per capita). Figures are in US dollars, for easy comparison between countries.

+ **PPP (Purchasing Power Parity)**: All GDP data are now PPP. This means that GDP is given in terms of what it will buy using local prices. A low-income country (like Malawi) will probably have low prices, so $1 will buy more there than in the UK.

+ **Life expectancy**: The average number of years a person can expect to live.

+ **Poverty line**: the minimum level of income required to meet a person's basic needs - the World Bank considers this to be $1.25 per person, per day.

+ **Dependency ratio**: The proportion of people who are too young (0-14) or too old (>65) to work. It's calculated by adding both groups together, and dividing that by the number aged 15-64 (the working population), multiplied by 100. The lower the number, the greater the number of people able to work.

+ **Infant mortality**: The number of children per 1000 live births who die before their first birthday.

+ **Maternal mortality**: The number of mothers per 100 000 who die in childbirth.

+ **Access to safe drinking water**: The percentage of the population with access to an improved water supply.

+ **Being underfed**: The percentage of the population who do not consume at least 2000 calories a day.

+ **Literacy rate**: The percentage of the population, aged over 15, who can read and write.

▲ Collecting water in rural Malawi.

your questions

1 **a** In pairs, choose another country in sub-Saharan Africa, and one in Europe. Then, use The *CIA World Factbook* online to research development data for your two countries. Compile a table which includes the development indicators shown opposite.

   b How far are your two countries similar to Malawi and the UK?

2 Make a large copy of the table below to explain the following about the world's poorest countries:

	Reasons for this	The effect of this on the country
a They have low life expectancy.		
b They have high maternal mortality.		
c They have low literacy rates.		
d They have a high dependency ratio.		

3 **Exam-style question** Describe two indicators which can be used to show a country's level of development. (4 marks)

➕ **In this section you'll learn how different indicators can give different patterns of development.**

And a special welcome to …

It's nearly midnight on 27 July 2012, at the Opening Ceremony for London's 2012 Olympic Games. The crowd cheers to welcome athletes from all over the world, but a rousing roar goes up for the Saudi Arabian team as they enter the stadium. For the first time, a woman is competing for the Saudi team. It's also the first time that the Olympics have included women from every country in the world.

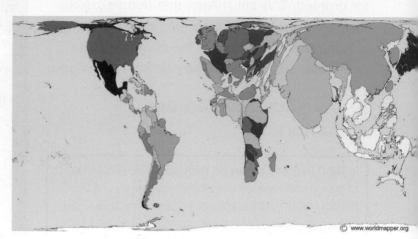

▲ A 'Worldmapper' map showing gender equality. Each country on the map has been located accurately, but it has been drawn in proportion to how far it offers equality for women – the larger it is, the more equality there is.

Corruption and development

One problem with trying to achieve economic development is corruption. The Corruption Perceptions Index (see below) was devised to help investors to work out where their money would be safest. This index uses a scale from 10 (honest) to 0 (very corrupt). In corrupt countries, invested money is more likely to be used to bribe officials, for private wealth, or to purchase weapons.

Key
Corruption Perceptions Index
- 9 – 10
- 8 – 8.9
- 7 – 7.9
- 6 – 6.9
- 5 – 5.9
- 4 – 4.9
- 3 – 3.9
- 2 – 2.9
- 1 – 1.9
- 0 – 0.9
- no information

▼ The Corruption Perceptions Index for most countries.

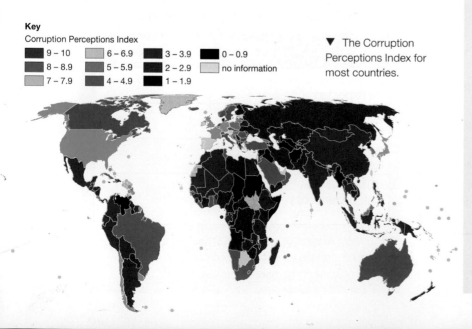

Women and development

Development is usually about economic indicators which often ignores the contribution of women. There is no country in the world where opportunities for women are as good as for men.

- Countries where women have the best opportunities are in Western Europe, and those with the worst are in the Middle East, e.g. Yemen and Saudi Arabia (see map above).
- Women account for more farm labour in developing countries, but the shift to manufacturing in those countries now offers more jobs for women.
- A few democracies (e.g. Lebanon) only allow women to vote if they are educated - a condition which does not exist for men. In 2015, women will be allowed to vote in Saudi Arabia for the first time.

The Human Development Index

Although GDP is most commonly used to show levels of development, some countries with a high GDP have a very unequal distribution of wealth. Qatar and the United Arab Emirates are in the top three countries for GDP, but are very unequal, with the wealth going into the hands of a few. Other countries spend what limited wealth they have on health and education, which benefits the whole population.

Because of this, the United Nations developed the **Human Development Index (HDI)** to measure development. It consists of a single figure per country, between 0 and 1 (the higher the figure the better). The HDI is calculated using an average of four indicators:

- Life expectancy
- Education – literacy
- Education – average length of schooling
- GDP per capita (using PPP$)

HDI gives a different rank order when compared with GDP (see tables).

However, there is still a close link between GDP and HDI – the poorest countries in the world for GDP also have the lowest HDI. It's still a matter of money – how can the poorest countries afford health care or education?

Many of the lowest countries on the HDI have also been involved in conflict, and all are in Africa. As most investors are unwilling to put money into countries where there may be unrest, there will be little wealth. Without wealth, health and education remain poor – and countries struggle to develop.

HDI – the top five

HDI rank	Country	HDI figure	GDP rank	Free democratic elections?
1	Norway	0.943	4	Yes
2	Australia	0.929	10	Yes
3	Netherlands	0.910	13	Yes
4	USA	0.910	12	Yes
5	New Zealand	0.908	9	Yes

HDI – the bottom five

HDI rank	Country	HDI figure	GDP rank	Free democratic elections?
183	Chad	0.328	165	Yes but unstable
184	Mozambique	0.322	177	Unstable with corrupt elections
185	Burundi	0.316	186	Yes but unstable
186	Niger	0.295	183	Yes but unstable
187	Democratic Republic of the Congo	0.286	187	Yes but only recently

your questions

1 Describe the strengths and weaknesses of **a** GDP and **b** HDI as an indicator of development.

2 Explain why:
  a wealth is needed to improve health and education
  b better health and education can improve a country's wealth.

3 **a** In pairs, use the *CIA World Factbook* online, and the information in this section, to research ONE of the following countries to find out its GDP, HDI, level of gender equality, level of perceived corruption, and any five other indicators: China, Russia, Brazil, Saudi Arabia, India.
  b Prepare a short 200-300 word report on your chosen country to describe how developed it is.

4 **Exam-style question**
  a Describe two features of the countries with the highest HDI. (4 marks)
  b Suggest one reason why countries with the lowest HDI score poorly on this indicator. (2 marks)

✚ **In this section you'll learn about the global development gap and how this has changed over time.**

Mind the gap

Imagine a report which highlights the poverty of the developing world, and compares it with the wealth of rich countries. It describes how people live in the world's poorest countries, and how the rich countries rely on the poor countries for their raw materials.

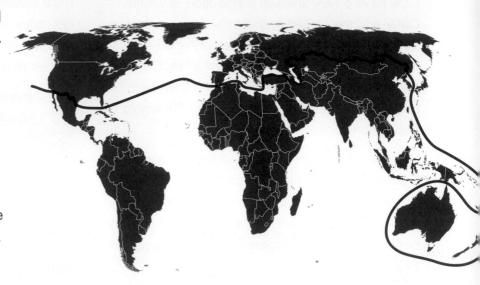

▲ The 'North-South Divide', from the Brandt Report of 1980.

Such a report does actually exist. It was written in 1980 by the then German Chancellor, Willie Brandt, and identified two major groups of countries:

- A group of wealthy countries, consisting of those in North America, Western Europe, Japan and Australasia. Because most were in the northern hemisphere, they were identified as the 'global north'. These were the world's **High Income Countries** (HICs).

- A second group of much poorer countries in South and Central (or Latin) America, almost all of Africa, and Asia. Because most were in the southern hemisphere, they were identified as the 'global south'. These were the world's **Low Income Countries** (LICs).

The Brandt Report identified the differences between these two groups of countries (in terms of both GDP and quality of life) as the 'North-South Divide', or 'Development Gap' (see map).

So what's changed?

Over three decades have passed since the Brandt Report, and world now looks very different:

- In the 1980s, rapid economic development took place in Latin America. This created a new group of countries known as **Middle Income Countries** (MICs) — including Brazil, Peru and Chile. Major resources of raw materials (such as iron ore) in these countries encouraged investment and development — especially

▲ Rio de Janeiro, Brazil.

in the major cities which experienced big increases in their populations (see Section 15.7).

- In the 1990s, equally rapid development took place in many of the countries of south-east Asia, such as Hong Kong, Singapore, Thailand, and Malaysia. Much of their growth was due to the relocation of manufacturing overseas by many US and European TNCs. Growth in south-east Asia was so aggressive that these countries became known as the 'Asian Tigers'. Their economic growth doubled the size of their economies between 1988 and 1996. Most of this region can now be classed as **Newly Industrialising Countries** (NICs).

- Even more astonishing, has been the rapid industrialisation of China and India in the 2000s. They tend to be referred to as **Rapidly Industrialising Countries** (RICs). Together with Brazil and another emerging giant, Russia, these countries are also known as the BRICs.

A new world order

Has Brandt's view of the world survived? The table shows how the world's ten largest economies, by 2015, have changed since the Brandt report was written. One thing has stayed the same though. In 1980, Brandt identified the world's poorest countries as almost entirely in Africa – the map on the right shows that this is still true.

Rising / Falling	1980 rank	2015 (est.) rank	The 10 largest economies
	1	1	USA
▲	9	2	China
▼	2	3	Japan
	4	4	Germany
▲	12	5	Brazil
	6	6	United Kingdom
▼	5	7	France
▲	13	8	India
▼	3	9	Russia (was the USSR in 1980, which was much larger!)
▼	7	10	Italy

▼ The ten poorest countries in the world in 2012, all of which were in Africa.

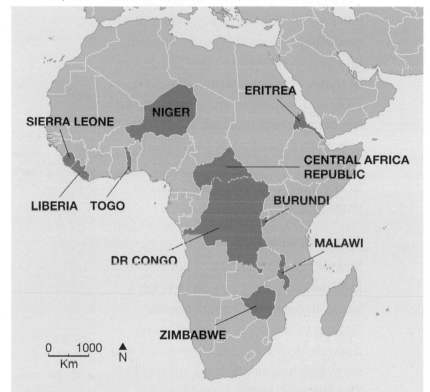

your questions

1 Explain the difference between the 'global north', 'global south', MICs, NICs, RICs and BRICs.

2 a Make a copy of the thick black 1980 North-South Divide line on a world map. Now use different colours to shade in the MICs, NICs, RICs and BRICs.

b Using your map and the table of the world's largest economies above, explain how the world has changed since 1980.

c How far would you say that there is still a 'North-South Divide'?

3 **Exam-style question** Using examples, explain what is meant by the global development gap. (6 marks)

✚ In this section you'll learn about recent development in Malawi, and barriers to its progress.

What's holding Malawi back?

In June 2012, Malawi got a new President, Joyce Banda. She has got her work cut out. Malawi faces increasing food prices, fuel shortages, and has 1.2 million people facing hunger. She believes that Malawi faces an uphill task in joining a globalised world, and that globalisation would bring investment and jobs in new industries. She is one of those who believe that without it, GDP and incomes remain low, and so the government receives little tax to spend on education or health care. If people are sick, they are unable to earn. It's a vicious cycle (see the diagram).

1 It's landlocked

Malawi has no coastline – it's **landlocked** – so it has no port from which to export or import goods. Getting to the coast involves going through Mozambique. There's just one slow, 800 km, single-track railway line. The line runs south, from Malawi's border with Zambia, to Nacala, a port in Mozambique (see map). Nearly all of Malawi's exports (e.g. tobacco, sugar, tea) are shipped out of Nacala. The trains return with imports (farm fertiliser, fuel, manufactured goods). But it's a slow, and expensive, process.

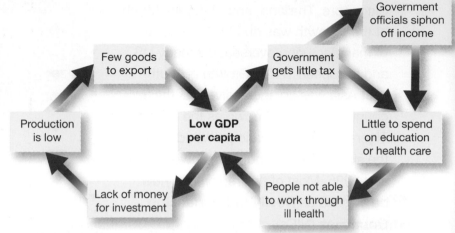

▲ The national economic cycle of poverty.

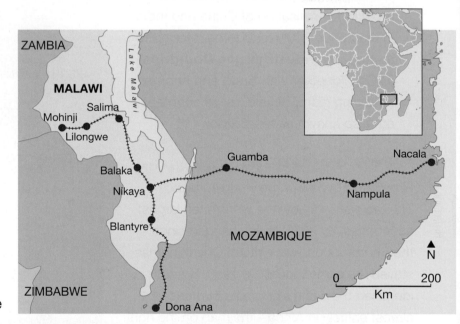

▲ Malawi is dependent upon its limited railway network to export its products.

Indicator	2010 data	Trend since 1998
Death rate per 1000	12.8	Down from 23
Infant mortality per 1000 live births	79	Down from 132
Life expectancy (years)	52.3	Up from 36.3
Malnutrition (% of under-5s underweight)	15.5	Down from 48
Maternal mortality per 100 000 births	460	Down from 1221
Male adult literacy rate (%)	81	Up from 72
Female adult literacy rate (%)	69	Up from 42

▲ Development progress in Malawi since 1998

Malawi desperately needs the transportation and communication networks (**infrastructure**) that allow businesses to thrive. But Malawi's government does not have enough money to deal with that, or other problems like an unreliable power supply, water shortages, and poor telecommunications.

2 The impacts of HIV/AIDS

Since HIV/AIDS reached Africa in the 1980s, 20% of adults in Malawi have become infected. Every year tens of thousands die from AIDS - the main cause of adult deaths and low life expectancy in Malawi. The result is poverty among Malawians. HIV/AIDS is also an economic problem. Those dying are mainly in their 20s and 30s – the age group which is normally the most economically important.

HIV/AIDS has a number of impacts on those affected:
- They become weaker and unable to work, so their income falls.
- The high costs of the drugs needed every day mean that many families cannot afford them. Only about a third of those with HIV are getting treatment.
- The death of a wage earner, plus the cost of the funeral, pushes many families further into poverty.
- By 2009, half a million orphans in Malawi had lost parents to AIDS. Orphans often live with grandparents, who are less able to work.

3 Trade

In the long run, Malawi is only likely to develop if it increases trade. Although the WTO (see Section 11.5) claims that it tries to make trade fairer to countries like Malawi, trade still presents problems. Malawi exports raw coffee beans. It could earn more by roasting them first. However, tariffs into the EU and the USA (allowed by the WTO) add 7.5% or more to the price of imported roasted coffee beans. It's in the interests of coffee-processing companies in the EU to roast their own beans as it's cheaper for them to sell to companies like Costa or Starbuck's.

▲ A train in Malawi. The single track is narrow gauge, which limits the speed and amount of weight on each train.

> We don't have machinery for farming, we only have manpower ... if we are sick, or spend our time looking after sick family members, we have no time to spend working in the fields.
> (Malawian Farmer)

your questions

1 a In pairs, draw up a spider diagram to show the problems facing Malawi over: infrastructure, HIV/AIDS, and trade.

  b Now decide which is the greatest of these three problems. Justify your reasons.

2 a Imagine that Malawi has investment and aid to pay for *either* a faster rail link to the coast, *or* a programme to treat everyone with HIV, *or* the building of factories to roast, and process, coffee beans. List the advantages and disadvantages of each project.

  b Which project would you choose? Justify your choice in 400 words.

3 **Exam-style question** Using examples from one country in sub-Saharan Africa, describe the barriers that prevent its progress. (6 marks)

In this section you'll learn about two theories of development regarding how societies develop over time.

Why are some countries poor?

Malawi (see Section 12.4) illustrates how a low-income country faces barriers to progress. These barriers may be economic (e.g. debt, poor infrastructure), social (e.g. HIV/AIDS), or environmental (e.g. drought). However, there are different theories about the causes of poverty in countries like Malawi. The first theory, by Rostow, sees a path to progress that Malawi and others simply have to follow. The second, by Frank, believes that Malawi's poverty is due to its past relationships with other countries.

▲ A market stall in Uganda - subsistence farmers selling cash crops. Rostow would say this is typical of a Stage 2 economy.

Rostow's theory

Walt Rostow was an American economist who worked in the US government after the end of the Second World War. He was anti-communist, and believed that poverty was the reason why China and other countries had overthrown their governments and become communist. In 1960, he published his theory (explained opposite) – usually called Rostow's Model – based on the experience of Europe, North America and Australasia. He believed that countries should pass through five stages of development:

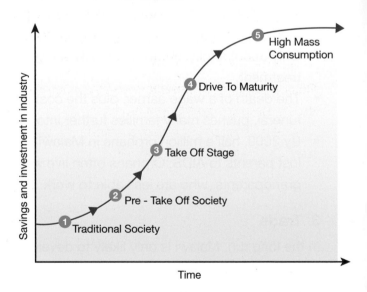

▲ Rostow's five-stage model of economic development.

▼ Five countries – but at which stage of Rostow's model are they?

Country	GDP (US$)	GDP growth rate (%)	Literacy (%)	Doctors per 100 000 people	People working in farming (%)	People working in manufacturing (%)
Australia	40 800	2.1	99	2991	3.6	21.1
Mali	1100	2.7	31	49	80.0	Very few
Singapore	59 700	4.9	93	1833	0.1	19.6
Chile	17 400	5.9	96	1090	13.2	23.0
Sri Lanka	5700	8.3	91	492	32.7	24.2

1 **Traditional society** – Most people work in agriculture, but produce little surplus (extra food which they could sell). This is a 'subsistence economy'.

2 **Pre-conditions** for take-off – there's a shift from farming to manufacturing. Trade increases profits, which are invested into new industries and infrastructure. Agriculture produces cash crops for sale.

3 **Take-off** – growth is rapid. Investment and technology create new manufacturing industries. Take-off requires investment from profits earned from overseas trade.

4 **Drive to maturity** – a period of growth. Technology is used throughout the economy. Industries produce consumer goods.

5 **Age of high mass consumption** – a period of comfort. Consumers enjoy a wide range of goods. Societies choose how to spend wealth, either on military strength, on education and welfare, or on luxuries for the wealthy.

Rostow's theory became known as modernisation theory, because he thought it was a way to 'modernise' – like the world's developed countries. All that countries had to do was follow his model!

> **+ Communism** is a system of government, based on the theories of Karl Marx, which believes in sharing wealth between all people. Communist countries include China and Cuba, and once included the USSR and most of Eastern Europe until 1991.

Frank's dependency theory

In the 1970s, dependency theory was developed by the economist André Frank – in opposition to Rostow's ideas. Frank believed that development was about two types of global region – core and periphery (see below). The core represents the developed, powerful nations of the world (i.e. North America, Europe and Australasia), and the periphery consists of 'other' areas, which produce raw materials to sell to the core. The periphery therefore depends on the core for its market.

In Frank's theory, low-value raw materials are traded between the periphery and the core. The core processes these into higher-value products, and becomes wealthy. Frank disagreed with Rostow, because he believed that historical trade was what had made countries poor in the first place. He believed that poorer countries aren't primitive versions of wealthier countries, but are weaker members of a global economy whose rules are decided by the wealthy.

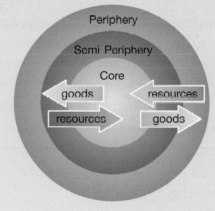

your questions

1 a Study the data in the table opposite. In pairs, decide which country belongs at which stage of Rostow's model. Justify your decisions.

b Make a copy of Rostow's model. Label the five countries where you think they should go, and annotate with reasons (e.g. high GDP).

2 Now make a copy of Frank's dependency model. On it, label where you think the five countries in the table belong.

3 Research each of the five countries in the table, using the *CIA World Factbook* online, to find out their main exports and imports. Label where these countries fit on Frank's model.

4 Who do you think put forward the better theory to explain development – Rostow or Frank? Explain your reasoning.

5 **Exam-style question** Describe Rostow's theory about how countries develop over time. (6 marks)

➕ In this section you'll learn how levels of development vary within India.

Booming India ... for some

Everything about India is huge!

- Its population (1.2 billion in 2012) is the world's second largest.
- Its workforce is 500 million people!
- Its economy is racing ahead, growing at an average of 7% a year!

But India's wealth is not evenly shared out. It varies between states – from the wealthiest (Maharashtra), to the poorest (Bihar).

Core and periphery regions

Much of India's economic growth has been around its ports and cities. As this growth happens, people migrate to these places for jobs. Then, as they earn money they spend it – boosting demand for housing and services, which creates even more jobs. This leads to an upward spiral that economists call the **multiplier effect** (see the diagram). Over time, the multiplier effect increases and a whole region develops – called a **core region**.

While Maharashtra booms, some states (like Bihar) receive less investment, have poor transport, and are distant from cities. Employers are not attracted to these places and there are few jobs, so young people have to leave to find work. These areas are called the periphery – regions on the outside (away from the core), which produce less wealth.

India – Factfile 2012

- Total population: 1.2 billion (72% is rural)
- Population annual growth rate: 1.3% (birth rate 20.6/1000, death rate 7.4/1000)
- Infant mortality rate (per 1000 live births): 46
- Life expectancy at birth (years): 67
- Population living below the national poverty line 25%
- School enrolment (primary) 99.2%
- Adult literacy rate (age 15 and above) 61%

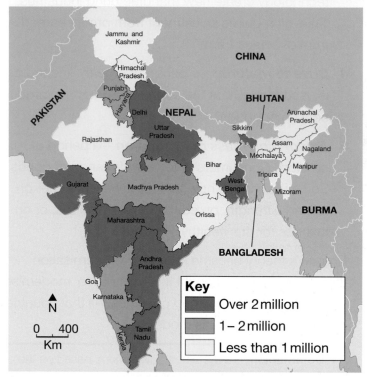

Key
- Over 2 million
- 1 – 2 million
- Less than 1 million

▲ The GDP of India's states (in million rupees).

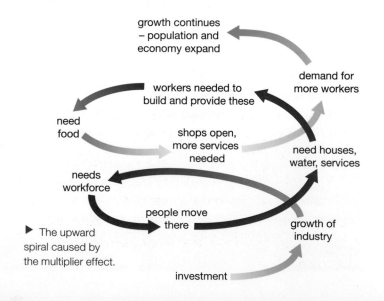

▶ The upward spiral caused by the multiplier effect.

growth continues – population and economy expand

demand for more workers

workers needed to build and provide these

need food

shops open, more services needed

need houses, water, services

needs workforce

people move there

growth of industry

investment

Maharashtra – the urban core

Maharashtra is India's richest core region, with the largest GDP (both in total and per capita). It contains India's largest city, Mumbai. Maharashtra's economic growth has come from:

- **services** – e.g. banking, IT, call centres. Graduates from Mumbai's universities often find jobs with large TNCs (e.g. BT).
- **manufacturing** – half of Mumbai's factory workers make clothing. Other industries include food processing, steel, and engineering.
- a **construction boom** – building factories and offices.
- **entertainment** – Mumbai has the world's largest film industry (Bollywood).

Bihar – the rural periphery

Bihar is India's poorest state. 86% of its population is rural, – mostly working in farming, where wages are low. Annual incomes in Bihar average 6000 rupees (£75) per person – 33% of India's average income, and 20% of Maharashtra's. Half of Bihar's households live below the poverty line, and 80% work in low-skilled jobs that pay little. Although this state has 100 million people, it gets little investment, because people can't afford to pay for basic services. Only 59% of Bihar's population has electricity.

- Bihar is also a **caste-based society**. Those in the higher castes are literate, whereas almost all of those in the lowest castes are illiterate.
- School attendance is poor. Only 35% of children in one part of Bihar go to primary school, and only 2% reach Years 12 and 13. Overall, literacy in Bihar is 47%.
- Women are the poorest in Bihar, because they have India's lowest literacy rates (33%). Uneducated women marry early and have babies before they are 20. They are poorer than men, rarely own land, and work as low-wage labourers.

Farming in the rural periphery

Less than half of India's rural population own land (most of it is owned by a wealthy minority). Most farmers rent small plots of land – paying for it by a system called 'sharecropping', where they pay half of their produce to the landowner. Every year they lose half of their earnings.

Most farmers work the land by hand, and few have any surplus income available (they live on what they grow, called **subsistence farming**). With no surplus produce to sell, there is no investment available for machinery or fertiliser. They become trapped in a **cycle of poverty** (see below).

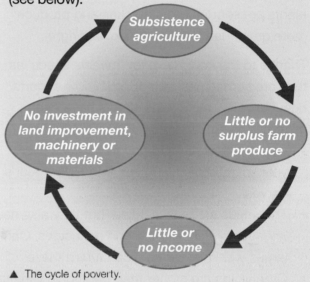

▲ The cycle of poverty.

1 Using data, make four statements to show that **a** Maharashtra is part of India's 'core', **b** Bihar is part of its periphery.
2 Copy the cycle of poverty above. Add arrows/labels to show what would happen if a family member **a** becomes ill, **b** gets into debt, **c** goes to Mumbai and sends wages home.
3 Design spider diagrams to explain **a** why land in Bihar produces less food, **b** why women are poorer than men, **c** why few children attend school.
4 **Exam-style question** Using examples, describe how levels of development may vary within a country. (6 marks)

+ In this section you'll learn about differences between top-down and bottom-up development.

The 'Green Revolution'

The Green Revolution of the 1970s changed rice growing forever. It offered HYV (High Yielding Variety) seeds, instead of the traditional lower-yielding varieties – which had sometimes not produced enough food and led to hunger. The HYVs were developed by scientists working for trans-national corporations (TNCs). Now the rice plants are shorter, grow quicker, and produce more grain than traditional rice (see right).

Thanks to the new HYVs (or 'miracle seeds' as they are sometimes called), India now exports rice. However, the overall effects have been mixed:

- Farmers now have to buy new seeds every year, instead of saving some from last year's harvest.
- The seeds are high yielding, but they also need irrigation water, fertiliser and pesticides. Only larger, wealthier farmers can afford these.
- Crop yields are higher than traditional varieties, so incomes have risen – for the wealthy.
- The over-use of chemicals to control pests, has reduced the resistance to pesticides.

Working from the bottom up

ASTRA (Application of Science and Technology in Rural Areas) is a recent development project in rural India. Local researchers found out what people's lives were like. They talked to families, recorded how they spent their time, and listened to their problems and needs. This is an example of how a **bottom-up development** project works.

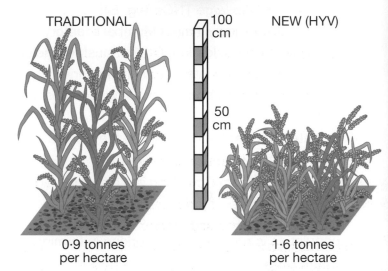

TRADITIONAL 100 cm 50 cm NEW (HYV)

0·9 tonnes per hectare

1·6 tonnes per hectare

▲ The differences between HYVs and traditional varieties of rice.

Top-down development

The Green Revolution is an example of '**top-down development**' – where decisions about development are made by governments or large private companies. These decisions are then imposed on people, because – supposedly – there will be benefits for them.

Top down development involves:

- Decision-makers looking at the 'big picture' to identify needs or opportunities, e.g. to establish national energy sources, food security, or better transport networks.
- Experts helping to plan the changes.
- Local people being told about them, but with no say in whether they will happen or not.

The argument goes that all people will benefit by a process called 'trickle down' – where jobs and therefore wealth, 'trickle down' to the poor.

The problem of time

ASTRA found that, for most rural families, the daily routine takes time – especially for women and girls. Cleaning, collecting fuel, preparing and cooking food, fetching water, tending sacred cows, looking after the vegetable patch – all before any paid work is done in the fields! Rural girls have little education and few complete primary school. Most time is spent collecting fuelwood. Every family needs 25-30kg of it every week, and it takes hours to collect. As population increases, it's in increasingly short supply.

Solution – think cow dung!

However, the answer could be right under their noses – cow dung! Cow dung is a highly valued resource, because it produces gas – called **biogas** – which is used for cooking by day, and powering an electricity generator at night. It is fed into a brick, clay or concrete-lined pit that forms part of a biogas plant (see diagram). The pit is sealed with a metal dome and the dung inside ferments to produce methane. As pressure builds up, the methane is piped into homes. It's a simple technique using local materials and is another example of **intermediate technology** (see Section 4.8).

Families gain because:

- less time is spent collecting fuelwood
- there's no ash, so less time is spent cleaning
- heat is instant, so cooking is quicker
- there's less smoke and fewer cases of lung disease.

> **+ Bottom-up development** means
> - experts work with communities to identify their real needs
> - giving local people control in improving their lives
> - experts assisting with progress

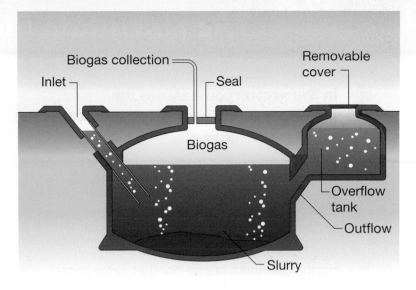

▲ A biogas plant

1 Explain the difference between 'top-down' and 'bottom-up' development projects.

2 a Copy and complete the table below to compare these 'top-down' (HYV) and 'bottom-up' (biogas) development projects.

	Top-down project	Bottom-up project
Size and scale		
Aims of the project		
Who pays for it		
Who makes the decisions about what's needed		
Raw materials and technology required		
Who benefits		

  b Use your completed table to decide which type of project is best for (i) national interests, (ii) local communities.

3 **Exam-style question** Using examples, describe the differences between top-down and bottom-up development. (6 marks)

A top-down project: The Narmada River Scheme

✚ In this section you'll assess the benefits and problems of a top-down development project.

Top-down – the government decides!

Over much of India, rainfall is seasonal and unevenly spread (see right). Parts of north-west India are so dry that semi-desert exists, which prevents people from making a decent living. Across the rest of India, between:

- May and September, the Indian monsoon brings huge falls of rain that are difficult to imagine – think of the heaviest rain you have ever seen, and then double it.
- November and March, almost no rain falls across large areas of India.

As India's population increases, and its economy booms, demand for water is rising. As a result, the Government decided that western India needed super dams to:

- encourage economic development, by providing drinking water and electricity for cities and industries
- open up dry lands for farming using **irrigation** to feed a growing population.

Building large dams makes it possible to store monsoon rains to use during the dry season. By 2008, the Indian Government had built over 4500 dams – 14 of which are super dams. Now the Narmada – one of Western India's major rivers (see right) – is being tackled with a series of 3000 dams (big and small). The scheme will take a 100 years to complete! But how well will it work for people and the environment?

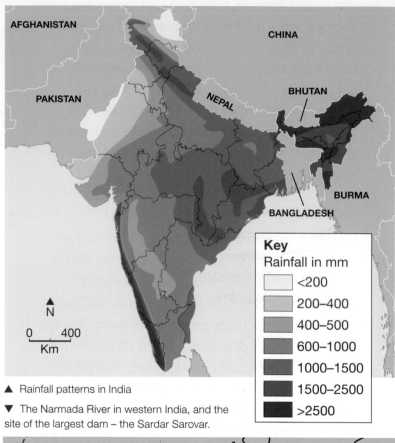

▲ Rainfall patterns in India

Key
Rainfall in mm
- <200
- 200–400
- 400–500
- 600–1000
- 1000–1500
- 1500–2500
- >2500

▼ The Narmada River in western India, and the site of the largest dam – the Sardar Sarovar.

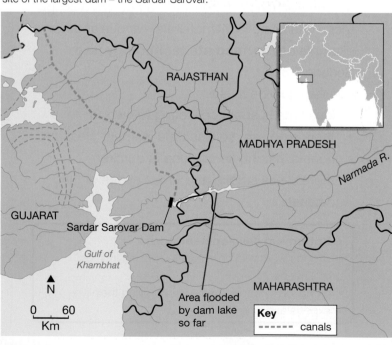

Key
- - - - - - canals

✚ **Irrigation** is taking water from areas that have it, to those that don't, in order to allow farming.

The Sardar Sarovar Dam

The Sardar Sarovar Dam, along the Narmada River (see photo), is already one of the world's largest dams. When completed, it will provide water all year round to areas of India that suffer drought.

Groups benefiting from the dam

Originally designed to be 80 metres high, the government now plans to raise the dam to 136.5 metres – to increase its capacity.

- **People in India's cities.** The dam is multi-purpose – it provides 3.5 billion litres of drinking water a day, and 1450 megawatts of hydroelectric power (HEP), which is more than 750 wind turbines!

- **Farmers in other parts of western India.** A series of canals distribute water to other states in India. When complete, they will **irrigate** 1.8 million hectares of farmland in the driest parts of Gujarat, Maharashtra, Rajasthan and Madhya Pradesh (see maps opposite). Gujarat and Madhya Pradesh suffer from drought and lose £20 billion in farm production each year.

What do you think?

+ Do you think governments have the right to develop 'top-down' schemes like this if they affect so many people?

Groups losing out from the dam

- **Local residents.** 234 villages have been drowned so far, forcing 320 000 people out. Few villages can afford the electricity generated by the dam – only the cities benefit.

- **Local farmers.** Good quality farmland has been submerged. Those gaining from irrigation will lose out from increased soil salinity, making the soil less usable. Damming the river means that fertile sediment, normally deposited on flood plains each year, will be lost.

- **Western India.** Religious and historic sites have been flooded by the dam. The silt brought down by feeder rivers will collect behind the dam and reduce the reservoir's capacity.

- **People downstream.** This area has a history of earthquake activity. Seismologists believe that the weight of large dams can trigger earthquakes, which could destroy the dam and cause massive loss of life.

your questions

1 a Copy and complete the following table about the economic, social and environmental benefits and problems of the Sardar Sarovar Dam.

	Benefits	Problems
Economic		
Social		
Environmental		

b Highlight in one colour those benefits or problems which are **local**, and in another those which are **further away**.

c Which are the greatest benefits – economic, social or environmental? Are they local or further away?

d Which are the greatest problems?

e Explain whether you think top-down schemes like this should be built if they cause such problems.

2 Exam-style question For a top-down development project that you have studied, explain its benefits and problems. (8 marks)

➕ In this section you'll learn about changes in primary and secondary employment in the UK.

> ➕ See Section 11.2 for definitions of **primary**, **secondary**, **tertiary** and **quaternary** employment.

UK plc

In a globalised world, most of the goods you use at home or at school are made overseas, including the clothes you're dressed in and the food you eat. Your clothes were probably made in China, Vietnam or Bangladesh, and there's a 50% chance that your food was produced overseas. If that's the case, how on earth does the UK plc make its money? This section ought to give you some answers!

Trouble in Dagenham

Dagenham in east London has a long industrial history. Shown in the film 'Made in Dagenham', it was where women working for Ford Motors in the 1960s went on strike for equal pay with men. At the time, Ford's car factory at Dagenham was its largest outside the USA – with 20 000 workers. Many jobs were skilled and well paid. But now Dagenham is suffering. Only car engines are made there now – by just 3000 workers. Like many manufacturers, Ford now produces its cars in Europe (where wage costs are lower). Between 1998 and 2005, the number of jobs in Dagenham fell by over 14%. Now there are few jobs in the area for young people.

▲ Ford's car factory at Dagenham was once the largest in Europe. Its gradual closure has resulted in thousands of job losses.

However, despite the downward trend at Dagenham, the UK's car industry isn't dead! Cars are heavy items, so it still makes sense to make them near where they will be sold. Over 70 000 people in the UK still assemble cars, and another 110 000 make and supply parts. There are still plenty of car manufacturers – in fact, the UK produces nearly as many cars now, as it did at its peak in the 1970s. It's just that the UK's car industry has changed:

- **Brand names** have changed. Most major British car manufacturers have either gone out of business or been taken over by overseas companies. The five biggest car manufacturers in the UK are, in order, Nissan UK (Japanese), Jaguar Land Rover (Indian-owned), Toyota UK (Japanese), Honda UK (Japanese), and Mini (German-owned).
- Cars are now made in **automated** factories, where robots and machinery assemble them with far fewer employees.

> *On your planet*
> ➕ In the nineteenth century, the UK used to be known as 'the workshop of the world', because so many of the world's manufactured goods were made here.

All change!

The table shows changes to the UK's employment structure since 1980. The last thirty years have seen an employment revolution:

- **Primary** employment has nearly halved. For example, over 250 000 people were employed in coalmines in the 1970s – but only 6000 by 2009.
- **Secondary** employment has also crashed.
- However, **tertiary/quaternary** employment has increased sharply (see Section 13.2).

Date	Primary	Secondary	Tertiary and quaternary
1980	0.89	8.8	17.6
1985	0.79	7.3	18.3
1990	0.70	7.5	21.0
1995	0.56	6.3	21.0
2000	0.49	6.2	23.0
2005	0.46	5.6	25.2
2010	0.49	5.0	25.9
Trend	**Down by 45%**	**Down by 43%**	**Up by 48%**

▲ Changes to primary, secondary and tertiary/quaternary employment in the UK between 1980 and 2010 (in millions of workers).

Meanwhile, at Burberry …

Burberry's brand (see right) includes bags, shoes and clothes. It's a British company selling expensive goods with the label 'Made in Britain' – a coat can cost over £1000! But, like many fashion companies, most of its products are now made in China. In 2007, its 300 employees at a factory in Treorchy, South Wales lost their jobs to China. In 2009, 300 jobs went from its Yorkshire factories. Luxury labels like Burberry have built their reputations on British-made goods – using skilled labour to justify their high prices. But the temptation to move production to places with cheaper labour is strong. From Asda's 'George' label – to brands like Burberry – most clothes are now made overseas. However, the label 'Made in Britain' survives! In Burberry's case, it kept two British factories open in Castleford and Keighley, which allows it to say 'Made in Britain' on those products. Products made overseas are simply not labelled!

On your planet

+ The fashion industry is one of the UK's biggest industries for both design and retail – but most goods are made overseas!

your questions

1 Why do some factories now produce as much as they did 30 years ago, but with fewer people?
2 What would the effect be on the community in Dagenham if **a** Ford reduces employment, **b** there are fewer jobs for young people.
3 In pairs, research **one** of the top five car companies making cars in the UK. Find out **a** where they make cars, **b** the number of employees, **c** the brands made, **d** why they locate in the UK.
4 Copy and complete the table below to show the advantages and disadvantages of clothing companies locating overseas for **a** the company, **b** its UK workers, and **c** consumers.

	Advantages	Disadvantages
For the company		
For UK workers		
For consumers		

5 **Exam-style question** Suggest reasons for the changes in primary, secondary and tertiary employment in the UK. (6 marks)

✚ In this section you'll learn about changes in the UK's tertiary and quaternary industries.

Up early – but it's worth it!

It's 6.30am at Canary Wharf in East London's Docklands. The working day has already started, with traders at their desks. By getting there early, they can trade with colleagues in Singapore and Hong Kong (which are seven hours ahead of the UK). Many traders put in long hours until late into the evening, so that they can also trade with colleagues in New York (which is five hours behind the UK).

100 000 people work in Canary Wharf, with most commuting in from all over London and south-east England. Their journey is worthwhile, because average salaries there are £100 000 a year. That's four times higher than the UK's average salary – and doesn't even include their annual bonus, which can be as much as a year's salary!

▲ Canary Wharf in London's Docklands

Working in Canary Wharf

What jobs are these people doing? Most companies in Canary Wharf are banks (e.g. HSBC and Barclays) and investment companies (e.g. J P Morgan). They make their profits by investing money for individuals, companies and pension funds – money that people save each month towards their pensions. Banks trade in currencies, stocks and shares, hoping that their value will rise. Each time they trade, they earn a commission. That's how investment banks make money.

London is one of a few major cities in the world – known as 'World Cities' – that trade and invest all over the world. Each year, half of all money in the world – several trillion pounds – passes through London at some stage! The growth of banks has led to related companies moving to Canary Wharf:

- Investment in different currencies requires lawyers, so there are law firms there.
- Insurance is big business. Shipping companies like Maersk (see Section 11.3) insure container ships travelling around the world.
- IT companies also work there, helping banks to trade instantly all over the world.

The knowledge economy

Canary Wharf didn't happen by chance. In the 1980's, the UK government, under Margaret Thatcher, planned it this way. Those working at Canary Wharf were part of a government policy to develop the **tertiary** and **quaternary** sectors (see Section 11.2). All offer a service (tertiary), but many are also highly specialised (quaternary). They all work in the **knowledge economy** – an economy based on knowledge and mental skills. Almost everyone has a university degree, plus specialised training after university (e.g. law or accountancy). The knowledge economy also includes creative industries (film, media, advertising, and marketing, also concentrated in south-east England.

	IT	Finance	Professional scientific & technical
1980	767 000	853 000	992 000
2010	1 133 000	1 127 000	2 367 000
Rise	48%	32%	139%

▲ The growth of quaternary employment in the UK.

The biotechnology boom

Do the names Johnson & Johnson, Pfizer, or GlaxoSmithKline mean anything to you? If not, that's not surprising, because their brands are better known than their company names (e.g. *Anadin* made by Pfizer). They're part of a huge biotechnology industry.

Biotechnology is not a well-known industry. Companies like this need land for offices and research laboratories, so they tend to locate away from city centres (see the map). But they remain close to major universities (for research) and airports (for business meetings). Some biotechnology research is sensitive and controversial (e.g. because it uses animal testing, or involves the development of genetically modified crops). However, it's a major industry in the UK, with the number of employees doubling to 2.3 million between 1985 and 2010.

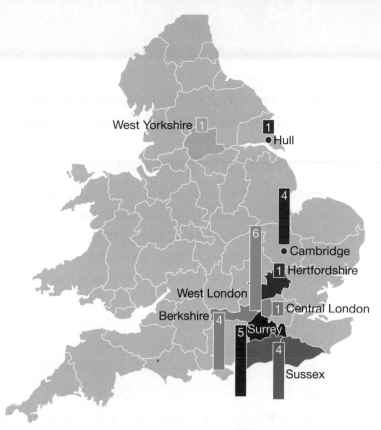

▲ The laboratory and office locations of the 27 largest biotechnology companies in the UK.

+ **Biotechnology** applies science to global problems. For example:

- drugs (or pharmaceuticals)
- medical biotechnology (using living cells to treat human diseases)
- medical technology (using technology in medicine, from sticking plasters to high-tech products, e.g. body scanners)
- industrial biotechnology (using natural biological processes for industrial purposes, e.g. enzymes which destroy dirt in washing powders).

your questions

1 Explain the difference between tertiary jobs, quaternary jobs, and the knowledge economy.
2 Give two specific examples of each of the different types of tertiary and quaternary jobs listed in the table.
3 Explain, with examples, why some knowledge-based industries may prefer to locate in city centres, while others prefer out-of-town locations.
4 In pairs, brainstorm and list the reasons why office workers in Canary Wharf often work long hours.
5 **Exam-style question** Using examples, describe changes in tertiary and quaternary employment in the UK. (6 marks)

➕ In this section you'll learn about classifying jobs, and why types of employment have changed.

All change in Dinnington

Dinnington is a small town of about 9000 people, located within the borough of Rotherham in South Yorkshire. It had one of the Yorkshire coalfield's biggest collieries until 1992, when it was closed. During the twentieth century, most jobs there were done by men – boys followed their fathers to work down the mine. Mining jobs were full-time and well paid, and the town's shops and pubs thrived – and so did the Dinnington Colliery Brass Band. Dinnington was a town based on primary sector employment.

The colliery band still exists – it's just won a £1 million recording contract! – but there are no miners left. The colliery site is now a large Business Park (see the photos), which created 1700 new jobs between 1998 and 2008. Most companies there are tertiary sector, such as home delivery or sales (e.g. a motorhome dealer). Like many places in the UK, Dinnington has seen a change from primary and secondary employment to tertiary.

Working in Dinnington now

Losing the mine is not the only employment change in Dinnington. Most people now have to commute, instead of working locally. In November 2012, of the jobs advertised locally in Dinnington:

- few were in Dinnington itself (most were in Sheffield, Rotherham, Doncaster or even further away)
- some were permanent, but many were temporary or contract posts for a few months only
- most were full-time, but some were part-time
- many were low wage. Even salaries for skilled workers (e.g. engineers) offered just £30 000 (15% above the UK average salary).

Classifying jobs

The change from primary to tertiary employment is important, but there are other equally useful ways of classifying jobs. As well as by job type, the UK government now classifies jobs depending on whether they:

- involve a level of skill and responsibility – for example, professional and managerial, skilled or unskilled
- are full-time or part-time
- are temporary or permanent.

These indicators determine income, but they also help to show changing employment patterns in the UK.

	Rotherham	London	UK
Level of skill and responsibility			
Professional and managerial	33%	50%	43%
Administrative and skilled	21%	23%	22%
Semi-skilled and unskilled	46%	27%	35%
Full- and part-time			
Full-time work	68%	78%	74%
Part-time work	32%	22%	26%

▲ Comparing the breakdown of job types in Rotherham with those in London and the UK as a whole.

The retail sector

Between Sheffield and Rotherham lies Meadowhall Shopping Centre. Located beside the M1, it attracts shoppers from both the Midlands and northern England. It's one of the UK's largest shopping centres (with stores such as M&S and Apple) and, for people in Dinnington, it provides a useful source of retail jobs.

- However, like tourism, the retail industry is seasonal – Meadowhall employs up to 7500 people at peak times (e.g. Christmas), but much less at other times. Therefore, many jobs are **temporary**.

- Trade even varies during the week. Most shops make over 60% of their sales at weekends. Therefore, many staff are employed **part-time** the majority of which are women.

These jobs can suit people looking for part-time hours – e.g. mothers choosing hours to match their children's school hours – but are not sufficient for people seeking full-time work.

your questions

1 Explain why London has more 'professional and managerial' jobs than Rotherham.
2 Why is it important whether a job is professional, skilled or semi-skilled?
3 Compare the jobs on offer at Dinnington and Canary Wharf by copying and completing the table below.

	Dinnington	Canary Wharf
Level of qualification and skill		
Full- or part-time		
Temporary or Permanent		
Salary levels		
Benefiting men or women?		

4 Why is Dinnington likely to find it hard to attract employers from the 'knowledge economy'?
5 Exam-style question Using the table of data on the opposite page, explain why average incomes in Rotherham are lower than in the rest of the UK. (4 marks)

Rotherham and the new economy

The 'new economy' is a service-sector economy. It's very different from the 'old economy', which was based on traditional industries like coal mining and engineering (which declined in the 1980s and 1990s, because of overseas competition).

Companies in the new economy are called 'footloose' (see Section 11.7), because they're not tied to location. But the 'new economy' has two sides (see table).

- At the high salary end, companies in the 'knowledge economy' need workers with high qualification and skill levels (found in large cities and science parks near university towns and research centres).

- However, the majority of jobs in the 'new economy' are at the lower salary end. Many people work on small industrial estates or business parks (as in Dinnington), or at retail centres such as Meadowhall.

Low salary 'new economy', e.g. Dinnington and Rotherham	High salary 'knowledge economy', e.g. London's Canary Wharf
Tertiary sector.	Quaternary sector.
Examples are jobs with delivery firms, in retail parks, or in shopping centres.	Examples are jobs in major cities with global companies. Jobs are advertised globally to get the right people.
Often located on the outskirts of towns – attracted by cheap land and local labour.	Attracted by highly skilled and educated staff nearby, and good electronic networks (e.g. high-speed broadband).
Many jobs are unskilled, needing few qualifications.	Jobs require a degree, e.g. law, finance, ICT, media, biotechnology.
A quarter of all jobs are now part time. Many jobs are also temporary – lasting just weeks or months (e.g. at Christmas).	Most jobs are full-time, with high salaries plus bonuses. Contract jobs are also available (often with very high daily rates of pay).
Wages are usually low (minimum wage or just above), with more for supervisor roles (e.g. in supermarkets).	Salaries depend on qualifications – even new graduates can earn salaries of up to £60 000.
Employees are a mix of male and female, but women are in the majority.	Employees are mostly male, especially in banks or biotechnology companies.

▲ Winners and losers in the new economy

✚ In this section you'll learn about the UK's North East region, its industry and employment.

Perhaps it's the accent – research says that Geordie is the most trusted accent from call centres! It might be the food (ever eaten a stotty?), Northumberland's stunning beaches, or the 'Angel of the North'. Whatever it is, the North East has a great sense of identity – and generates great loyalty from the 2.6 million people living there. Its largest city, Newcastle upon Tyne, buzzes with life (even when it's freezing in January). But the North East is facing serious problems with unemployment.

Deprivation in the North East

The North East is one of the UK's 12 standard Economic Regions (see the lower map on the right). It's England's lowest-income region, with an average **household income** of £26 000 per year. It's also England's most-deprived region, with high unemployment, poor health, and overcrowded housing:

- In 2012, unemployment there was over 11% (one of the highest rates in the UK).
- In 2011, a quarter of all jobs in the North East were in the **public sector** (local and national government organisations, such as the NHS). Public-sector jobs are being cut, due to reduced government spending (especially in Newcastle and Middlesbrough, where one third of jobs are in the public sector). Between 2010 and 2011, 32 000 jobs were lost from the North East's public sector.
- The North East has the worst indicators in England for deaths from smoking and early deaths from heart disease, strokes and cancer, as well as the lowest percentage of adults who eat healthily.

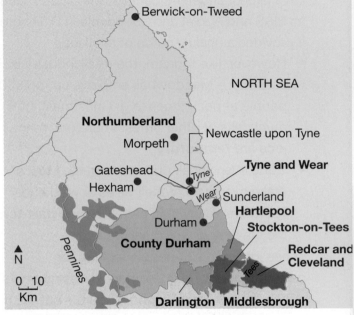

▲ The North East region.

✚ **Household income** is the total income earned by everyone in a household, after paying tax. The average UK household has two income earners.

▲ The UK's 12 standard economic regions.

Understanding the 'domino effect'

In the 1960s, the North East employed 55 000 men in coal mining. Much of the coal went to Redcar's steelworks, or to an aluminium smelter at Lynemouth. The local engineering industry made ships' engines, and the Tyne and Wear estuaries were lined with shipyards building ships to carry the world's trade. These 'heavy' industries were all linked like a series of dominoes – coal was needed to make steel, which was used for engineering, and to make the steel plates that were eventually turned into ships.

However, each industry suffered problems:

- The coal was expensive to mine, because it was very deep down. Hardly any coal is produced in the region now.
- The steel industry suffered from globalisation and cheap overseas competition.
- Shipbuilding and engineering collapsed when Asian countries began to build larger ships with cheaper labour. Shipbuilding on the Tyne finally ended in 2007, when Swan Hunter sold all its cranes and equipment to a shipyard in India.
- In 2012, the aluminium smelter closed in Lynemouth.

What happened is a process called the 'domino effect' – when one industry collapses, it leads to the collapse of others. When this happens, it also damages other local businesses and services. Like Glasgow (see Section 13.6), this led to deindustrialisation in the North East.

▲ Shipyards lining the River Tyne in the 1970s.

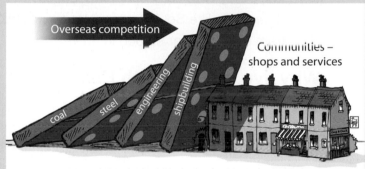

▲ The 'dominoes' of the coal, steel, engineering and shipbuilding industries

So what's left?

As in the rest of the UK, manufacturing in the North East has declined and the tertiary sector has increased. There's still some manufacturing – Nissan in Sunderland is Europe's largest car factory (it moved there in the 1980s, with financial help from the British government), but most employment is tertiary. As well as the public sector, private companies include:

- the transport companies DB Arriva and Go-Ahead
- call centres (e.g. Tesco, Orange, Npower), because wage rates are 40% lower than in London
- Virgin Money, which took over much of Northern Rock.

Part of the region's problem is a lack of high-salary jobs in the quaternary sector. The Sage Group is the first major software business to locate there, but there are few others.

your questions

1 Outline **a** three pieces of evidence to show that the North East is one of the UK's most deprived regions, **b** why health can suffer as a result of deprivation.

2 **a** Make a copy of the diagram showing the dominoes of the coal, steel, engineering and shipbuilding industries. Label your diagram to show how these four industries are linked.

   b Now draw a second diagram, with labels, to show what happens when one of the dominoes falls to (i) each domino, (ii) the local community.

3 In pairs, discuss whether the British government should help with the costs of companies re-locating to the North East. List the pros and cons.

4 **Exam-style question** Describe the impacts of changing employment trends on one UK region you have studied. (6 marks)

✚ In this section you'll learn about employment in the South East, and how it differs from the North East.

On the right is a photo of Hart (a Hampshire district in South East England). It's a pretty area, with typical 'English' small towns and villages in rolling countryside. In a 2011 Halifax report, Hart was considered to be the UK's most desirable place to live. The report used indicators like: jobs, housing, health, crime, weather, and broadband access to make its evaluation. 95% of Hart's residents were considered to be in good health and affluent (well off), with incomes above the UK average. Crime levels were low there. With easy commuter access to London (and other cities like Southampton) Hart's population has risen from 9000 to over 90 000 in 30 years.

Why live in the South East?

In spite of the fact that – with 8.7 million people – the South East is the UK's most densely populated region, much of it is open countryside. Even though it's close to London, Surrey has a high percentage of woodland, and the Sussex coast (as the driest and warmest part of the country) is one of the UK's most popular retirement and holiday spots. Outside London, South East England is the most expensive area in the UK for house and land prices. Many people live in commuter towns outside London, so good rail links to London are vital.

Life in the South East

London salaries boost average household earnings in this region, because many people commute. It's not surprising that people here have the UK's highest spending power. The average household income of South East residents is the UK's second highest (£35 200) – over 35% higher than in the North East. Household surveys show that people in the South East spend the most on healthy foods, e.g. fish, fresh fruit and vegetables, and they also live longer. They also eat out more – up to three times a week. For many, this region offers a high quality of life.

On your planet

✚ Someone living in South East England has an average life expectancy that is 11 years longer than in some parts of Northern England.

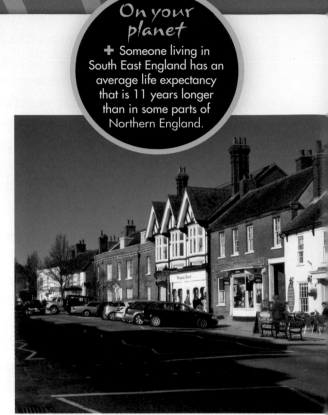

▲ The affluent district of Hart in Hampshire.

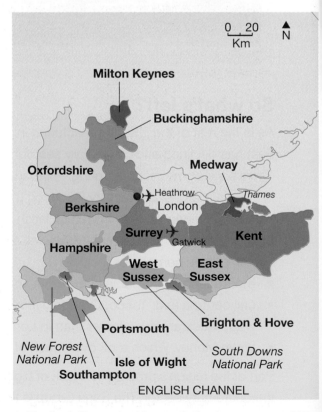

▲ The South East region (also see the lower map on page 224)

Jobs in London

The South East makes the second largest contribution to the UK's GDP, after London. Many workers living in the South East commute to London daily. It's very time consuming, but salaries in London are higher – there are many well-paid tertiary and quaternary jobs in the capital. In the public sector, civil-service jobs in government departments are better paid, with more opportunities for promotion. North East England (see Section 13.4) lost public-sector jobs between 2010 and 2011, but the South East gained them. In the private sector, knowledge-economy jobs (see section 13.2) and company headquarters in London provide senior, highly paid jobs.

Jobs in the South East

The South East also has many job opportunities. In 2011, its unemployment rate was 6.3% (when the UK average was 8.4%). Being the UK region closest to Europe, the South East contains two of the UK's largest ports (Southampton and Dover), which provide thousands of jobs. It also has many significant employers in the knowledge economy:

- Many biotechnology companies are based in South East England, rather than in London. Between them, the 'Top 20' companies have 13 research laboratories and Head Offices in Surrey, Sussex and Berkshire.
- Many top IT companies (e.g. Microsoft and Oracle) are located in the 'M4 corridor', between Heathrow Airport and Reading.

Between them, businesses in the region spent nearly £6 billion on research and development – 22% of the UK's total and the highest of any UK region.

▲ The Oracle headquarters at Thames Valley Park in Reading.

	South East England			London
	Primary	Secondary	Tertiary & quaternary	Tertiary & quaternary
1996	1.4%	19.4%	79.2%	87.3%
2001	1.6%	17.8%	80.6%	89.0%
2006	1.1%	15.6%	83.3%	90.3%
2011	0.8%	14.4%	84.8%	91.6%

▲ Data showing changes in employment in South East England and London, 1996 2011.

your questions

1 Using the data in this section, and Section 13.4, describe the evidence to show that the South East region is more affluent than the North East.

2 Study the upward spiral diagram in Section 12.6 about the multiplier effect. Make a copy of the diagram on page 212 to show why there are many jobs and more wealth in South East England.

3 a Explain the benefits that companies in South East England might get from moving to North East England.

   b Using the evidence in Section 13.4, and this section, suggest possible reasons why very few companies actually do move away from South East England.

4 **Exam-style question** Explain the employment trends in one UK region you have studied. (8 marks)

In this section you'll learn about the impacts of deindustrialisation and economic change in Glasgow.

Glasgow's decline

The Glasgow shipyard in the photo lies empty – a victim of deindustrialisation (see Section 11.6) caused by overseas competition. The shipbuilding industry created close communities that were devastated by decline throughout the twentieth century, but particularly in the 1980s. Deindustrialisation had enormous economic impacts, especially on cities that depended on traditional 'heavy' industries, such as steel making, shipbuilding and coal mining. The decline of shipbuilding in Glasgow caused a decline in steel making (to make the ships) and coal mining (to make the steel). The economic impacts included:

- a loss of income for the workers
- a loss of income for local shops and services (as people had less money to spend)
- a loss of government tax income
- a rising demand for income-support benefits (which meant that government spending was rising, just as its income was falling)
- migration away from the city, as skilled workers moved to find employment elsewhere.

There were also serious social impacts. In the 1980s, Glasgow contained some of the most deprived areas in the UK. There were widespread problems with family breakdown, alcoholism, and crime. The area developed a poor image that put off new investors, which led to a further cycle of decline (see right).

Environmental impacts

A range of environmental impacts can also be associated with deindustrialisation. Some are positive and some are negative (see right).

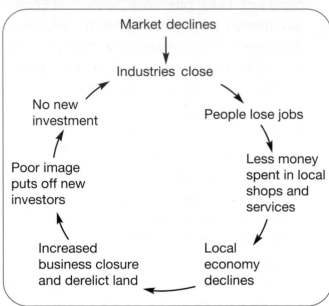

▲ The cycle of decline.

Market declines
↓
Industries close
→ People lose jobs
→ Less money spent in local shops and services
→ Local economy declines
→ Increased business closure and derelict land
→ Poor image puts off new investors
→ No new investment
→ Industries close

Positive impacts of deindustrialisation	Negative impacts of deindustrialisation
• More available land for future development (e.g. housing) • Less water used in industrial processes • Less energy required to operate machines (reduces CO_2 emissions) • Reduced traffic congestion near factory • Reduced noise and air pollution	• Derelict land looks unsightly • Deteriorating infrastructure (e.g. lack of investment in roads) • Empty factory buildings (and associated litter/ vandalism) • Manufacturing goods further away leads to greater transport problems and pollution elsewhere

On balance, deindustrialisation has had some very positive effects on the local and regional environment of developed countries like Scotland. But, globally, the picture is different. As manufacturing has moved overseas, ships and aircraft criss-cross the oceans and skies transporting raw materials and finished goods from place to place. In doing so, they are pumping out carbon dioxide and increasing air pollution, and are likely to be contributing to an increase in climate change. Overseas factories also generate pollution, so the environmental problem has just moved overseas.

Glasgow's alive again!

One of Glasgow's problems was that all of its main industries were linked. So, a programme of **diversification** has now taken place – using Scottish, UK, and EU government grants to kick-start the process. Private industries by themselves rarely invest without a good return on their money, so the government invests first to try to create a cycle of growth (see right).

As in the rest of the UK, manufacturing has declined in Glasgow (although there is still some shipbuilding). However, tertiary and quaternary industries have expanded in the following areas:

- **Arts, culture and tourism.** Glasgow has a wealth of architecture from famous architects, such as Charles Rennie Mackintosh. The UK and Scottish governments have invested in the new Burrell Collection, to create an internationally famous art museum that attracts tourists. They've also invested in the Scottish Exhibition and Conference Centre, the Glasgow Science Centre, and the Riverside Museum of Travel and Transport. Tourism has brought increased employment in hotels and restaurants.
- **Mixed-use developments.** Investment from private property developers has led to the development of riverside flats along the River Clyde, together with shops and restaurants.
- **Media.** The new BBC Headquarters for Scotland's TV and radio broadcasting opened in 2007. Commercial broadcaster STV is also located nearby.

> **+ Diversification** means creating varied industries and economic activities. Then, if one fails, it won't affect the others (as happened in Glasgow).

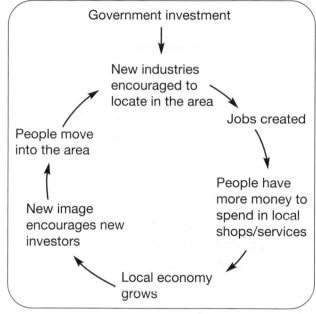

▲ The cycle of growth.

your questions

1 Explain why deindustrialisation can lead to social problems, such as family breakdown and crime.
2 Copy the scales diagram opposite, but rewrite it for the economic effects of deindustrialisation and investment in Glasgow.
3 Design a poster showing the appearance before and after deindustrialisation of Glasgow's docks and riverside. Use the Internet to find some images to use.
4 On balance, is deindustrialisation in the 1980s a price worth paying for the improved environment and employment opportunities in Glasgow now?
5 Exam-style question Using examples, explain the environmental impacts of deindustrialisation and economic change. (8 marks)

✚ In this section you'll learn about the costs and benefits of greenfield development versus brownfield regeneration.

Brownfield sites

In Glasgow (Section 13.6), one of the economic impacts of deindustrialisation and the closure of industry was unemployment. However, an important environmental impact of this process was the creation of derelict land and **brownfield sites**.

In Chesterfield, derelict brownfield sites (like the one in the photo) were once used for the city's coal and engineering industries. Like Glasgow, Chesterfield is on a large coalfield, which created a demand for mining equipment. When the mines closed in the 1980s and 1990s, they left a legacy of derelict land contaminated with pollution from tar and other coal products.

However, redeveloping brownfield sites (e.g. for housing) can have disadvantages:

- It can be more expensive than developing **greenfield sites**, because of the clean-up costs involved.
- Most brownfield land in former industrial areas is far away from the areas where housing demand is highest – in South East England.
- Some brownfield and derelict land provides important local wildlife habitats or public green spaces in cities.

Fort Dunlop in Birmingham

At its height, over 12 000 people worked at Fort Dunlop – a tyre-storage facility on the outskirts of Birmingham. Built in 1916, it closed in the 1980s (when Dunlop's manufacturing and storage facilities were moved overseas). It has become one of Birmingham's major brownfield sites.

Sites like Fort Dunlop are important for future development. They provide a number of regeneration options (see the diagram).

✚ A **brownfield site** is an area of land that has been built on before and is suitable for redevelopment.

✚ A **greenfield site** is an area of land that has not previously been built on.

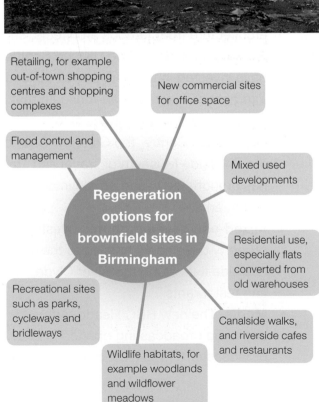

Retailing, for example out-of-town shopping centres and shopping complexes

New commercial sites for office space

Flood control and management

Mixed used developments

Regeneration options for brownfield sites in Birmingham

Residential use, especially flats converted from old warehouses

Recreational sites such as parks, cycleways and bridleways

Canalside walks, and riverside cafes and restaurants

Wildlife habitats, for example woodlands and wildflower meadows

In 2002, Fort Dunlop received planning permission for a new sustainable 24-hour community. Opened in 2006, it now includes a 100-bed hotel, a business park with office and retail space, as well as places to eat and drink. It's a good example of how a brownfield site can be regenerated to provide both employment and leisure, and improve the local environment. Its tenants include architects, designers, accountants, and independent retailers (including a coffee company and children's activity centre).

▲ The empty Fort Dunlop factory before regeneration.

Greenfield sites in Solihull

Few new developments in the West Midlands are on greenfield sites, as there is brownfield land available. However, Solihull is desperately short of housing and in 2010 wanted to build 10 500 homes on greenfield sites. In addition, it has a social housing waiting list of over 10 000.

Greenfield sites have advantages, with no clean-up costs. It's easier to plan – greenfield land is a blank canvas where developers can plan what they wish without considering what's already there (e.g. roads).

But there are also disadvantages:

- Valuable farmland is lost. The total greenfield area lost to urban development in the UK between 1945 and 2010 was 750 000 hectares (more than the size of Greater London, Berkshire, Herefordshire and Oxfordshire combined).
- The negative impact on the rural landscape. The Council for the Protection of Rural England (CPRE) believes that this is as serious as the loss of farmland.
- Greenfield sites have no existing roads, water or energy infrastructures, so developers must pay the costs of installation.
- Many greenfield sites close to cities (which would be ideal for housing) are on protected greenbelt land (see Section 14.8).

Some loss of greenfield land is almost certain. The demand for housing in Solihull and the West Midlands can only be met by developing some large greenfield sites.

▲ The revamped Fort Dunlop building now has many uses.

your questions

1. Based on what you have learned about Glasgow in Section 13.6, outline the likely problems faced by Birmingham after the collapse of its traditional industries.
2. In pairs, read the spider diagram of possible uses for brownfield sites. Select the three that you think would bring the greatest benefits to Birmingham, and present your ideas to the class.
3. Make a list of the options from the spider diagram and tick which ones the Fort Dunlop scheme has done well.
4. Summarise in a table the advantages and disadvantages of using **a** brownfield sites, **b** greenfield sites for new development.
5. Exam-style question Using examples, compare the benefits of brownfield and greenfield urban development. (8 marks)

✚ **In this section you'll learn about three areas of possible future growth for the UK economy.**

In the 1960s, Harold Wilson made a speech outlining a vision of the future. He spoke about the 'white heat of technology' and its part in Britain's future. In the 1980s, Margaret Thatcher looked forward to the knowledge economy. This section asks where the UK economy might grow in the next 20 years.

1 The digital economy

A digital economy is based on digital technology. With rapidly growing online shopping and banking, the UK leads the world in digital spending per person. The IT, software and digital industries were worth £100 billion to the UK economy in 2011, and it's estimated that the digital economy will generate 10% of the UK's GDP by 2015. However, there's further to go, such as expanding superfast broadband and cable and Internet TV and radio. Already, leading academics and businesses are working in new technologies, especially Internet security.

Where might the digital economy expand next?

- More home Internet use for shopping online, paying bills, managing heating, home security, etc.
- Environmental management, e.g. flood warnings, traffic flow in cities.
- Health care – Will patients in future visit the doctor online, via Skype?
- Education – Could university teaching be delivered online in rural areas, or at home, as university fees increase?
- Tele-working – Will more people work remotely from home (see Section 13.9)?

On your planet
✚ Some people fear that the digital economy, and the boom in online shopping, will kill off the traditional town and city 'high street'.

2 Education and research

A knowledge economy needs an educated workforce. The expansion of Higher Education (universities) means that in the world's wealthiest countries the percentage of young people going to university doubled from 20% to 40% between 1995 and 2008. The UK is starting to fall behind other countries – with about 30% of men and 40% of women in their early 20s having degrees. Greater expansion of Higher Education is needed (e.g. in 2012, ten UK colleges became universities).

Within companies and universities, research and development (R&D) benefits the UK economy hugely. Without it, where would the next generation of TVs come from, or drug treatments, or new seeds for growing larger amounts of food? Many people object to genetically modified (GM) seeds, but scientific research has been the key to feeding the growing human population. The biggest R&D sectors in the UK are pharmaceuticals and biotechnology, aerospace and defence, software, the car industry, and telecommunications.

3 Green employment

The 'green' sector is attempting to improve air and water quality, recycle and reduce waste, and improve the environment. It includes 'green':

- products made from renewable materials, or recycled goods (e.g. fleeces made from old plastic bottles)
- buildings that use less energy, recycle water, and are built from natural materials (e.g. buildings in London's Olympic Park)
- tertiary industries (e.g. eco-tourism)
- quaternary services (e.g. architects designing 'green buildings').

The UK government estimated that the world market for green goods and services was £350 billion in 2010. In the next decade, the renewable energy market has the biggest potential.

Who's doing the work?

Who will work in these three growing areas – employees from the UK or from overseas? In 2012, 580 jobs a day in the UK went to overseas workers. Economically it makes sense, because they:

- meet skills shortages
- help to balance the UK's ageing population.

Overseas migrants are mostly young and skilled. Many companies would rather recruit ready-trained foreign workers than pay to train UK youngsters themselves. Companies don't want controls on immigration – a 2012 survey by Manpower showed that globalisation provides real benefits from a mobile labour force. UK employers use foreign labour to meet shortages in jobs that are hard to fill, such as labourers, skilled trades, accountancy, doctors and nurses. Equally, young graduates from the UK migrate to Australia and the EU to fill skills shortages there.

On your planet
+ London's Olympic Park consists of low-carbon, low-energy venues – to help promote east London as a centre for a new 'green economy'.

▲ Constructing solar panels – is this an industry where there's future growth?

your questions

1 **a** In pairs, draw a large spider diagram called 'The future economy'. Draw three arms coming from it (one for each of the three growth areas). Brainstorm and list possible jobs that people might do for each area. Share your ideas in class.

   b Now classify the jobs into primary, secondary, tertiary and quaternary. Where will most growth come from?

2 In pairs, discuss and summarise the arguments **a** in favour and **b** against companies employing highly skilled people from overseas.

3 In Finland, 80% of young people go to university. Draw up a table showing the advantages and disadvantages for the UK of aiming for the same percentage.

4 **Exam-style question** Using examples, outline two areas where economic growth might take place in the UK. (6 marks)

✚ **In this section you'll learn how work is changing and how people might work in future.**

The UK has undergone a dramatic change from a secondary to a tertiary economy since the mid-twentieth century. Now the UK economy is shifting again – towards a knowledge-based (quaternary), rather than a product- or service-based economy. There are now more people employed in knowledge-based jobs in London (e.g. R&D and consultancy), rather than traditional City jobs (e.g. banking). There has also been an end to the 'job for life', as employers look for a workforce that's flexible (that is, to meet demand when they're busy, but to save costs when things slow down). So, how does all this affect the ways in which people are working now – and might work in the future?

IT and tele-working

It's difficult to predict the future. People writing about this subject in the 1980s predicted a shorter working week, because they thought that computers in the future would do much of the work then done by people. This has happened in some jobs (e.g. robot production lines in car factories), but many people in the UK actually work longer weeks than in the 1980s!

Computers are now a feature of every desk. They are vital to most jobs and the ways in which people work. Email is almost essential. This book was written and produced mainly on a computer! Laptops and WiFi access now enable work to take place almost anywhere – on trains, in coffee shops, in hotels or in isolated rural areas. And working from home (**tele-working**) is increasingly common – often by people living in rural locations. Many new-build homes contain studies or home offices, and public buildings such as train stations are being 'wired', so that tele-workers can work via the Internet from almost anywhere.

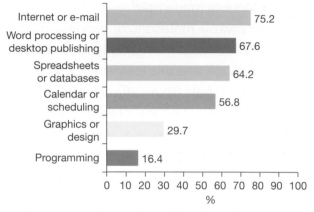

	%
Internet or e-mail	75.2
Word processing or desktop publishing	67.6
Spreadsheets or databases	64.2
Calendar or scheduling	56.8
Graphics or design	29.7
Programming	16.4

▲ The bar chart shows how we use computers in the workplace (over 75% of workers use computers for the Internet or e-mail).

▲ WiFi is often available in coffee shops.

On your planet

✚ Intel (the computer chip company) has manufacturing and processing operations around the world. It expects its managers to take part in teleconferencing outside normal working hours, because of the different 'time zones'.

Flexible working

New technology allows employees to work **flexibly**, which means that:

- those who use IT can work anywhere and at any time
- they can work with colleagues or customers and clients on the other side of the world, who they've never even met
- companies don't have to pay for expensive city-centre offices, where rents can be over £5000 a year for an office the size of a small bedroom!

So the rigid pattern of working fixed hours is disappearing for some. However, this change has its advantages and its disadvantages (see right).

Self-employment

The 'job for life' has ended for many people. By building up contacts and a range of skills, it's increasingly common for tele-workers to become self-employed by **freelancing** (taking on a single job without any commitment to further work once it's finished). Self-employment is one of the fastest-growing employment sectors. It's common in a range of jobs, such as:

- the construction trades (e.g. joiners, builders)
- knowledge-workers (e.g. IT analysts, accountants, investment analysts)
- services (e.g. teaching professionals and social work contractors).

Self-employment is a risk, because money only comes in when work's available. But many people choose it because they like the variety, independence and flexibility. Work they're doing now isn't what they'll be doing in five years' time. But it's also less secure, because there's no sick pay, pension or health-care package. They have to save for their own future. Many employers like it, because they can out-source work when there's too much to do, without taking on anyone full-time (or having to pay things like pension contributions or employer's national insurance).

Advantages of flexible working

- Better health – people can take breaks during the day
- Less stress from commuting
- Less absenteeism and sickness
- Lower staff turnover, so people with key skills are kept
- Parents can work from home, saving money on childcare
- It suits disabled people, who don't have to travel
- It allows those who wish to work part-time to do so easily
- Better productivity – people work longer, instead of wasting time commuting
- Less traffic congestion and pollution because people aren't travelling

Disadvantages of flexible working

- Lower wages (possibly) – less contact with your boss, so you miss out on promotion or a pay rise
- Isolation from colleagues
- It's sometimes difficult to motivate and organise home-workers
- Work never disappears – when you're working from home, it's always around you

your questions

1 **a** Explain the differences between: tele-working, self-employed, out-sourcing, and flexible working.

 b Give two examples of jobs done for each of these terms.

2 Copy and complete the table below showing the advantages and disadvantages of different ways of working (be sure to say whether these are for employees or employers):

	Advantages	*Disadvantages*
Having a job for life		
Tele-working		
Flexible working		
Being self-employed		

3 **Exam-style question** Using examples, explain the impacts of changing working practices on different groups of people. (8 marks)

✚ In this section you'll investigate the reasons why some urban areas grow, while others decline.

Boom time for London

While London basked in the success of the 2012 Olympic and Paralympic Games, the results of the 2011 Census were announced. They showed that, after 50 years of decline, London's population had grown by 12% since 2001.

There are a number of **demographic** (population) reasons why London's population is growing. The two main ones are:

- migration from *within the UK* (in 2009, 178 000 people moved to London from other places in the UK)

- migration from *overseas* (in 2009, 154 000 people arrived in London from overseas – about 2% of the population).

In both cases, many migrants are in their 20s and 30s — and, are in time, starting families. So a third reason for London's population growth has recently emerged – a rapidly rising birth rate. The number of children under five increased by 24% between 2000 and 2011.

Ealing, in west London, is London's third largest borough in terms of population — with a much higher than average percentage of people in the 25-44 age group (see bar chart). It is this age group which is driving both population, and economic, growth. Most of these people are well educated and skilled, able to compete in London's jobs market.

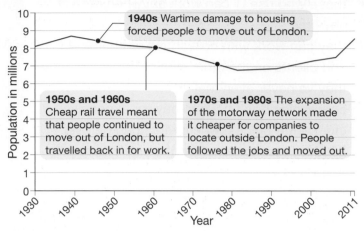

▲ In 2011, London's population was just over 8 million — not quite the peak of the 8.6 million in 1939.

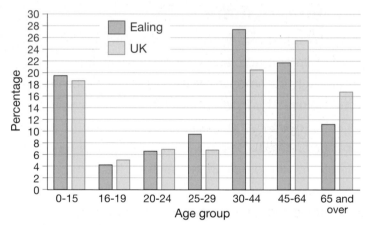

▲ Population breakdown – Ealing and the UK, 2011.

London acts as an **economic** hub — the sixth largest city economy in the world. In 2008, London's GDP was $565 billion — more than the GDP of some entire countries.

Over 3 million people work in service industries in Greater London.

- 120 000 civil servants work in central government departments.
- More than half of the UK's top 100 companies have their headquarters in central London.
- Media companies (TV, advertising, and publishing) are concentrated there.
- Tourism and entertainment are also big employers.

Liverpool in decline

Liverpool was booming in the first half of the twentieth century — when 40% of the world's trade passed through its port. In 1937, the city's population was 867 000 (compared to under 500 000 today). In the 1950s and 60s engineering, manufacturing, banking and insurance were all important industries. However, by the 1970s things began to change — 10 000 people were leaving Liverpool every year.

Social and political processes	• In the 1950s and 60s the government wanted slums cleared and houses modernised. In 1955, over 88 000 houses in Liverpool were considered unfit to live in. • Nineteenth-century terraced housing was demolished and replaced with high-rise flats in the inner city, and housing estates on the outskirts.
Economic processes	• Liverpool was badly hit by the recession in the 1970s and 80s. Manufacturing declined — often moving overseas instead where labour was cheaper — and fewer ships passed through the port. • Goods began to be transported in containers needing less labour — so jobs were lost at the port. • People left the city because of the decline in jobs.

Liverpool today

Liverpool is still an important port, despite its decline in the 1970s. But the UK now relies on Europe for 70% of its trade, rather than America. Therefore, a lot of that trade now goes through southern ports, like Harwich.

- Compared with other large urban areas, Liverpool's unemployment is high.
- Liverpool has a low level of new business start-ups (two-thirds the rate of London).
- Like London, Liverpool is dominated by service sector employment, but with lower levels of pay, and more people working part-time, than in London.
- Liverpool has five of the ten poorest communities in England.

Since 2004, Liverpool's population has started to rise again — following a regeneration of the city centre and the Kings Waterfront (a new concert arena, restaurants and apartments) — especially attracting people in their 20s.

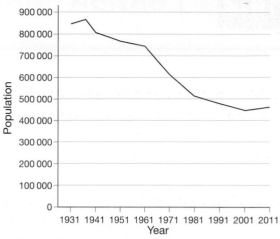

▲ Liverpool's population, 1931-2011

▲ Anfield, in Liverpool, is one of the poorest communities in England. 43% of children live in poverty, 30% of adults are unemployed, and the population is declining.

your questions

1 In pairs, discuss and design a spider diagram to show the reasons why London's population is rising rapidly. Use different colours to show economic, social, political and demographic reasons.
2 Why should well-qualified young people be in demand for jobs in London?
3 Now produce a second spider diagram to explain why Liverpool is not booming in the same way as London.
4 **Exam-style question** Outline one economic and one demographic process that has led to population increase in a city you have studied. (4 marks)

✚ In this section you'll look at how different areas of the same city can vary.

Deprived London

Compared with Liverpool, London appears to be booming. But if you dig a little deeper, you'll find that some parts of London remain quite deprived. Deprivation has several different causes. The government gathers together information about jobs, health, education, housing, income and services to produce an **Index of Multiple Deprivation** (see map).

Canning Town, Newham

Canning Town is in Newham – one of London's most deprived boroughs (see photo).

- Housing in Newham is too expensive for people on low incomes – and far more expensive than the UK average (see table).
- In 2001, nearly half (43%) of working-age adults in Canning Town had no qualifications at all. That means they tend to work in low paid, unskilled jobs.
- In 2001, 21% of the population had poor health.
- Canning Town has a high percentage of minority ethnic residents who were born outside the UK. If their English is poor, it can prevent them from getting good jobs. Also, the qualifications they gained in their home country might not be recognised in the UK. So, someone who trained as a doctor at home might end up doing a poorly paid job.

However, Newham is home to the recently transformed Queen Elizabeth Olympic Park which includes a whole range of sports facilities.

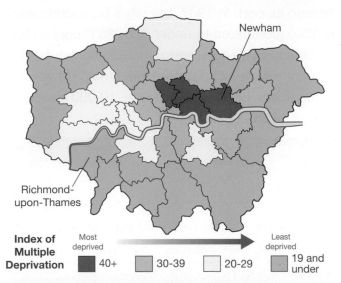

Newham

Richmond-upon-Thames

Index of Multiple Deprivation	Most deprived			Least deprived
	■ 40+	30-39	20-29	19 and under

▲ The pattern of deprivation in London's boroughs in 2008.

Type of property	Richmond upon Thames	Newham (including Canning Town)	UK average
Detached house	£809 770	£300 594	£264 707
Terraced house	£472 860	£231 741	£136 706
Flat/maisonette	£307 925	£278 870	£162 616

▲ Average house prices in 2010.

Richmond upon Thames

The borough of Richmond upon Thames in south-west London is one of the city's least deprived areas.

- The average annual income in 2009 was £46 415 (four times that of Canning Town in Newham). This is because Richmond has a high level of very skilled and highly qualified workers.
- Richmond has more open space, per person, than any other London borough – it includes Kew Gardens and Richmond Park.
- Richmond does not have any NHS hospitals, but nearby Hounslow and Kingston upon Thames have large teaching hospitals.

Richmond does have some problems though:

- 18% of the population has poor health.
- Richmond's health and social services may struggle in future, with the expected increase in people aged over 85.
- Richmond has some very high-quality housing, but it's also very expensive. It has one of the smallest percentages of affordable housing in Greater London. So many people who work in Richmond, live outside the borough.

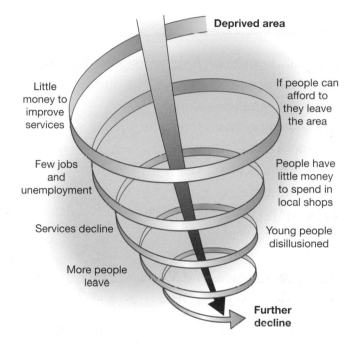

▲ The problems that places like Canning Town face can lead to a spiral of decline or **negative multiplier** — where one problem leads to another.

✚ In this section you'll look at the impacts of the rising demand for more residential areas.

The Thames Gateway

London's population is growing faster than at any time in its history. One of the problems with London's expansion is that – as the city grows – more and more of the surrounding countryside is being lost (see diagram).

The Thames Gateway project is one example of how London's growth is eating up the countryside. The Thames Gateway is an area of land stretching 70 km (43 miles) east from inner east London – on both sides of the Thames and its estuary. The project has been planned to develop **brownfield sites**, drain marshland for new housing, provide employment opportunities and improve transport links.

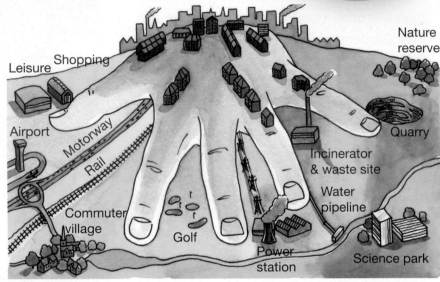

▲ London is eating up the countryside.

London's expansion is driving up the demand for, and price of, land for both housing and business. As land and housing become more expensive, people who work in London have to make a choice: *either* to live there and accept the high cost of housing, *or* to move out of London (e.g. to places like Chelmsford) for cheaper housing – but then spend more time and money travelling back into London for work.

> ✚ **Brownfield sites** are areas of land that have been built on before and can be built on again.

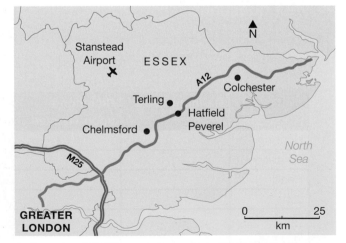

▲ Every day, 650 000 people leave places like Chelmsford to go to work in London.

Chelmsford's growth

In 1971, 58 000 people lived in Chelmsford. Today, 120 000 live there.

- Most people who live in Chelmsford live on recently built estates — preferring to buy, rather than rent their homes.
- The average income in Chelmsford is high — £30 000 in 2009 — when the average for the UK was £24 300.
- Many Chelmsford residents work in professional, service industries, and commute to central London and London Docklands. Unemployment is therefore low.
- People are attracted to Chelmsford because of its excellent transport links to central London and Stansted airport.

As urban areas, like London and Chelmsford, **sprawl** — because of the increasing demand for more housing — the **rural-urban fringe** comes under more pressure:

- Open spaces, fields and woodlands, disappear — habitats are destroyed.
- Traffic and pollution increases.
- There is more pressure on water resources for domestic use.
- Quarrying increases to provide bricks and cement, for house and road building.
- House prices increase — local people can no longer afford to live there.
- As more of the people living in the rural-urban fringe become workers from the larger towns and cities, the smaller towns and villages start to lose their facilities, like shops, pubs, post offices and bus services.
- People become more dependent on cars. If they can't afford one, they lose out.

> + **Urban sprawl** — the growth of towns and cities into the surrounding countryside.
>
> + **Rural-urban fringe** — the area where the town/city meets the countryside.

▲ A modern housing estate on the outskirts of Chelmsford.

your questions

1 Explain these terms in your own words: urban sprawl, rural-urban fringe.

2 Copy and complete the table to compare the advantages and disadvantages of living in London and Chelmsford.

	Advantages	Disadvantages
Living in London, working in London		
Living in Chelmsford, working in London		

3 Draw a Venn Diagram to show the environmental, social and economic impacts of the rising demand for more homes around London.

4 Exam-style question.

a Describe the effects that rising demand for housing can have on people and the environment. (4 marks)

b Using examples, explain how the rising demand for residential areas has affected one urban area. (6 marks)

14.4 Improving urban areas

In this section you'll evaluate how successful the 2012 Olympic Games were in improving Newham.

Walk in the Park ...

... the Olympic Park, that is. During the 2012 Olympic Games, visitors were invited to take a walk in the Park – to see it from a different angle – with its meadows, trees, flowers and wildlife. Back in 2005, when London won the bid to host the 2012 Olympics, much of the site was rundown and derelict. The Lea Valley (site of the Olympic Park) was, for many years, the industrial centre of London. But, as London's docks declined and manufacturing moved overseas during the 1980s, the area changed.

Improving urban areas

In 2005, the London borough of Newham (see Section 14.2) was one of London's most deprived areas. By choosing Newham as the location for the Olympic Park, the intention was that it would help to **regenerate** the area and create an upward spiral of improvement – or **positive multiplier effect** (see the diagram). The regeneration plan also hoped to **rebrand** the area.

▲ The Olympic Park in 2012.

> + **Positive multiplier** is where one improvement leads to another, (see diagram below).
>
> + **Quality of life** includes housing, employment, quality of the environment, and access to services.

A sustainable games

London's 2012 Games were planned with various sustainable principles in mind, including to:

- build on brownfield sites
- minimise waste, energy use, water use and pollution
- create affordable housing
- benefit people from all communities
- help people have a good quality of life, by including features such as parks.

> + **Regeneration** is the ways in which an existing area of housing or industry is improved or renewed.
>
> + **Rebranding** is improving and positively changing the image that people have of a place.

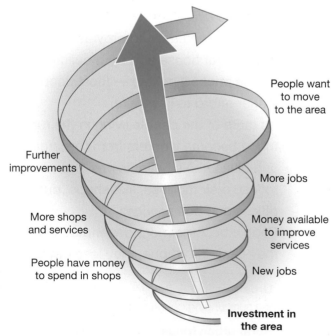

People want to move to the area

More jobs

Money available to improve services

New jobs

Investment in the area

People have money to spend in shops

More shops and services

Further improvements

▲ A positive multiplier – an upward spiral of improvement.

What problems does Newham have?

Newham's problems are to do with things like low income levels, a high percentage of people with no educational qualifications, problems with housing and so on.

- Jobs — more people work in lower-paid, semi-skilled or unskilled jobs than in better-paid managerial or professional jobs (see opposite). Unemployment is higher than other parts of London or the UK.

- Housing — as Newham's house prices and rents have increased, incomes have stayed the same (or gone down). Newham, therefore, experiences more overcrowding than other parts of the UK. The council believes it needs more family housing.

- Environment — there were also a number of environmental problems, before the land was cleared to create the Olympic Park. Abandoned land was used as a dumping ground. Land, soil and waterways were polluted with chemicals and waste, and ugly electricity pylons were built on open spaces.

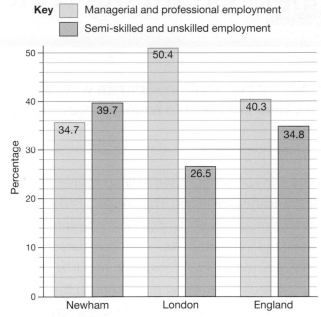

Key — Managerial and professional employment — Semi-skilled and unskilled employment

▲ Types of employment 2001

What improvements have the Olympics brought to Newham?

- About 7000 temporary jobs were created to develop the Olympic Park, with a further 8000 new jobs planned for its transformation.

- Five new neighbourhoods, with up to 8000 new homes, will be built in the Park. This includes family homes — and up to 35% will be affordable housing.

- The polluted land and waterways were cleaned up — over 4000 trees were planted, and 45 hectares of wildlife habitat were created.

your questions

1 Your task is to evaluate how successful the 2012 Olympics have been in regenerating Newham.

  a Use the information on these two pages to score the improvements to Newham in the table below. Add up your total score.

  b Write a short report (400 words maximum) on how successful you think the Olympics were in regenerating Newham. Include the reasons for the scores you gave in the table.

  c Write a conclusion to your report to suggest how you think further improvements could be made to areas of east London.

Improvements in:	Score +2 = Big improvement; +1 = Small improvement; 0 = Not sure; −1 = A bit worse; −2 = Much worse
jobs	
housing	
the environment	
TOTAL (max = +6: min = −6)	

2 **Exam-style question** Using examples, explain how urban areas can be improved. (6 marks)

14.5 Changing rural settlements

In this section you'll investigate why, and how, rural settlements have changed.

Rural living

Villages in rural areas have existed for hundreds, if not thousands, of years. They originally developed because most people worked in farming and associated industries. However, compared with 50 years ago, farming now employs very few people — only 1.4% of the working population in the UK.

Rural settlements in the UK have seen some big changes. In smaller towns and villages, close to cities:

- there has been an increase in **counter-urbanisation**
- as they have grown in size, they have become **suburbanised**.

In more remote rural areas:

- There has been an increase in the number of houses used as **second homes** and holiday lets (particularly in tourist areas like Cornwall); this tends to increase house prices.
- Schools, pubs, shops and other services have closed, because there aren't enough people living there year round to support them. Local people are using the services in nearby towns instead.

All rural areas have seen a change of building use — with old farm buildings, village schools and pubs, often being converted into housing.

+ **Counter-urbanisation** is the movement of people out of cities to smaller towns and villages.

+ **Suburbanised villages** are villages that grow to look like the suburbs of towns or cities.

+ **Second homes** are owned by people who have their main home somewhere else (usually in an urban area).

▲ Many village pubs and services are closing down.

As you move away from large cities and towns, you see a change in the type and nature of settlements and land use. Geographers refer to this as the **rural-urban continuum**. The diagram shows how it works.

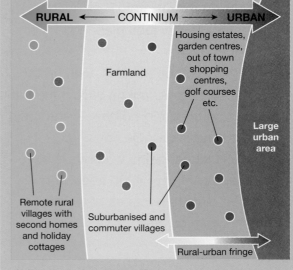

▲ The rural-urban continuum

Terling – a commuter village

Terling, in Essex, is a traditional-looking English village near Chelmsford, with a population of about 1600. People who lived in the village used to work on local farms, but now the farms employ contractors.

Because of its location, Terling has become a **commuter village**. The nearby railway station, at Hatfield Peverel, has half-hourly trains to London. Terling is also close to the A12 – the main road from London to Colchester.

However, Terling's accessibility to other towns and cities, is causing problems. Many of Terling's residents work, and shop, elsewhere — so local services are struggling.

The village shop has fewer customers; the doctor's surgery only opens 5 hours a week; the bus only runs twice a week; and the village pub has closed. Houses in the village have also become expensive.

▶ Terling – a twenty-first century commuter village

Roseland Parc – a retirement village

Roseland Parc is a new type of 'village'. It's a retirement village – built *within* the village of Tregony in south Cornwall (see map). It consists of one and two-bedroom flats that are 'sheltered' (looked after by a warden).

The Parc includes a restaurant, a gym, a library, and a care home with full nursing facilities.

However, it isn't cheap – a flat costs about £300 000, and care home provision can cost over £4 000 per month. Some people in Tregony also criticise it, because the residents don't take part in many village events, or spend money at the local shop.

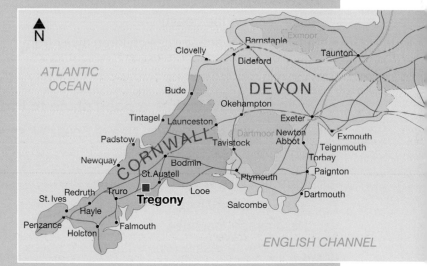

your questions

1 In pairs, identify the advantages and disadvantages of living in Terling, compared to London or Chelmsford (see Section 14.3).

2 In pairs, research a retirement villages such as Roseland Parc. Prepare a presentation (4-5 slides) using these headings

- What is its location like?
- What does it offer people?
- What are the disadvantages of places like this?

3 Exam-style question Using examples, describe the reasons why two types of rural settlement have developed. (6 marks)

➕ **In this section you'll investigate two contrasting rural regions.**

South-west England

South-west England (which includes the counties of Devon and Cornwall) is the UK's most popular tourist region. The city of Exeter, in Devon is fast becoming a centre for professional, financial and property services, and is only two hours by train from London – close enough to attract companies (like the Met Office) with its low land prices and good quality of life. South-west England also has one of the UK's fastest-growing populations (10% between 2001 and 2011). This is due to inward migration, especially from:

- people over 50, who want to retire there because of the quality of the environment (especially on the coast)
- families who want to raise children away from cities.

▲ Every year over 4 million people visit south-west England for its scenery and beaches.

Devon and Cornwall — a good place to live?

In spite of its popularity, this area of South-West England, faces problems:

- **Incomes are low**. In 2011, Cornwall's average annual income was 25% below the UK average, because:
  - its remoteness prevents large companies from locating there
  - full-time jobs in farming, fishing and mining have declined
  - jobs in tourism are seasonal, often part-time, and low paid.
- **Young people leave** because housing is expensive, due to the demand from retired people and for holiday cottages.

- **Access to services is poor**. Tourism keeps some rural shops and pubs open, but they are under threat. Having a car is essential to get to doctors' surgeries, hospitals etc. Education and health services suffer because:
  - few secondary schools in Cornwall have a sixth form, so students have to travel long distances to Truro, or stay there during the week
  - the three largest hospitals are in Exeter, Plymouth and Truro — so many people live at least an hour away from urgent care.

The Scottish Highlands

The Scottish Highlands have one of the lowest population densities in the UK, averaging just 8 people per km² (compared with 66 per km² for Scotland). This falls to 2 people per km² in some supersparse areas!

Kinloch Rannoch – a remote upland village

Kinloch Rannoch lies at the eastern end of Loch Rannoch in the Grampian Mountains (see photo). The road through the village continues west for 18 miles and then ends abruptly at Rannoch station! Kinloch Rannoch faces the same problems as many Scottish upland villages:

- Isolation - It is remote. There is a small primary school, but secondary school students face a 22-mile journey, to Aberfeldy.
- Trains are infrequent and expensive which makes it difficult to commute to work.
- Services - The village has a small population of about 300 residents, so it lacks basic services.
- Jobs – mainly in tourism, forestry and farming, but jobs are low paid and often seasonal. There are few jobs for young people.

▲ Kinlock Rannoch in the Scottish Highlands — its remoteness causes many economic and social problems.

The Highlands, like south-west England, suffer from deprivation. Some indicators of deprivation have been grouped to produce a map (see right), to identify 'fragile areas', in danger of decline. The indicators include:

- Population decline – due to an ageing population, and young people moving away to towns — often for university or jobs.
- High long-term unemployment.
- Distance from services – such as a post office, primary school, shop, and doctor's surgery.

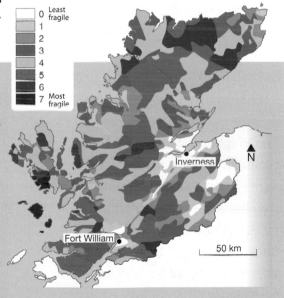

0	Least fragile
1	
2	
3	
4	
5	
6	
7	Most fragile

your questions

1 Draw a spider diagram to show why quality of life can be low in rural areas.
2 Explain the reasons for each of the following trends in remote rural areas, and the problems that this might cause.
  a Most young people leave at age 18.
  b There are few jobs.
  c There are many second homes.
3 Exam-style question
  a Describe the distribution of 'fragile areas' shown on the map. (3 marks)
  b Using examples, explain why remote rural areas can be described as 'fragile'. (4 marks)

In this section you'll look at two rural development projects in south-west England, and assess, their success.

Improving Cornwall

In the last section you found out that parts of south-west England are in crisis. Young people leave because they can't find jobs, and housing is expensive. The local economy isn't helped by Cornwall's remote location either – or the fact that a lot of tourism-related jobs are seasonal.

However, two **sustainable development** projects are now helping to improve life in Cornwall for thousands of people.

Eden Project, Cornwall

The Eden Project opened in 2001. Built in the bottom of an old clay pit, it has transformed the environment and local area in many ways. It consists of two 'biomes', which house collections of plants from around the world – ranging from banana trees to Mediterranean herbs. The Eden Project offers a range of activities throughout the year, including music and art events, ice-skating, rock-climbing and story-telling. It is mainly an indoor attraction, so people can visit at any time of the year.

+ **Sustainable development** is about meeting the needs of people now and in the future, limiting harm to the environment, and avoiding excessive use of limited resources.

▲ Inside the Rainforest biome

Eden Project – the benefits

The table below summarises some of the benefits that the Eden Project has brought to Cornwall.

Visitor numbers	• 13 million people have visited since the Eden Project opened.
Visitor impacts	• People who visit the Eden Project for a day, tend to stay in Cornwall for an average of five days. • As well as spending money at the Eden Project, visitors spend money locally on a range of things from B&Bs to ice cream and car parks.
Impact on local suppliers	• Since opening, 2500 local businesses supplying the Eden Project have benefited (89% of catering supplies come from local suppliers).
Impact on the local economy	• The Eden Project has earned over £1.1 billion for the local economy. • It has drawn attention to Cornwall's other big attraction for older tourists – its Gardens. Tourists now come in spring and autumn to see them. • It has enabled some hotels and B&Bs to stay open for most of the year, when in the past they had to close in the winter.
Jobs	• In March 2011 the Eden Project employed 574 people (nearly 400 on a permanent basis). • Since opening in 2001, 3000 jobs have been sustained locally as a result of the Eden Project.

Combined Universities in Cornwall (CUC)

CUC is a partnership of six universities and colleges, working together to give students a chance to study in higher education in Cornwall. Part-funded by the EU, CUC was created:

- because the nearest universities were at Plymouth and Exeter, in Devon, so young people were forced to leave Cornwall to study at higher level
- to help transform the local economy into a **knowledge economy** to build on the future prosperity of the South West.

Has CUC achieved its aims?

Cornwall now has 7700 students in higher education at CUC. This has brought a boom to Falmouth, particularly to landlords renting rooms and houses, and to local pubs and clubs. The pie chart shows that, in 2010, 38% of CUC graduates found jobs in Cornwall — and a further 22% in south-west England. A project called 'Unlocking Cornish Potential' tries to find graduate placements with local businesses. Over three years, 500 graduate jobs were created, and 76% of graduates stayed with the company after their placement finished.

But, there's still a problem. Although some CUC graduates stay in Cornwall, many of them can't afford to settle down there and buy a house — because prices are high thanks to tourism, second homes, and the demand for rented accommodation from students. So, many young, educated people are continuing to leave Cornwall.

> + **The knowledge economy** is one based on specialist knowledge (rather than goods) to produce economic benefits such as jobs.

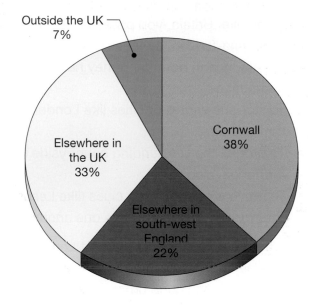

Outside the UK 7%

Elsewhere in the UK 33%

Cornwall 38%

Elsewhere in south-west England 22%

▲ Where graduates from CUC found work in 2010.

It is possible to measure the value something adds to the economy. This is known as **GVA** – Gross Value Added. The total value of CUC's GVA, up to 2011 was over £70 million. This reflects people receiving paid employment who didn't have jobs before, increased spending in local shops, and so on.

your questions

1 Copy and complete the following table to compare the **benefits** brought by the Eden Project and CUC.

	Eden Project	CUC
Economic		
Social		
Environmental		

2 How far has each project helped to create sustainable development in south-west England?

3 **Exam-style question** Using examples, describe how a rural development project can help to bring economic benefits to an area. (4 marks).

✚ In this section you'll evaluate how successful green belts and National Parks are in making rural regions sustainable.

Green belts

If you fly over Britain, you pass over settlements that are 'ring-fenced' by areas of countryside. These are '**green belts**' and they have three main functions:

- To stop the sprawl of cities like London and Birmingham.
- To protect the surrounding countryside from further development.
- To prevent neighbouring cities (like Leeds and Bradford) from merging into one another.

But, do green belts work? Without London's green belt, much of south-east England could have become one huge urban area. Although our green belts have mostly remained intact, they are under constant threat:

- Over 11 km² of green belt land has been lost to motorways and housing every year since 1997.
- Nearly 50 000 houses (equivalent to a city the size of Bath) have been built on green belts since 1997.
- They cause 'leap-frogging' – where towns like Chelmsford in Essex (see Section 14.3) grow outside the green belt. And commuters then travel back across the green belt to work in London.

Some people are concerned that blocking developments on green belt land will harm the economy, but the Campaign to Protect Rural England (CPRE) believes that there are enough brownfield sites to build new offices, factories and over 1.5 million houses.

▲ London's green belt is clearly visible from above.

✚ **Green belts** are areas of open land around cities, where development is restricted.

▲ England's green belts cover 13% of its land area.

The Cairngorms National Park

All of the UK's National Parks share the same aims:

- To conserve and enhance the natural and cultural heritage of the area.
- To promote sustainable use of the natural resources of the area.
- To promote understanding and enjoyment of the area by the public.
- To promote sustainable economic and social development of the area's communities.

▲ Some of Scotland's most remote communities are found in the Cairngorms National Park.

The Cairngorms National Park is home to three of Scotland's five ski centres – CairnGorm Mountain, Glenshee and The Lecht. So, is it possible to allow economic development, and conserve the landscape, at the same time?

In 2004, the Glenshee resort was sold to new managers after making losses over the previous two winter seasons. They wanted to ensure that skiing and snowboarding would continue. In the winter, Glenshee can suffer if the weather is too warm and wet, so the resort aims to become more of a year-round attraction.

- the Cairnwell chairlift is open in July and August to take visitors up Cairnwell Mountain
- a 3.2 km mountain bike track was opened in 2010.

The resort has also widened its winter activities by:

- hosting a Highlander freestyle competition for skiers and snowboarders
- holding a World Snow Day to encourage families to explore what is available at the resort.

A wide range of businesses support the ski resorts in the Cairngorms, including: ski and snowboard hire shops, and companies offering outdoor activities such as rafting.

The environment at Glenshee ski resort is protected by the fact that it's part of a National Park. But there is also a Nature Reserve, two SSSIs, as well as a Special Protection Area for birds, and a Special Area of Conservation for vegetation.

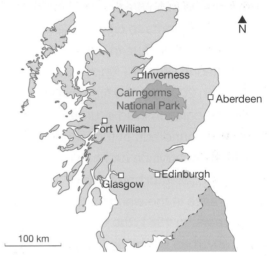

▲ Cairngorms National Park

your questions

1 Copy and complete the table below. In each box:
   a give a score between 1 and 5 (where 1 is very poor and 5 is very successful)
   b explain your reason for each score.

	Green belts	The Cairngorms National Park
How well they conserve landscapes	Score: Reason:	Score: Reason:
How well they allow economic development	Score: Reason:	Score: Reason:

2 Use your table to decide how successful green belts and National Parks are in making rural regions sustainable. Explain your answer.
3 **Exam-style question** Explain how either green belts or National Parks can help to make rural areas sustainable. (6 marks)

In this section you'll find out what urbanisation is, how it varies around the world, and why cities grow.

More urban than rural

On the right is Sunita. In 2007, her parents brought her to Mumbai in India. She now lives in slum housing in an area called Dharavi (shown opposite). Migrants to cities in developing countries often end up in places like this. Sunita – or someone like her – tipped the balance, because in 2007 (for the first time) more of the world's population lived in urban areas than rural. Every year, hundreds of thousands of people leave the countryside and move to cities. This process is called **urbanisation**.

▲ The Dharavi area in Mumbai, India.

The increase in the world's urban population has been going on for many years. In 1861, the UK was the first country in the world to reach an urban population of 50% – as a result of people moving to cities during the Industrial Revolution in the eighteenth and nineteenth centuries. Other countries in Europe and North America had similar increases in their urban populations.

▼ The world's urban population and largest cities.

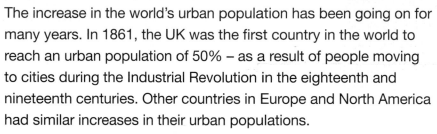

Key
Percentage urban population
0-15%	50-75%
25-50%	75-100%

City population
- 1-5 million
- 5-10 million
- 10 million or more

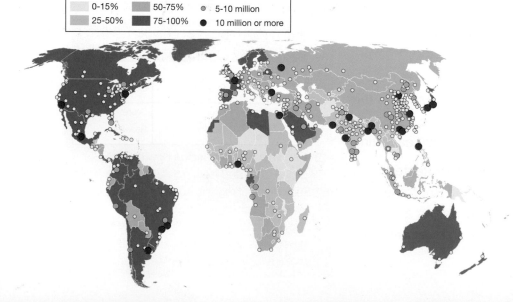

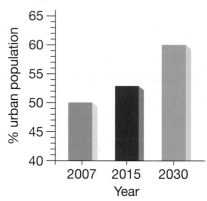

▲ The increase in the world's urban population, 2007-2030 (projected).

+ **Urbanisation** is the rise in the percentage of people living in urban areas, in comparison with rural areas.

The growth of urban populations in developing countries has mainly occurred since the 1950s. However, in those countries, the urban population is increasing more rapidly than it did in developed countries.

How does urbanisation vary in different regions?

Guangzhou in China was the world's fastest-growing city at the beginning of this century. Its population grew by over 300 000 people *a year* between 2000 and 2010 – a total of 3.3 million people! It's not surprising then that Asia, along with Africa, has one of the fastest rates of urban population growth (see below).

- Asia's urban population is expected to be about 64% by the middle of this century.
- Africa's urban population has grown from about 14% in 1950, to 40% today. It's expected to increase rapidly to 58% by 2050 (but Africa will still have the lowest urban population percentage, compared with the rest of the world).

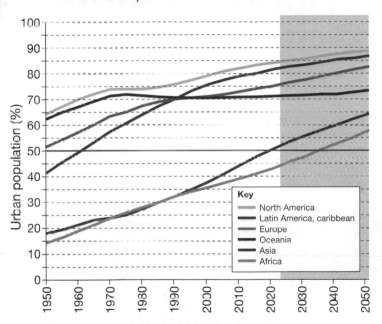

▲ The increasing urban population in different world regions, 1950-2050 (projected).

Why do cities grow?

Mumbai is home to over 20 million people, and around 1000 new migrants – people like Sunita and her family – arrive every day. By 2025 its population is likely to be over 26 million.

Cities grow for two main reasons:

- rural-urban migration
- natural increase, i.e. the number of births minus deaths

Rural-urban migration

Rural areas, especially in developing countries, have few jobs – apart from working on the land – and people often live in poverty. Factors like this help to 'push' people away from the countryside.

In the cities there are more jobs, better educational and health facilities, more entertainment options, and some people are better off. These factors help to 'pull' people to the cities. This movement of people from the countryside to the cities is called **rural-urban migration**.

Natural increase

One of the reasons why London has grown is that, like Mumbai, people have migrated there. People migrating to the city tend to be in their 20s and 30s, and they often start families once they are settled. So, London's birth rate is rising rapidly – the number of children under 5 increased by 24% from 2000-2011. It is this **natural increase** – sometimes called **internal growth** – which helps to drive up the population of the world's cities.

your questions

1 Use the map opposite to describe the distribution of countries with an urban population of **a** over 50%, **b** under 50%.
2 Explain why Sunita is likely to stay in Mumbai, and not return to a rural area.
3 Explain why industrial growth usually leads to **a** urban growth, **b** rural-urban migration.

4 Explain why **a** most migrants are young, **b** they get married in a city rather than at home, **c** they affect the rate of population increase in cities.
5 Exam-style question Using the graph above, describe the rate of urbanisation in different regions of the world. (4 marks)

✚ In this section you'll look at the differences (contrast) between megacities in the developed and developing worlds.

Population of megacities

The photo below shows one of the world's largest cities, but where do you think it is? New York, Tokyo, Buenos Aires? It's actually Mumbai, India, and one of the world's **megacities** – they're cities with a population of over 10 million people.

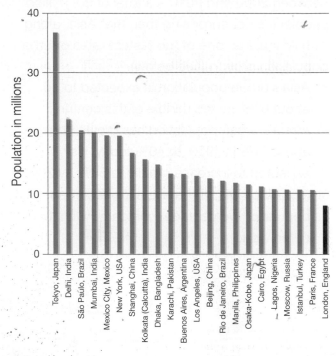

▲ The world's megacities in 2010 with London for comparison. The world's first megacities were New York and Tokyo.

◄ Mumbai at night.

▼ There are significant differences in megacities in different parts of the world, as the table below shows.

Developing world megacities	Developed world megacities
Spatial growth As cities grow in population, they also grow in space – **spatial growth**. Developing world cities are growing rapidly (e.g. Karachi in Pakistan is growing by 4.9% a year) due to industrialisation. Industries are often attracted by low tax rates. But with little tax income, governments don't have the money to provide essential services. Many people end up living in overcrowded, unplanned slums and shanty towns (like Dharavi in Section 15.1) often found on land that no-one else wants.	**Spatial growth** Megacities in developed countries are growing much more slowly (e.g. Tokyo in Japan is growing by just 0.6% a year). They have grown recently as a result of: • *merging* with other growing cities nearby to form **conurbations** (e.g. Tokyo's population now includes Yokohama and Kawasaki). • *sprawling*, which results in low population densities (e.g. Los Angeles). Most of the sprawl is caused by people moving out to the suburbs, where housing is cheaper and there's more space.
Economic activity All megacities act as service centres, but those in the developing world are often also important manufacturing centres. In developing-world megacities, thousands of people work in the **informal economy**.	**Economic activity** Most people in developed countries work in the **formal** economy, with relatively few people working in the **informal** economy.

Factfile: Mumbai (India) – *developing world megacity*	*Factfile: Los Angeles (USA) –* *developed world megacity*
Population	**Population**
• 20 million (predicted to become the world's largest city by 2050) • Population growth = 2.9% a year – an example of **hyperurbanisation** • 60% of the population live in poverty (in slums like Dharavi)	• 12.9 million (17.6 million total population of the Greater Los Angeles area) • Population growth = 1.1% a year – faster than most other megacities in developed countries, but more slowly than those in developing countries.
Size and spatial growth	**Size and spatial growth**
• 603 km² • 1000 people arrive in Mumbai every day • Navi Mumbai (New Mumbai) has been built on the mainland opposite Mumbai. Over 1 million people live there – mostly middle-class people moving out of Mumbai	• 12 520 km² while Greater Los Angeles covers a staggering 87 945 km²! • Los Angeles grew rapidly in the 1960s and 70s, because people from other parts of the USA were attracted by California's climate and lifestyle.
Economic activities	**Economic activities**
Mumbai is the commercial capital of India. Economic activities include: • services (e.g. banking, IT and call centres) • manufacturing (e.g. textiles and engineering) • construction • entertainment and leisure (e.g Bollywood).	The economy of Los Angeles is based on: • trade (the ports of Los Angeles and Long Beach are the fifth busiest in the world) • entertainment (Hollywood, TV, music) • aerospace and technology, oil, fashion and clothing, tourism

+ **The informal economy** refers to jobs that have little, or no, job security (e.g. street trading). No tax is paid.

+ The **formal economy** refers to jobs with contracts of employment and more job security. People pay tax.

+ **Hyperurbanisation** is where the urban population is growing so fast that the city can't cope with people's needs.

▲ Los Angeles, California, is the second most populated city in the USA.

your questions

1 a Use an atlas to plot the megacities from the bar chart opposite on a world map. Use one colour dot for developed world cities, and a second for developing world.

  b Describe what your map shows about the distribution of the world's megacities.

2 Draw two spider diagrams to contrast the economic activities, growth and population of **a** developed, and **b** developing world megacities (remember – contrast means you need to look for differences).

3 Exam-style question Using examples, explain how economic activities can lead to the population growth and spatial growth of 'megacities'. (8 marks)

In this section you'll look at some of the challenges faced by cities in the developed world.

At night you can see London from space; it's a city that never sleeps. Like all major cities, London consumes vast amounts of resources every day. Office buildings can look great at night, but think of the energy they use!

From local to global

The early development of most cities depended on the ability of surrounding rural areas to provide their needs, such as food and fuel for warmth and cooking. As cities grew, and technology developed, that close link between the city and rural area was broken – people began to look further away for exotic foods and energy supplies from overseas (e.g. oil). The people who live and work in cities now consume vast amounts of:

- food from all over the world
- energy at home, at work, and in getting to work.

Cities have become places of **concentrated resource consumption**.

▲ London as seen from space.

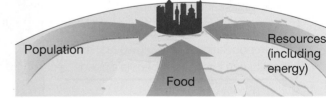

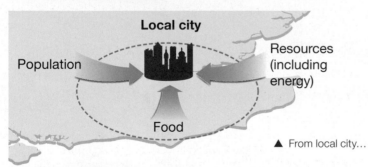

▲ From local city...   ▲ ... to global city

Eco-footprints

An **eco-footprint** measures the area of land needed to:
- provide all the resources and services consumed,
- absorb all the waste produced.

It is measured in **global hectares per person (gha)**. London's eco-footprint is 4.54 gha, almost twice the global average. London needs an area over 200 times the size of the city itself to support it.

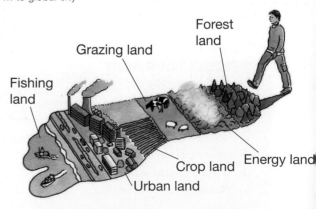

▲ Our eco-footprints extend far beyond the places where we live.

Inputs and outputs

London's growing population (8.2 million in 2011) creates challenges in terms of the city's food and energy consumption, its use of transport, and the waste it produces (see right).

Dealing with waste

London's waste costs £580 million per year to collect, transport, treat, and dispose of. Half of it is sent outside the city to Essex as landfill. However, the Environment Agency predicts that by 2013 all landfill sites in south-east England will be full. The remaining waste that isn't recycled is incinerated (burnt) within the city. How should London solve its waste problem?

Prevention – Waste increases as cities grow and people become wealthier. London now produces 10% less household waste than in 2000 – as a result of things like an increase in the use of e-mail and a reduction in packaging, etc.

Recycling – London is not as good at recycling as the rest of the UK. Only 32% of London's waste is recycled (the national average is 39%) – and 27 out of 33 of London's Borough Councils were below this level.

London's resources used (INPUTS)

Food
- 6.9 million tonnes a year – equivalent to 8 billion meals.
- 81% comes from outside the UK.

Energy
- All of London's energy is imported.
- 13.2 million tonnes of oil equivalent (i.e. equal to the amount of energy released by burning 13.2 million tonnes of oil). London's energy is made up of:
  21% electricity, 23% liquids, 55% gas, <1% renewable.
- Despite a rising population, the use of energy in London's homes since 2000 has stayed constant at 42% of total energy used. Total energy use has fallen by 10% since 2000.

Transport
- 64 billion passenger kilometres are travelled each year.
- Use of public transport is rising. It now accounts for 41% of all London journeys, compared with 37% for private transport (cars).

Materials consumed
- 49 million tonnes.

London's waste produced (OUTPUTS)

- 44.7 million tonnes of CO_2 (in 2008) – the equivalent of 5.9 tonnes per person per year. (It has gone down by 11% since 2000.)
- 20 million tonnes of waste a year.
- Organic waste is released into rivers.
- 49% of all waste is sent to landfill (down from 72% in 2000).
- 32% of waste is recycled or composted (up from 8% in 2000), but still the lowest rate in England.

your questions

1 Use a search engine to find an eco-footprint calculator and calculate your own eco-footprint. Briefly explain your score.
2 Explain, using examples, why cities consume more resources than rural areas.
3 In pairs, design a questionnaire to find out what items people recycle in **a** your school, **b** your local area. Carry out the survey and describe the results.
4 Based on your findings to question 3, decide on a programme of changes that need to take place to reduce eco-footprints in **a** your local area, **b** London.
5 Exam-style question Using examples, explain some of the challenges faced by cities in the developed world. (8 marks)

+ In this section you'll investigate why eco-footprints vary, and how London is reducing its eco-footprint.

The UK's eco-footprints

A 2007 report, produced for the WWF, showed that Winchester was the British city with the largest eco-footprint per person (6.52 gha), while Salisbury, Newport and Plymouth had the joint smallest (5.01 gha). Winchester's eco-footprint is the equivalent of 3.62 planets.

Why do eco-footprints vary?

Salisbury and Winchester are only 35 km apart, with similar populations of about 43 000, but Salisbury has one of the UK's smallest eco-footprints, while Winchester has the largest. The graph on the right shows that Winchester's impacts are greater in all categories, but why?

- Incomes in Winchester are higher than in Salisbury, so people in Winchester consume more goods and energy – all of which increase their eco-footprint.

- Both cities have high employment rates, but people commute further from Winchester than from Salisbury. This increases Winchester's carbon footprint. A higher percentage of Salisbury's workers walk, cycle or use public transport to get to work – generating less CO_2.

City	Planets	Eco-footprint (gha)
Winchester	3.62	6.52
St Albans	3.51	6.31
Chichester	3.49	6.28
Brighton and Hove	3.47	6.25
Canterbury	3.40	6.12

▲ The British cities with the largest eco-footprints per person

City	Planets	Eco-footprint (gha)
Gloucester	2.81	5.06
Stoke-on-Trent	2.79	5.03
Kingston-upon-Hull	2.79	5.02
Salisbury	2.79	5.01
Plymouth	2.78	5.01
Newport	2.78	5.01

▲ The British cities with the smallest eco-footprints per person

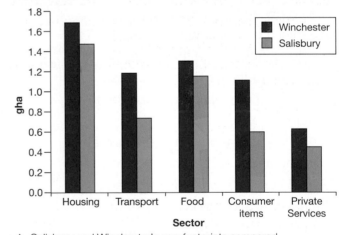

▲ Salisbury and Winchester's eco-footprints compared.

Where does London fit in?

You might think that London would be up with Winchester and St Albans in terms of its eco-footprint. However, it only came forty-fourth out of sixty cities in the WWF list. So, why isn't it higher?

- London's eco-footprint is higher than the UK average in terms of housing, food, and private services.

- BUT it's lower than average for transport – many people use public transport rather than cars. London also has policies to discourage high-polluting vehicles (see Section 15.5). This helps greatly to reduce the overall size of London's eco-footprint.

How is London reducing its eco-footprint?

1 Reducing energy consumption

London's energy strategy is focused on lowering CO_2 emissions and reducing energy use. Nearly 80% of London's CO_2 emissions originate from energy supply to, and use in, buildings, while transport accounts for the rest. In order to combat climate change and reduce energy consumption, the city is planning:

- major **retro-fitting** programmes:
- ten low-carbon zones to reduce energy emissions – including fitting solar panels and insulating buildings.
- a new low-carbon energy supply for London.

2 Reducing waste generation

By 2020, London plans to reduce the amount of household waste produced by 10%, as well as reducing municipal waste. It plans to achieve this by:

- **re-using waste** – 20 000 tonnes of municipal waste by 2020 (30 000 by 2031)
- providing cheap and accessible **recycling and composting** services
- providing recycling bins all over the city
- developing waste-burning power stations to generate heat and power.

> **+ Retro-fitting** in this case means adding new energy-saving and energy-efficiency features to existing homes and public buildings.

BedZED

BedZED (Beddington Zero Energy Development) in Sutton, south London, opened in 2002 it is an example of a sustainable community that promotes energy conservation. There are nearly 100 apartments and houses, as well as offices and workplaces.

BedZED's homes use 81% less energy for heating, 45% less electricity, and 58% less water than an average British home. They also recycle 60% of their waste.

▶ The new energy centre in the Olympic Park. A third of the fuel burned here is household waste, which is used to heat water to generate electricity for the whole Olympic Park development.

your questions

1 Use the graph opposite to suggest possible reasons why Winchester's eco-footprint is higher than Salisbury's.

2 In pairs, use an internet search engine to research the 'BedZED' scheme in Sutton, south London. Explain how energy consumption there has been reduced.

3 Discuss these statements as a class.
   a People should be fined for not recycling.
   b Every house should be built to BedZED standards.

4 **Exam-style question** Using examples, describe how one urban area is reducing its eco-footprint. (6 marks)

In this section you'll look at how London is reducing its eco-footprint by developing a more-sustainable transport system.

Monday 5 November 2012. London's 'jam cams' showed 49 traffic incidents around the city. Accidents, traffic-light failures, and a burst water main all led to hours of delays for people driving through the city. For some – the emergency services, taxis and delivery drivers – their jobs depend on their ability to move freely around the city. However many others could get out of their cars, free up the roads and help to reduce the city's eco-footprint – all at the same time!

Making London's transport more sustainable

London plans to make transport in the city more sustainable by changing the ways in which people travel, and by using new technology. Tackling transport is one of the main ways in which London aims to reduce its greenhouse gas emissions by 60% (from 1990 levels) by 2025. And, by reducing greenhouse gas emissions, London's eco-footprint will be further reduced.

Encouraging clean technology

Red buses go green – All new London buses were planned to be hybrid from 2012. This means that a conventional engine is combined with an electric motor – making the buses quieter, cleaner and more fuel-efficient (see photo below).

Electric vehicles – 'Source London' was launched in 2011 – the UK's first citywide electric vehicle charging point network. There are 1300 charging points throughout the city – more than the number of petrol stations in London. The intention is to increase the 17 000 hybrid and electric vehicles currently being used in Greater London to 100 000 within a few years.

Discouraging high-polluting vehicles

In 2008, in an attempt to improve London's air quality, the Greater London **Low Emission Zone** was set up. This meant that the worst polluting diesel vehicles either have to meet minimum emissions standards, or pay a daily charge to drive in London. Since it began, the scheme has been expanded to cover more vehicles, and there are plans to develop it even further.

◄ London's new twenty-first century Routemaster bus uses hybrid technology. It produces 40% less CO_2 and is 40% more fuel-efficient than the old bus.

Reducing congestion and pollution

London's main response to congestion and pollution was to introduce a **congestion charge** in 2003. The congestion charge operates in Central London from 7.00am to 6.00pm, Monday to Friday. It has resulted in:

- a 6% increase in bus passengers during charging hours
- the raising of money to invest in improving London's transport (£148 million in 2009/10).

Initially, there was a fall in traffic in the charging zone, but congestion is now back to the level it was before the charge was introduced. However, without the charge, congestion would be much worse.

Key

~~~ Low emmission zone

▨ Congestion charge zone

▲ London's congestion charge zone and Low Emission Zone.

▲ London's distinctive bikes for hire have become a familiar sight around the city.

Encouraging cycling

The **Barclay's Cycle Hire Scheme** was launched in London in 2010, with 6000 new bikes for hire, from 400 docking stations. Cycling is a green and healthy way to get around the city, and the hire scheme is part of a plan to increase cycling by 400% by 2026 (compared with 2001).

Barclay's Cycle Superhighways are being built to improve cycling in London.

- Four routes have already been launched, and eight more will follow by 2015.
- The superhighways are at least 1.5 metres wide, and have a blue surface so motorists can spot them easily.
- They will provide fast, direct cycle routes into, and out of, central London.

your questions

1 What are the benefits for **a** people, **b** the environment, and **c** the economy, if London makes its transport system more sustainable?

2 **a** Complete a table showing the advantages and disadvantages of each of the schemes being used to make London's transport system more sustainable.

   b Which scheme do you think has the most advantages? Which has the most disadvantages?

3 In pairs or threes, discuss and present your ideas on a spider diagram for the following:

   a How the schemes outlined here could be developed further.

   b Whether your local town or city should develop a congestion charge.

4 Exam-style question Using examples from a named city in the developed world, describe the possibilities for making transport more sustainable. (8 marks)

Urban challenges in the developing world

✚ In this section you'll examine some of the challenges that can lead to a low quality of life in the developing world.

> ✚ **Quality of life** is about more than how much money people have. It can include things like housing, employment, the environment, access to services and so on.

Welcome to India

- Rajesh and his wife Sevita live on a Mumbai beach. They support their children by running their home as an informal pub for other beach residents. But the council constantly threatens to bulldoze their illegal shelter.
- Kanye uses a handheld blowtorch to cut up ships discarded by the rest of the world. His dangerous job is paying for the education of his three daughters, and will hopefully provide his ticket to a brighter future.

- Johora started out as a rag-picker, but she has now built up a bottle-recycling business on a railway embankment. She has big plans for her seven children.
- Ashik buys beef fat from the abattoir and renders it down. It looks disgusting – but his thrifty use of waste could be destined for your soap or cosmetics.

(Adapted from the BBC TV series 'Welcome to India', broadcast in October 2012.)

Informal economy

Walk through the streets of any city in a developing country and you'll see people working in the **informal economy** (see Section 15.2). Up to 60% of the workers work in the informal economy – selling food from roadside stalls, cleaning car windscreens and so on.

These workers don't do jobs that earn a regular wage. They make and sell goods and services unofficially – often on a 'cash-in-hand' basis, and without a contract or job security. They also don't have any health and safety protection, or pension scheme. If they can't work, they won't earn anything. But on the positive side, they don't pay taxes either!

Tax is part of India's problem. It's not just the people in the informal economy who don't pay tax. India has attracted many large companies to operate in tax-free zones, which means that they don't pay tax to the government either. Without tax revenue, the city authorities can't provide services like clean water, sewage pipes, or electricity supplies.

▼ 'Rag-picking' (making money from other people's rubbish) is an important part of the informal economy in Mumbai and other cities in the developing world.

Slum housing

Quality of life for people like Rajesh, Kanye, Johora and Ashik is low, and poor housing contributes to that.

Over 1 million people live in Dharavi – about the same number as the entire city of Birmingham in the UK. Dharavi is a well-established slum that has been there for decades.

It lies between two railway routes in Mumbai. A lot of the homes are fairly solid – made from brick, wood and steel, and many have electricity. Dharavi has well-established communities that provide self-help clinics, food halls and meeting places.

Elsewhere, more-temporary slum housing is built from anything that's available – wood, cardboard, plastic sheeting, etc. The slums:

- are usually over-crowded
- lack proper sanitation or clean water
- experience pollution and disease
- are often built on marginal land.

Urban pollution

Rapid urbanisation and industrialisation in developing countries has created big environmental problems including **water pollution**. For a long time, the Mithi River (which flows through Mumbai) has been used by factories as a dump for untreated industrial waste – and by the airport dumping untreated oil. 800 million litres of untreated sewage also go straight into the river – every day!

Air pollution can be a major problem in developing world cities. In Mumbai, exhaust gases from vehicles, and smoke from burning rubbish and factory chimneys, pollute the air. As the Indian economy grows, more and more electricity is needed – most of which is generated by burning fossil fuels like coal. As a result, large quantities of greenhouse gases (including carbon dioxide) are released into the air.

▲ Temporary slum housing (top) and the heavily polluted Mithi River in Mumbai (bottom).

your questions

1 a In pairs, list the challenges faced by Mumbai residents.
  b Draw a Venn diagram and classify the challenges as economic, environmental, or social.
  c Explain in 100 words whether Mumbai's problems are mainly economic, social or environmental.
2 Explain why cities, such as Mumbai, experience pollution.
3 In pairs, discuss and present your ideas about the following:
  a India's government should increase the tax on companies, to pay for improvements to cities.
  b How far local people should club together to start city improvement schemes.
4 **Exam-style question** Describe two differences between the formal and informal economy. (4 marks)

In this section you'll look at how a self-help scheme has improved the quality of life in a developing world city.

In August 2012, when the London Olympics closed, Rio de Janeiro (in Brazil) was handed the Olympic torch as the host for the 2016 Games – the first Olympics to be held in South America. Before that, Brazil also has the Football World Cup to host in 2014. Rio has got its work cut out. It has to build sports venues, develop its transport infrastructure, and get to grips with its appalling crime rate.

Welcome to Rocinha

Rocinha, Brazil's largest **favela**, is built on a steep hillside overlooking Rio de Janeiro. It is home to as many as 210 000 people. Rocinha began to develop in the 1950s, when Rio grew rapidly as a result of rural-urban migration and natural increase. Demand for cheap housing led to the development of over-crowded favelas, which at first were homes made from scrap materials. They lacked services such as sanitation, water and electricity. Today, Rio has over 600 favelas – housing 19% of the city's population.

One of the major issues in the favelas today is crime. Rio's favelas tend to be controlled by drug lords and gangs – around 6000 people die each year in Rio as a result of drug related crimes. That's nearly 12 times the number of murders in the whole of the UK each year.

▲ Brazil is determined to prove that its Olympics in 2016 will be even better than the London Olympics.

> **+ Favela** is the word used in Brazil for shanty town (area of slum housing).

BRAZIL

● Rio de Janeiro

◄ The favela of Rocinha overlooks Rio de Janeiro, the second largest city in Brazil.

Self-help in Rocinha

The authorities in Rio set up self-help schemes to improve the favelas. They provided residents with building materials to construct permanent homes, and the residents provided the labour. In return for the payment of bills, the authorities also provided electricity and clean water supplies. People's quality of life in Rocinha has improved:

- They have better overall living conditions.
- They are healthier and suffer less disease (due to clean water and sanitation).
- They are better educated (schools have been set up), so they can get better jobs.

However, one of Rocinha's biggest challenges – that of controlling crime – was too big an issue for self-help, and is being tackled by the government.

Dealing with crime

In 2008, Police Pacification Units (PPUs) began to be established in Rio's favelas, so that the authorities could start to take control of neighbourhoods previously controlled by drug traffickers and gangs. Once a neighbourhood is considered to be under police control, officers trained in community policing replace the PPUs. In favelas where the police have regained control, there has been a fall in crime.

But, Rocinha is a headache for the authorities in Rio. It sits close to the main site for the 2016 Olympics. The authorities want to be able to provide a secure ring around the areas that will host most of the Olympic and World Cup events. A Police Pacification Unit was sent into Rocinha in late 2011, but it seems likely that it will take some time to bring Rocinha under control.

▲ Rocinha is built on a steep hillside overlooking Rio de Janeiro. Most of the houses are now built from brick and concrete.

▲ Police Pacification Units are attempting to reduce crime in Rio's favelas.

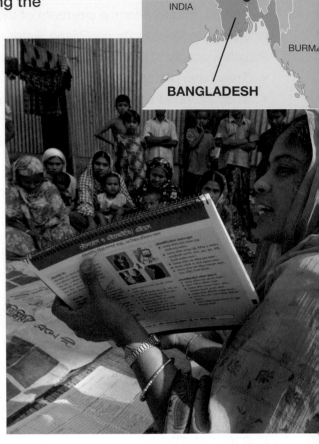

NEPAL   BHUTAN

BURMA

Dhaka

INDIA

BURM

BANGLADESH

+ In this section you'll look at different ways of improving the quality of life in developing world cities.

Dhaka, Bangladesh

Dhaka in Bangladesh teems with life. 14 million people live there – 40% of them in slums. Land is scarce and expensive, and slum dwellers are increasingly pushed out to the edge of the city. The government refuses even to acknowledge the slums, so it won't provide basic services like clean water, electricity and healthcare.

The Manoshi project

The Manoshi project, run by the Bangladeshi **NGO** Brac, was launched in 2007. Brac has 6000 volunteer healthcare workers (see photo on the right), who are helping 3 million female slum dwellers with family planning, ante-natal care and new-born care.

The project has almost halved the number of **maternal deaths** among its patients to 135 per 100 000 live births, but faces further challenges of undernourished mothers and babies becoming ill due to poor sanitation.

Old Zimkhana

Old Zimkhana, a slum community built on the site of a disused railway station in Dhaka, had no safe water or toilets. But with the help of an organisation called Prodiplan (one of WaterAid's partners), water sanitation and hygiene education projects have improved life for the local community.

Six **tube wells** have been constructed – saving people time and energy in collecting water. Tube wells are built where the water table is too deep to be reached by a hand-dug well. Two new sanitation blocks also provide toilets and water for washing. The result is improved health, which means that people are able to work more, earn more, and begin to move out of poverty.

> + **NGOs (non-governmental organisations)** develop small-scale sustainable solutions to local problems in developing countries e.g. Oxfam, the Red Cross, and WaterAid are examples of NGOs.
> + **Maternal death** is the death of a woman during pregnancy or shortly after childbirth.

▶ Tube well in Bangladesh.

Curitiba, Brazil

Small-scale solutions are fine for community-based schemes, like those described in Dhaka. But what if a whole city wants to tackle its quality of life? Curitiba is a city of 1.6 million people in south-east Brazil. Rapid growth and industrialisation meant that Curitiba faced the same environmental and social problems as other developing world cities – unemployment, poor housing, pollution and congestion. Now the city's government has improved the quality of life there as a result of **urban planning**.

Transport

- Curitiba developed a transport system based on buses. Routes cross the city in all directions, with timetabled inter-changes to provide a fast service (see photo).
- The start and finish times of businesses, services and schools were adjusted to spread the rush hour and avoid traffic jams.
- Affordable bus fares were introduced – over 80% of the city's population now use the buses. Traffic congestion and pollution have been reduced.

Recycling

In 1989, Curitiba became the first city in Brazil to introduce recycling of domestic waste. Now 70% of the city's waste is processed, creating jobs for those sorting the waste. Recovered materials are sold to local industries, and the money earned is used to fund social programmes.

A green city

In 1970, each of Curitiba's residents had less than 1m² of green space. Today that figure is 52m², thanks to the development of 30 parks and forested areas. 'Greening' the city improves the quality of life. Most of the parks are on riverbanks and in valley bottoms. They help to prevent the development of favelas and the use of land for landfill sites.

BRAZIL

Curitiba

▲ Curitiba has one of the most popular, yet inexpensive bus systems in the world.

your questions

1 Your task is to judge the success of attempts to improve the quality of life for people living in Rocinha (Section 15.7), Dhaka and Curitiba.

   a Copy and complete the following table for each project

Project name:		
Likely to ...	**Mark out of 5**	**Reasons for your score**
... improve health		
... improve living conditions		
... reduce poverty		
... improve the environment		
... benefit the economy		

   b Now write a short report (400 words maximum) on which project you think has been the most successful in improving the quality of life of those affected.

   c Describe and explain two targets that you would give each project to work on in the next ten years.

2 **Exam-style question** Using examples of projects you have studied, explain how far they have improved the quality of life in cities in the developing world. (8 marks)

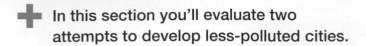

USA

MEXICO

Mexico City

+ In this section you'll evaluate two attempts to develop less-polluted cities.

The world's most polluted city?

Mexico City is one of the world's biggest megacities, with a population of about 20 million. It grew rapidly from the 1960s through to the 1980s, and is now home to 20% of Mexico's population. However, its rapid growth and enormous size have created some major environmental problems – particularly air and water pollution.

In 1992, Mexico City 'won' the title of the most polluted city on the planet. Since then it has made massive efforts to improve its environment.

Managing Mexico City's pollution

In the 1990s, in an attempt to reduce air pollution, a large proportion of Mexico City's heavy industry was moved out of the city. Mexico City is still an important industrial hub, but now cleaner service industries dominate economic activity.

One of Mexico City's achievements has been the implementation of the Green Plan. It was launched in 2007, and will run until 2021. The Plan covers things like water supply, transport, climate, and waste management. But, for the government, reducing air pollution and improving air quality is one of the highest priorities. The changes to transport (see the table) have been key to reducing air pollution and driving down CO_2 emissions.

▲ Mexico City, with the volcano Popocatépetl in the background. In 1990, air pollution was so bad that often you couldn't see Popocatépetl at all.

Action	Impacts
A bus rapid transit system has been developed, using cleaner fuel.	Bus passenger numbers have increased, and emissions have been reduced.
Mexico City's subway system (Metro) is being expanded to encourage more people to use public transport.	The subway now connects with bus stations and has higher passenger numbers.
Fuel-efficient vehicles have replaced 75 000 taxis.	CO_2 emissions have been reduced.
A bike rental scheme (called Ecobici) has been introduced. Ecobici has 4000 bikes available at 275 bike stations.	The bikes are expected to be used up to 30 000 times a day.
It is now compulsory for school children to use public transport to get to school.	The number of cars on the road has been reduced.

▲ Mexico City – transport improvements.

Is Mexico City's plan working?

Mexico City is no longer the world's most polluted city. The transport changes have led to a huge reduction in CO_2 emissions, but – compared with other cities in Latin America – it has below-average air quality. Cars still remain a major form of transport.

And tackling pollution doesn't come cheap. 8% of the city's annual budget is spent on the Green Plan – that's US$1 billion.

Masdar City, UAE

Although the United Arab Emirates (UAE) is one of the world's wealthiest nations (see Section 12.2), its oil wealth is not shared evenly – most belongs to a super-rich elite. The UAE's rapid development means that it's very much a developing country, but with a difference.

Masdar City is a totally new planned settlement being built in the country. It's located 17 km outside the city of Abu Dhabi, near the airport, and has been designed on sustainable principles – it's planned to be free of carbon, waste and pollution. Building work started in 2006, and it will eventually cover 6 km² and be home to about 50 000 people and 1500 businesses.

Masdar City has been designed with the hot desert in mind:

- A large wall will keep out the hot desert wind and airport noise.
- To reduce the need for air conditioning, wind towers will draw colder air into homes from the roof – forcing warmer air out.
- Fountains will also help to cool the air.

Most of the transport will be provided by a rapid transit system – running underground and above street level (the original plan was to ban cars). This will leave the streets free for pedestrians. Energy will be provided by solar, wind, geothermal, and hydrogen sources.

What are the issues for Masdar City?

- Building Masdar City is hugely expensive (up to US$20 billion).
- Scarce water is being used to cool the air in a desert!
- The environmental organisation, Greenpeace, supports the project, *but* says that there should be more focus on making existing cities sustainable, rather than building new ones.
- There is concern that only the wealthy will benefit from the new city.

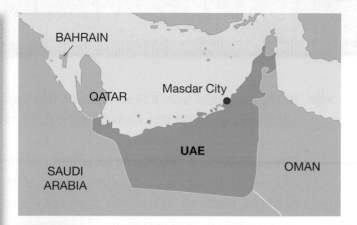

▲ Masdar City, Abu Dhabi, eco-city in the United Arab Emirates.

your questions

1 a Complete a table to show the advantages and disadvantages of Mexico City's attempts to reduce pollution.
   b Now complete a similar table for Masdar City.
   c Which city seems to you to be more sustainable? Explain your answer.

2 **Exam-style question** Using examples, explain how far one city you have studied has been successful in trying to reduce pollution. (8 marks)

✚ **In this section you'll learn about the varied rural economy in the Lake District.**

The Lake District in Cumbria (see right), is one of the UK's most popular National Parks, and home to nearly 45 000 people. National Parks are specially designated areas of outstanding landscape. The Cumbrian mountain scenery is the highest in England, and its lakes have inspired poets like Wordsworth. In 2011, this spectacular landscape had nearly 16 million visitor days (1 visitor spending 1 day = 1 visitor day).

Working in the Lake District

The Lake District is dominated by tertiary jobs (see section 11.2). There is some manufacturing (e.g. nuclear re-processing at Sellafield), but it's mostly outside the National Park. Tourism is the biggest industry (11% of jobs are in hotels and restaurants). When you add in the people who supply food and goods to hotels, shops and restaurants, tourism provides 33% of all jobs in the Lake District. However, only 1% of jobs are quaternary, e.g. business and financial services, so few jobs pay high salaries.

Commercial farming

Langdale (see right) is typical of the Lake District landscape tourists come to see. It's a landscape dominated by farming. Only 2% of people work in farming, but – without farms – the landscape would look very different. Most are small commercial sheep farms, worked by families who rent or own them. There are few job opportunities in farming now, because it's cheaper to hire temporary contractors to do jobs like shearing.

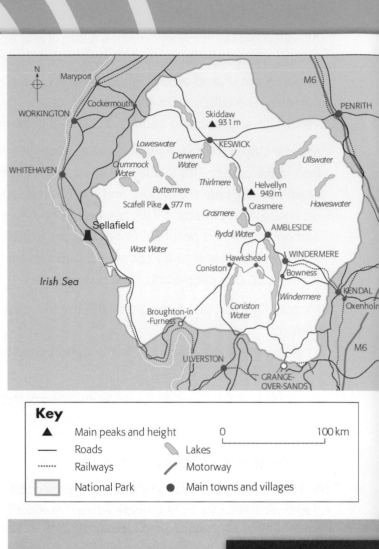

Key

▲	Main peaks and height	0	100 km
—	Roads	╲	Lakes
····	Railways	╱	Motorway
☐	National Park	●	Main towns and villages

1 Fell tops over 600m (60% of all land). Used for sheep grazing in summer.

2 Lower slopes (30% of all land). Fields separated by dry stone walls, used for raising sheep (for wool and lamb) between autumn and spring.

3 Flat valley floor (10% of all land). The most sheltered and fertile land, used for growing winter feed crops (e.g. hay), for jobs such as shearing, and for keeping a few cattle (for beef and milk).

Farm systems

A farm system is a simple structure – what goes in determines what comes out. It has three parts: inputs, processes and outputs (see the diagram). Think of it simply as buy seed, plant seed, harvest crop!

- **Inputs** are what a farmer invests in a farm. They also include physical inputs.
- **Processes** are the jobs carried out to turn inputs into outputs, and the challenges that may prevent or encourage this.
- **Outputs** are farm products and what happens to them.

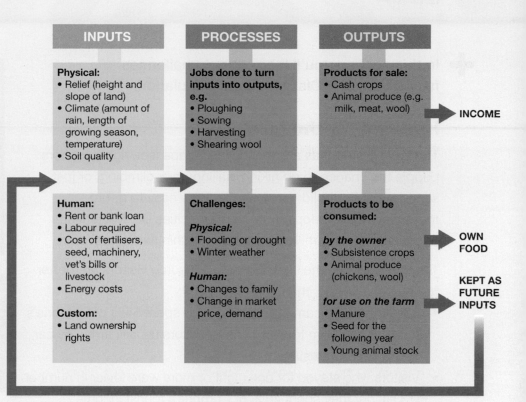

INPUTS	PROCESSES	OUTPUTS
Physical: • Relief (height and slope of land) • Climate (amount of rain, length of growing season, temperature) • Soil quality	**Jobs done to turn inputs into outputs, e.g.** • Ploughing • Sowing • Harvesting • Shearing wool	**Products for sale:** • Cash crops • Animal produce (e.g. milk, meat, wool) → **INCOME**
Human: • Rent or bank loan • Labour required • Cost of fertilisers, seed, machinery, vet's bills or livestock • Energy costs **Custom:** • Land ownership rights	**Challenges:** *Physical:* • Flooding or drought • Winter weather *Human:* • Changes to family • Change in market price, demand	**Products to be consumed:** *by the owner* • Subsistence crops • Animal produce (chickens, wool) → **OWN FOOD** *for use on the farm* • Manure • Seed for the following year • Young animal stock → **KEPT AS FUTURE INPUTS**

Challenges to farming

The Lake District's landscape is challenging for farmers. Apart from the valley floors, most land is too steep to be ploughed. Soils are thin, infertile and rocky.

The climate is also tough:

- Temperatures on the hills fall by 1°C for every 150m of height gained. There are 75 days of air frost a year above 300m (compared to just 20 on the coast).
- In winter, snow only lies occasionally on the coast, but over the hills it can occur as early as October and last until May.
- There is less sunshine on the hills (1200 hours a year, on average) than on the coast (1500 hours).
- Relief rainfall over the hills produces over 3000mm of rain a year, compared to about 1500mm on the coast.

Small wonder, then, that many young people don't want to farm here and leave.

your questions

1. In pairs, research and produce a presentation of six slides to show images of the Lake District. Finish by showing the advantages and disadvantages of the Lake District as a place to live.
2. Suggest reasons why **a** there is little manufacturing, **b** there are few quaternary jobs in the Lake District.
3. Make a copy of the farm system diagram above, and complete it for a commercial sheep farmer to show details of the inputs, processes and outputs described in this section.
4. Outline two reasons why there has been a decline in farm employment in the Lake District.
5. **Exam-style question** Using examples, describe how the rural economy can vary in the developed world. (6 marks)

➕ **In this section you'll learn about challenges facing the Lake District's rural population.**

Heading for the hills!

The Lake District has a dramatic landscape (see right)! It offers people the chance to go hiking, sailing, rock climbing, or just sightseeing. For holidaymakers, it's fairly easy to get to – motorways bring people from northern cities like Manchester and Leeds. Over 5 million people live within two hours' drive.

Tourism is vital to the Lake District. But some visitors are better for the Lake District than others. In 2009:

- 87% of visitors came for the day, and spent 54% of Cumbria's total income from tourism. These visitors usually arrive by car, and cause congestion on the roads.
- Visitors who stayed for one night or more were fewer in number (13%), but they spent 46% of Cumbria's tourist income.
- The vast majority (92%) were from the UK, and 8% from overseas. However, overseas visitors stay longer and spend even more!

Tourism and employment

Tourists bring many benefits to the Lake District:

- 55% of jobs in Windermere and Keswick come from tourism.
- Without tourists, services like the railway to Windermere, or local buses, would probably stop.
- Specific tourist shops (e.g. souvenirs and cafes) and others (e.g. outdoor clothing) couldn't survive without tourists.
- Car parks generate income for local councils to spend on schools and services.
- Farmers earn extra income by selling local foods, running campsites, or converting barns for holiday accommodation.
- Tourism brings investment in businesses, e.g. local pubs.

However, unemployment is a problem, because jobs in tourism are mostly seasonal, low paid and part-time. The number of full-time jobs is small.

Sector	% Expenditure
Accommodation	40.1
Food and drink	24.4
Recreation	8.0
Shopping	12.4
Transport	15.1

▲ How tourists spend their money

On your planet

➕ Most visitors to the Lake District come by car to see the scenery, get some clean air, and peace and quiet. The thing they dislike most is the traffic!

▶ This field has been turned into a caravan park to bring in extra tourist income for the farmer.

Rural isolation and changing services

The Lake District is isolated and it's difficult to live there without a car.

- It's far from cities, beyond daily commuting distance to Manchester or Leeds.
- Weather events create isolation. Floods and freezing winter weather in 2009-10 and 2010-11 cut off communities.
- Most local roads are winding, steep, and narrow, with summer congestion caused by tourists.
- Children in remote farms and villages face a long journey to school.

Some services are declining too:

- Bus services are reducing due to financial cutbacks.
- Many village shops face competition from supermarkets e.g. in Kendal; post offices and pubs are closing.
- Health centres are concentrating in larger settlements such as Ambleside, and hospitals in large towns such as Penrith and Kendal. Getting to hospital is a problem in summer traffic.

Counter-urbanisation and second homes

The Lake District's population is rising and ageing (Cumbria has more people aged over 45 than under). The typical newcomer has retired there from a city – an example of counter-urbanisation (see Section 14.5). This can cause problems:

- Property demands from tourism (e.g. holiday lets), inward migration and second homes, have led to rapidly rising prices. In 2012, house prices were 12 times local people's average income, and were unaffordable.
- 40% of the properties in Ambleside and Patterdale are second homes. In winter, pubs are quiet and businesses close. This can destroy communities, even though many second homeowners spend a lot with local builders and suppliers.

▲ Traffic congestion in Ambleside is serious for local people trying to get to work, and medical staff can't get through when accidents occur.

On your planet

✚ In a 2010 survey by Cumbria County Council, local businesses said their biggest problems were:
- local transport networks
- traffic congestion
- a lack of parking spaces
- a lack of skilled people
- poor bus services.

your questions

1   a   Draw up a table entitled 'The benefits and problems of tourism in the Lake District'. List as many as you can find using this section and the results of your own research.
    b   Now go through each, awarding it marks out of 5 to show how strong a benefit or problem it is.
    c   Add up the totals. Do benefits outweigh problems for the Lake District?
2   Discuss with a partner – 'Counter-urbanisation, good or bad for the Lake District?' Draw a spider diagram to show your ideas.
3   Exam-style question Using examples, explain the challenges facing rural areas in a developed country.
    (8 marks)

✚ In this section you'll learn how farmers in the developed world often diversify to earn extra income.

Too much reliance on too little

All businesses face risks, e.g. hill farming (Section 16.1) is not always profitable because the land is poor. However, farmers can sometimes find their whole income threatened by sudden events:

- Weather is a constant battle. Hay prices trebled in the snowy winters of 2009 and 2010, when livestock couldn't graze normally. £5 for a bale of hay doesn't seem much – until you see how quickly a few sheep or cattle can eat it!

- In 1986, after the nuclear explosion at Chernobyl in Ukraine, weather patterns brought rain containing radioactive waste to the Lake District. The grass on many Lake District hillsides was contaminated for over 20 years, so that sheep farmers had to sell their meat to the government (which destroyed it because of the risks to health). This only came to an end in 2012, and farmers can now sell their meat normally again.

▲ Hill farmers in the winter of 2010 had to battle deep snow to feed their sheep.

Diversification

How do farmers reduce the risk to their whole income caused by sudden events like those above? Many expand their businesses to include a second income (e.g. from a campsite). The EU offers grants to farmers to encourage them to **diversify** (widen the range of goods and services they offer). These new activities can be either farm-based or not.

Farm-based activities

These are based around the farm business. They include possible changes (e.g. to organic farming), or rearing other animals (e.g. deer for venison). Farm shops also allow farmers to sell their produce at retail price, rather than at the lower price offered to them by supermarkets.

▲ Quad-biking on the farm – a money-earner for farmers.

Low Sizergh Barn (farm-based)

Low Sizergh Barn is part of Low Sizergh Farm (a dairy farm in the Lake District). The barn is designed to encourage visitors to increase their understanding of the countryside, their knowledge of where food comes from, and how it's produced. A farm trail has been designed to show visitors the farm's livestock – as well as help them appreciate the wildlife and ecology of the farm. There's also a gift shop and café. The Farm Shop not only sells the farm's own meat and vegetables, but also pre-packed ready meals made on the farm. Like any manufacturing industry, cooking adds value (and therefore profit) to raw foods.

Holmescales (non-farm-based)

This business near Kendal hosts corporate events for businesses wanting to develop team skills out of the office. It offers companies a range of team-building activities, such as:

- team games like paintballing (which consists of role plays between teams on a mission)
- adventure challenges (e.g. fell walking, mountain biking, rock climbing, or canoeing)

It also offers family events like off-roading, archery skills – and even a tank with a paintballing gun!

Non-farm-based activities

These activities use farmland, but for different purposes. They include campsites and other tourist facilities (e.g. barn conversions for holiday rental). But many farmers are expanding beyond these into other activities to draw people to the Lake District, including:

- day activities such as paintballing, outdoor pursuits (e.g. mountain biking) and off-road driving on forest or farm tracks (see Holmescales above).
- specialist holidays, such as learning arts and crafts.

Even then, some farms are unable to survive, so their only option is to convert and sell off some of the farm's buildings – often as second homes.

your questions

1. List all the problems that farmers in the Lake District face.
2. **a** In pairs, come up with a list of things that a farm would have to do in order to develop a business like Low Sizergh Barn.
   b Put together a plan for your local bank to try to persuade them to lend you money for this business. Include **(i)** what you want to achieve, **(ii)** what you need to do, **(iii)** how you think this business will raise your income.
3. **a** Explain the benefits that Low Sizergh Barn and Holmescales can bring to **(i)** the farmer, **(ii)** the local community, **(iii)** the Lake District.
   b Which of the two businesses do you think best suits the Lake District and the people who live there? Give reasons.
4. **Exam-style question** Using examples, explain how farmers in the developed world can diversify to generate new income. (8 marks)

✚ In this section you'll learn how rural areas in northern Malawi face different challenges.

The changing farm economy

Mozesi Katsonga is a 35-year-old farmer with 2 hectares of land near Mzimba in northern Malawi. Mozesi works mainly with hand tools (see photo) and grows crops like maize to feed his family He also sells some tobacco and has a second job as housekeeper and cook at a nearby guesthouse. Neither job is enough by itself to support his family. Mozesi is affected by the same problems as many other farmers in Malawi:

- Rising fertiliser costs — these have become unaffordable for most farmers. The global price of oil has risen sharply since 2005. Many fertilisers are made from oil products, so fertiliser prices have also soared. A 50kg bag of fertiliser cost MWK 5000 (£10) in 2010, rising to MWK 15 000 (£30) by the end of 2012. Without fertiliser, farmers produce – and earn – less. For many, this might mean poverty.

- Falling tobacco prices — in 2009-10 the price of tobacco collapsed at auction (the price of 1kg of tobacco fell below 3 pence). Some farmers (including Mozesi) lost all their money, and could not grow tobacco the following season.

▶ Oxfam agrees with most Malawian farmers – that farming will get drier with climate change, and that this could create more poverty.

Living with hazards – drought

As well as rising prices, Mozesi is also concerned about climate change. Oxfam agrees with him and published a report in 2005, called '*Africa - up in smoke*' which outlined its belief that climate change will affect Africa more than any other continent, causing:

- water shortages — as average temperatures rise (which increases evaporation)
- food shortages — caused by more variable rainfall and increased drought
- increased desertification (see Section 2.9)

When compared to the last three decades of the twentieth century, recent rainfall in Malawi has been much lower. Also, the rains often don't last long, so streams and rivers dry up — destroying crops due to drought. On the other hand, very heavy rains in 2012 reduced Malawi's maize harvest by 7% and left over 10 000 families' homeless, due to flooding.

Rural-urban migration

Many families in northern Malawi are trapped in poverty. Their lives are limited by low crop yields, from small plots, or low wages on plantations (see Section 16.5). As well as having to pay for medical care and school fees:

- Rural areas have few healthcare facilities, so rural families have poorer health than families in urban areas.
- Primary schooling is free, but fees for one pupil at secondary school can cost a family its yearly income. So, what happens if they can't pay the fees or if they want to send a second child to secondary school?

The solution for many families is for one or more members to find paying work elsewhere. Male family members often move to a town or city in search for work (known as rural-urban migration). If they are able get a job, they then send home money (known as **remittance payments**), to increase the family's income. This can help to pay for medical and school fees.

However, there are also disadvantages:

- The loss of men from villages makes some problems worse. Women, young children and the elderly are left behind. With fewer hands to work the land, it's harder to farm. Yields fall and the family may have less to eat or sell.
- A lot of land can be under-used, and families are less likely to afford new farming techniques, such as higher-yielding seeds (see Section 12.7).

Rural isolation in northern Malawi

Much of rural northern Malawi is very isolated. It has poor **infrastructure**, e.g. dirt roads, so it can take several hours to travel 20km to local market towns during the wet season. When roads flood, rural areas can be cut off.

▲ A Malawian road in the rainy season – it's not easy for farmers to get to market in these conditions.

Rural telecommunications also vary. Services are slow, with congested telephone lines. In 2011, Malawi had 175 000 telephone landlines (1 per 90 people), and 700 000 Internet users (1 in every 23 people). But landlines are expensive to develop, so Malawi is expanding its mobile-phone coverage instead. Mobile-phone ownership is growing rapidly – from 1 million in 2007 to 4 million by 2011 (out of a population of 16.3 million), but rural coverage is still weak.

your questions

1. Explain why **a** good roads and **b** improved telecommunications, are important if rural Malawi is to develop.
2. Study the farm system diagram in Section 16.1. List the ways in which Mozesi's farm is different in its **a** inputs, **b** processes, and **c** outputs.
3. Draw and complete a table to show the potential **a** economic, **b** social and **c** environmental impacts of climate change on small farmers in northern Malawi.
4. List the possible benefits and problems for a rural family if a husband/father leaves to work in the city.
5. **Exam-style question** Outline one challenge faced by farmers in a developing country. (3 marks)

16.5 Malawi's rural economy

+ **In this section you'll learn how the rural economy varies in Malawi.**

Farming is critical to Malawi. Over 80% of its population works in farming, and the country depends on cash crops for its exports. But Malawi's farmers are finding it tough, because – on top of rising fertiliser prices (see Section 16.1) – global food prices constantly change, so they never know what price they'll get. There are three types of farmer in Malawi, as outlined below.

1 Cash crop plantations

The British colonised Malawi in the nineteenth century. They took over land and developed the plantations that produce most of Malawi's exports – tea, coffee and tobacco. Many estates remain in British ownership, including some by large TNCs (e.g. Unilever, which produces PG Tips tea). Large estates can afford large inputs (see Section 16.1), such as irrigation, fertiliser, storage, and transport to global markets. They hire local landless labourers, or subsistence farmers seeking extra income. The workers only get paid about 1p per kg of tea leaves or coffee cherries picked – although the estate owners argue that they also get housing, water, firewood and a daily lunch. The tea is sold in Malawi onto the global market at about £2 per kg. Therefore, most of the profit goes to tea dealers and retailers in the world's developed countries.

2 Tobacco tenants

Tobacco earns 10% of Malawi's GDP, and 2 million Malawians depend on it for their income. Almost all of the 900 000 adult tobacco growers work as tobacco tenants (see opposite) for tobacco companies like British American Tobacco (Malawi). However, many tobacco workers are children (see right).

▲ A tea estate in Malawi.

On your planet

+ Processed leaf tea was sold at £8 per kg in Tesco in 2012 – 800 times the price paid to farmers, and 4 times the price earned by Malawi.

The other face of tobacco

Chisomo was 9 when she first worked in the fields. Estate owners took her and her parents 1000 km from their village (with 45 other families) to tobacco fields in northern Malawi. Her mother says that they left home for a better life, but four years later, the whole family is still poor. *'My daughter has to work as hard as us if we are to afford necessities. The money that my husband and I get from the estate is not enough.'*

Now 13, Chisomo has never been to school. Her health is failing and she cannot work as hard. She is malnourished and gets sick easily. The family often goes without proper meals for 3 days. In the past 2 months, Chisomo has had malaria, diarrhoea and pneumonia, and her parents are scared that she may not survive.

Malawi has 1.4 million child labourers – the highest level of child labour in southern Africa. The Catholic organisation, Centre for Social Concern, estimates that 78 000 children are working full- or part-time in the tobacco fields – 45% aged 10-14, and 55% aged 7-9.

Tobacco estate owners allocate tenant farmers a plot to produce a specific amount of tobacco. The owners lend the farmers seed and fertiliser, and deduct the costs from any future profits. The owners are also supposed to supply food, but supplies sometimes run out. Many tenants lack medication, proper housing, and safe drinking water. They are amongst the poorest people in Malawi. The Tobacco Control Commission (TCC), a Malawian government watchdog, estimates that it costs US$1 for workers to grow 1kg of tobacco, which they then sell for 70c (losing 30c per kg). The tenants cannot make a living unless they send their children to work as well.

What do you think?

+ Can child labour ever be justified?

▼ Child workers drying tobacco leaves in the sun in Malawi.

3 Smallholders

Smallholders make up the majority of farmers in most areas. Across Malawi, 1.8 million families occupy 1.8 million hectares of land, so that's an average of 1 hectare per family. Between them, they produce 80% of Malawi's food. However, half of these smallholders actually farm less than half a hectare per family – only enough for subsistence farming (see Section 11.1). They grow maize, rice and groundnuts to feed the family. There is little income to invest in inputs like fertiliser to improve the soil, and their seed is whatever they've been able to save from the previous year's harvest. Those with larger plots produce small quantities of cash crops, like coffee or tea, but the income from these is normally used to pay for school fees or medical bills.

The collapse of Malawi's currency has made fertiliser and oil expensive, so inputs and transport costs have increased. Until food prices rose in 2007-8 there was little point in travelling to market to sell crops for an uncertain price.

your questions

1 **a** In pairs, think about the three types of farmer described in this unit. Then draw three large concentric circles and label them as follows:
   - In the inner circle – write down the **problems** that each farmer faces (use different colour pens for each farmer).
   - In the middle circle – **what decisions** are being made that affect these people?
   - In the outer circle – **who** makes these decisions, and **where** are they made?

   b In what ways do these three farmers' lives link to people's lives in the UK?

2 **a** Research child labour in Malawi by logging on to corpwatch.org.

   b Using your research, write a letter to tell shareholders in a large company like British American Tobacco (Malawi) how you feel about the use of child labour by their company.

3 **Exam-style question** Using examples, describe how the rural economy can vary in a developing country you have studied. (6 marks)

In this section you'll learn about different groups involved in rural development projects.

If you're like Mozesi, in Section 16.4, it's hard to make progress. Scraping a living from the land means that you need another job just to make ends meet. Now, as if he needs anything else, global anti-smoking campaigns have reduced the demand for his tobacco. This section looks at how two kinds of organisations – governments and non-governmental organisations – can kick-start rural development projects to help people like Mozesi.

The magic of mushrooms

Malawi's government has introduced a Sustainable Livelihoods Programme, funded by the UN Development Programme (UNDP). It's promoting mushroom growing as a new source of income for villagers.

Mushroom growing in Ndawambe village is making a real difference to the poorest villagers. Malawi's government (using UNDP money) has helped them to begin growing mushrooms by:

- providing **training** from government experts
- offering **micro-loans** (small loans which are repaid in a few months) to help buy the first mushroom compost
- paying for an **advertising programme** to promote mushrooms as a food with high protein value and medicinal properties.

One big advantage of this scheme is that it's cheap and sustainable, because the farmers can use waste from their land (e.g. maize stalks) as the raw material to make new mushroom compost.

The scheme has been a success, with Ndawambe's farmers only just meeting the demand. The extra income earned is now being used to fund other developments. For example, one family now has 900 chickens and has been able to build the first two-storey home in the village; other farmers have started bakeries, or now produce fruit juice, vegetable oil, and honey.

▲ Mushroom growing in Ndawambe – part of the Sustainable Livelihoods Programme.

Know your development organisations!

Development projects are normally funded by:

- **Governments** (both regional and national)
- **International or inter-governmental organisations** (IGOs), e.g. the partnership between Malawi's government and Japan (see Section 16.7). It can also involve global organisations (e.g. the UN Development Programme – UNDP) providing funding and expertise.
- **Non-governmental organisations** (NGOs) – see section 15.8, which are mostly charities, e.g. Oxfam and World Vision.

Fish farming

Families affected by HIV/AIDS often fall into poverty, because they are less able to work. The NGO 'World Vision' supports a project to develop fish farming with rural families affected by HIV/AIDS in 15 locations in Malawi. They help farmers by digging small, rain-fed ponds, measuring 20m by 10m. The ponds are designed for species such as tilapia (see the photo), which feed on aquatic plants and farm and kitchen waste. Children and the elderly can help out, which makes it easier to manage for households with few fit adults.

The benefits of the project include the following:

* It's cheap. The costs are low, because the fish are fed on waste.
* It provides a regular income. The ponds produce up to 1.5 tonnes of fish per hectare each year.
* It provides protein, calcium, vitamin A and nutrients – critical to those with HIV/AIDS. Good nourishment can extend the lives of HIV/AIDS patients by up to 8 years.
* It has led to a fall in child malnutrition in the area from 45% to 15%.
* It provides water for crops during droughts. Farmers with ponds produce 20% more crops than those without.
* Sediment from the bottom of the ponds makes excellent fertiliser. Some farmers grow bananas and guava on the pond edges, using water seeping into the soil.

The project has:

* doubled the income of 1200 households
* increased fish and vegetable consumption
* quadrupled the number of fish farmers in Malawi
* benefited women, who form 30% of those taking part.

▲ Tilapia – the species for which the ponds have been designed. They are common freshwater fish, which grow rapidly and taste good!

On your planet

✚ What feeds on living or dead plant or animal material? Answer – chickens, mushrooms and tilapia!

your questions

1 a Make a large copy of the table below and complete it using the details in this section.

	Mushroom farming	Fish farming
Type of project (e.g. NGO or government)		
How is the project financed?		
Economic benefits		
Social benefits		
Environmental benefits		
Spin-off effects (i.e. those which have arisen because of the projects)		

  b Which of the projects would you most recommend for **(i)** solving rural poverty, **(ii)** families affected by HIV/AIDS? Explain your reasons.

2 **Exam-style question** Using examples, explain how one organisation has helped to develop rural areas. (8 marks)

+ In this section you'll learn how farming can benefit from Fairtrade schemes and intermediate technology.

Middle Shire is a district in southern Malawi. Its name comes from the local river, the Shire, which provides almost all of Malawi's hydroelectric power (HEP). Most families there grow maize, cassava or rice, while others work on sugar cane or cotton plantations. But, as the photo shows, Middle Shire is now losing its soil to erosion.

The problem of soil erosion

Middle Shire's population has doubled since 1980, so farmers have to grow more food. Also, falling global tobacco prices mean that – to earn the same income – farmers need to grow more tobacco as well. These two pressures have led farmers to clear forested land to grow more crops. Between 1990 and 2005, Malawi lost 13% of its forest. This deforestation has had the following impacts:

- Heavy rain on the cleared slopes has led to surface runoff, causing **soil erosion**.
- Rivers and lakes have silted up with the eroded soil. Silting behind the HEP dams has reduced the water flow, which has lowered the HEP output.
- Malawi's potential for eco-tourism has also been threatened. Malawi has over 650 bird species, as well as mammals (including big game), reptiles, and nearly 4000 plant species – all of which can potentially attract tourists.

COVAMS – saving the soil

Using Japanese aid, the Malawian and Japanese governments have formed an **inter-governmental organisation (IGO)**. They're working on a project called Community Vitalization and Afforestation in Middle Shire (COVAMS), which involves local people conserving the soil and re-planting trees (see opposite).

▲ Severe soil erosion, due to the loss of forest on the slopes.

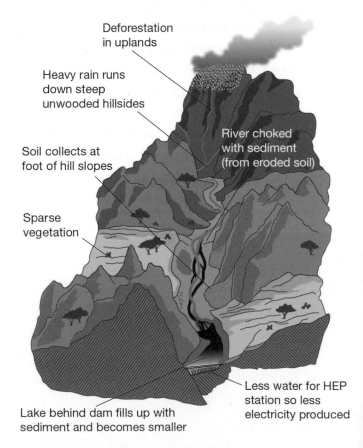

Deforestation in uplands

Heavy rain runs down steep unwooded hillsides

Soil collects at foot of hill slopes

River choked with sediment (from eroded soil)

Sparse vegetation

Lake behind dam fills up with sediment and becomes smaller

Less water for HEP station so less electricity produced

▲ The effects of soil erosion on Middle Shire.

COVAMS is tackling the causes of the problem using the following methods:

- Teams build rock, wood and bamboo barriers across streams (to prevent soil loss during rainy seasons).
- Villagers receive training in conserving soil and tree planting. Their job is then to train thousands of others.
- Farmers are trained to 'contour plough' around hillsides before planting crops, rather than up and down them, preventing surface runoff eroding the soil. (see below).
- Steep hillsides are 'stepped' to reduce surface runoff.
- Local tree nurseries grow fast-growing species, so that re-afforestation is quick.

There is nothing high-tech about this – it uses **intermediate technology**, where local materials and low-tech solutions prevent a problem from getting worse.

The COVAMS project has had three main impacts:

- In 2011, 183 households out of 247 in Middle Shire were taking part, and the project was planned to spread to 100 000 people.
- Crop yields have improved dramatically (as the quote below shows).
- With less soil erosion and silt, water quality has improved. This allows the dams to produce more HEP.

▶ Agnes Robben, a farmer in Middle Shire

> We can now sometimes harvest 15 bags of maize on one plot, which previously yielded only three bags. Instead of spending all of our money on food, we can now buy other things – including our children's education.

The Kasinthula Cane Growers Project

This project was set up in 1996 by Malawi's state-owned sugar corporation and a South African-owned sugar mill. It began by dividing up 1200 hectares of poor land into 2-3 hectare sugar cane plots, in order to provide income for subsistence farmers in three villages. It was certified for Fairtrade in 2002, and now supplies sugar to the UK (e.g. the Co-op), Europe and the USA. 30% of its production is sold to Fairtrade buyers.

Fairtrade sales earn an extra US$60 per tonne. This money has been spent on the following community development projects:

- Providing access to clean water. Villagers previously had to collect water from the Shire River – risking both water-borne diseases (e.g. bilharzia) and crocodile attacks. All three villages had boreholes by 2008, and taps are being installed in farmers' homes.
- Providing electricity via a government electrification project.
- Building a secondary school and health clinic. The new health clinic opened in 2010. Four bicycle ambulances have also been purchased.

There has also been enough income to provide cash payments to:

- increase farmers' incomes
- replace old sugar cane plants and machinery
- provide emergency food during a drought in 2005.

✚ **In this section you'll learn about development projects which are designed to improve the quality of life.**

Malawi is not the only country struggling to develop. Elsewhere, India is starting big development projects, such as the Sardar Sarovar Dam (see Section 12.8). These projects can bring big benefits, but also big problems. What poor communities often need are small, affordable projects over which they have some control. This section explores three projects designed to help such communities.

1 The benefits of biogas in India

By 2010, four million cattle dung biogas plants (see Section 12.7) had been built in India. These had created 200 000 permanent jobs (mostly in rural areas). Biogas has also brought other benefits to villages throughout India:

- Cooking with gas produces smoke-free kitchens (unlike firewood), so there are fewer lung infections.
- Women and children no longer have to spend time gathering wood or dung, so girls in particular now have more time available to go to school.
- Cattle are now kept in the family compound, making dung collection easier. Previously, cattle would graze the forest – eating saplings and preventing woodland from regenerating.
- When cattle dung is fed into the biogas digester, the micro-organisms that cause disease are destroyed as the dung ferments.
- After digestion, the sludge is richer in nutrients than raw manure, so it makes a better fertiliser.
- Many villages now use biogas to power electricity generators that provide light at night and allow water to be pumped from underground. Farmers now get up to three crops of vegetables a year by using irrigation (with pumped water).

▲ A village biogas plant with its dried cow dung fuel stacked in the background.

▲ Cooking indoors using a new, clean biogas stove, instead of a traditional stove which burned wood or dung and caused health problems.

2 Bangladesh's Grameen Bank

The Grameen Bank is a micro-finance bank in Bangladesh. It began in 1974 when its founder lent a total of US$27 to a group of 42 families to help them create a business making items to sell. The scheme grew and the Grameen Bank became a formal bank in 1983. It now makes small loans (grameencredit) to the rural poor. By 2009, it had lent US$8.7 billion to the poor, and had 2100 branches.

The Grameen Bank is based on the following principles:

- Everyone has potential – especially the poor.
- Charity doesn't solve poverty – it creates dependency.
- Developing business skills can lift people out of poverty.
- Loans should be small and repayable within a short period (e.g. 6 months).
- Once a loan has been repaid, a person can borrow again.

The bank concentrates on lending to:

- women – who form 98% of Grameen's borrowers. They are rarely able to raise loans with traditional banks, because few of them own land (so they have no security to back up their repayments).
- communities – for wells, farm equipment and livestock.

▼ A Grameen Bank borrower and her livestock – bought with the support of the bank.

3 Mobile health clinics in South Africa

John Taolo Gaetsewe is a district in South Africa's Northern Cape province. It faces major health challenges, with some of South Africa's highest rates for infant mortality, maternal deaths, and HIV/AIDS infection. But access to healthcare is limited. For those who can afford it, there is a taxi service 130 km to the nearest hospital. For those who can't, it's a donkey cart or nothing.

In 2011, a medical team working for Anglo-American (a TNC that mines iron ore in the area) decided to help. So they developed the Batho Pele health units. These are mobile clinics that bring health services to remote rural areas. There are nine units, which are taken by 4wd vehicles to four sites in the district for a week at a time. A free bus then collects patients from nearby villages. The services include screening for infection, eye testing, dental care and surgery – all free of charge. Anglo-American has said that it will pay for the service for two years, and then gradually hand over responsibility to the government's Department of Health. 2000 patients were treated in the first month.

your questions

1 **a** Copy and complete the following table to show the benefits of **each** of the three projects for communities, and whether the benefits are short, medium or long term.

Impact	Short term (immediate or in a few months)	Medium term (over a period up to a year)	Long term (over a few years)
Social			
Economic			
Environmental			

   b Which project produces the greatest benefits (**i**) for the local economy, (**ii**) for people and communities, (**iii**) overall?

   c Which of the projects would you recommend as the best for long-term development? Justify your choice.

   d In pairs, decide how far these three projects would suit Malawi. Use evidence from Sections 16.4-16.7 to support your ideas.

2 **Exam-style question** Using examples, explain how a rural development project can improve opportunities and the quality of life. (8 marks)

Unit 3 Making geographical decisions

In this section you'll learn about the six 'key ideas' which you may be asked to explore within Unit 3.

What is Unit 3 about?

Unit 3 is very different from Units 1 and 2. There is no particular content that you can learn or revise for! The exam will provide you with resources that you won't have seen before. It will ask you questions, and you'll need to understand what the resources are about.

How can you prepare for an exam whose content you can never know in advance? In fact, if you've covered Units 1 and 2 you'll be perfectly prepared to manage this exam. Unit 3 is about key ideas which are woven into the whole course. You'll probably be able to think of examples you have studied once you read through the key ideas below.

▲ Sustainable development in Bangladesh (see Section 15.8)

Key Idea	Ways in which you could study this
1 Sustainable development is an important concept.	• Investigate the 'Brundtland' definition of sustainable development – whether current social and economic needs can be met while protecting the environment and its resources for future generations. • Ways of judging whether development is sustainable socially, economically or environmentally by comparing small scale bottom-up projects using intermediate technology with large scale, top-down approaches.

Examples in this book which could help you prepare for this include:
• Section 3.5 – Threats to the Biosphere
• Sections 4.7 and 4.8 – Solutions to the water crisis in developed and developing countries.
• Section 12.7 – Biogas and sustainable development in rural India

▲ New technology could help to provide sustainable solutions to environmental problems (see Section 10.10).

Key Idea	Ways in which you could study this
2 Since the 1990s 'environmental sustainability' has become increasingly important.	• Investigate attitudes towards environmental sustainability e.g. those of TNCs, governments, NGOs and pressure groups such as Greenpeace. • Explore reasons why these organisations have different attitudes towards environmental sustainability and adopt different polices, e.g. no-growth and switching to renewable resources.

Examples in this book which could help you prepare for this include:
• Section 3.4 – conflicts of interest in St Lucia
• Section 11.6 – who wins and loses from Globalisation?
• Section 12.7 – the debate about GM crops and sustainable development

Key Idea	Ways in which you could study this
3 Demand for resources is rising globally but resource supply is often finite which may lead to conflict.	• Investigate how pressure on resources can lead to environmental and social problems, at a range of scales, e.g. the exploitation of forests, energy and water resources. • Consider how pressures on resources are likely to increase, due to population growth and affluence through development and globalisation and how this leads to conflict between individuals and organisations, e.g. oil drilling in Nigeria and conflicts between TNCs, governments, local people and NGOs.

▲ Could eco-cities reduce the pressure upon the worlds' limited resources? (see Section 15.9).

Examples in this book which could help you prepare for this include:
- Section 2.5 – how human activity can change the atmosphere and may be a cause of climate change
- Section 9.1 – the rising global population
- Section 10.1 and 10.2 – increasing global demands on resources.

Key Idea	Ways in which you could study this
4 Balancing the needs of economic development and conservation is a difficult challenge.	• Investigate how governments try to meet economic and social needs but also protect the environment, e.g. conservation areas and greenbelts versus urban and industrial development. • Investigate how global organisations (e.g. the United Nations) have become more important in managing environmental threats, and why national governments have differing attitudes to global agreements e.g. Kyoto Protocol.

▲ Big dams can help to meet the economic needs of countries, but with social and environmental consequences. (see Section 4.7)

Examples in this book which could help you prepare for this include:
- Section 3.7 – ways of conserving the biosphere e.g. RAMSAR
- Section 4.8 – the development of small-scale solutions to problems of water supply
- Section 10.7 – how governments encourage the development of renewable energies.

Key Idea	Ways in which you could study this
5 Achieving sustainable development requires funding, management and leadership.	• Examine the management and funding challenges for governments trying to achieve sustainable development both locally and nationally, e.g. renewable national energy targets and recycling. • Investigate the role of NGOs in achieving sustainable development, e.g. the impact of environmental groups on deforestation or the campaign to promote fair trade.

▲ Tree planting programmes in Cameroon have required the involvement of local, national and international communities (see Section 3.8).

Examples in this book which could help you prepare for this include:
- Section 2.9 – how Egypt is threatened by climate change
- Section 3.8 – encouraging the sustainable management of ecosystems
- Section 10.10 – the development of renewable energies and technologies.

Key Idea	Ways in which you could study this
6 Physical processes and environmental changes increasingly put people at risk.	• Examine trends in population and urbanisation to understand why increasing numbers of people, property and livelihoods are vulnerable to tectonic hazards and the impacts of climate change. • Investigate why managing risks is challenging due to the rising demand for places to live and the uncertain and unpredictable nature of risks.

▲ Coastal areas are more at risk of flooding as sea levels rise (see Section 5.5).

Examples in this book which could help you prepare for this include:
- Sections 1.7 and 1.8 – Living with the threat of earthquake hazards and how these can be managed
- Section 2.8 and 2.9 – the changing climate, and risks posed to the UK and to the African continent
- Section 4.4 – living with chronic water shortage in the Sahel.

In this section you'll learn how to prepare for Unit 3, the decision-making examination (DME).

The Resource Booklet

The Resource Booklet is a short booklet containing geographical resources that you won't have seen before. It will be about an issue somewhere in the world, and it's likely that you won't know much about this place, if anything. Don't worry – it isn't your knowledge of the place that's being tested, but your ability to understand the issue. To help you understand, the Resource Booklet will contain maps, photos, diagrams, text, and data. The exam is about your ability to study resources, and to make sense of them.

When you read the booklet, you'll see that it contains words and concepts that you'll understand from Units 1 and 2. That's intentional – examiners will set a Resource Booklet and exam to help you make links between topics you've studied.

Chapter 17.2 gives you an idea of what you can expect in a Resource Booklet. The booklet is likely to be organised into five sections:

- **Section 1** will introduce you to the place. The example in this chapter is Christchurch in New Zealand.
- **Sections 2 to 5** will explore an issue. You'll see that Christchurch was hit by a series of earthquakes in 2010 and 2011. These had major impacts. The city now faces an issue – what kind of city should be rebuilt? The Resource Booklet will take you through this, stage by stage.

Remember
+ It's your ability to understand the issue that's important.

The Exam

The exam lasts for 1 hour 30 minutes for both Foundation and Higher Tiers. It has 53 marks, 3 of which are for Spelling, Punctuation and Grammar. It sounds like a lot of time, considering the number of marks – but remember you'll need time to study the Resource Booklet. The exam will advise you to make only rough notes for the first 30 minutes. You don't have to do this – but it's good advice and can help you understand the whole picture before you start answering the questions properly.

▲ Tourism in Antarctica. Should Antarctica be left alone or should it be used for human activity, such as tourism? In the DME, you decide what's best!

What will the topics be?

The topic for the DME will be different each year. You won't know until the exam is given out. It will usually be about some aspect of sustainability, so you'll need to understand what sustainability means – based on the key ideas (see pages 286-287) and your learning from Units 1 and 2. It could be a large-scale or a small-scale issue. Examples of issues include:

• Continental scale – e.g. should Antarctica be used for human activity?

• National scale – e.g. Australia - How many people can it support?

• Regional scale – e.g. Christchurch in New Zealand, the example for this chapter.

It will also link one core topic from Unit 1 and one from Unit 2. In this case, the Christchurch earthquake links 'Restless Earth' (Chapter 1) and 'Development Dilemmas' (Chapter 12).

Why is it called a decision-making exam?

The Resource Booklet helps you to understand the sort of decision that will need to be made. For example, the resources here are about Christchurch, and who should make decisions about re-building the city after the earthquakes.

In the last question of the exam, you'll be given some options for the future. You could be asked to analyse these or perhaps even choose one which you think is best. The options will be real ones. There won't be a 'right' one — similarly, there'll be no 'wrong' answers. Section 17.3 provides questions on Christchurch to show you what the exam could be like.

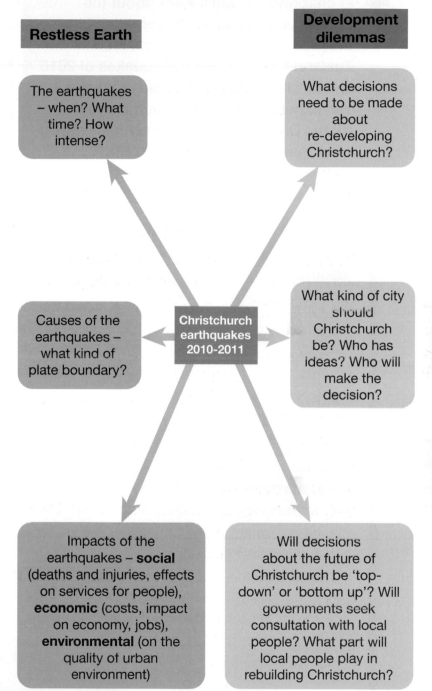

▲ Diagram showing how the study of Christchurch links 'Restless Earth' and 'Development Dilemmas'

In chapter 17.2 you'll learn about the decisions which need to be made about the rebuilding of Christchurch, New Zealand following the earthquakes of 2010 and 2011. This is an example of what you can expect to find in a resource booklet for the DME. Chapter 17.3 then shows you what questions you would be asked.

The problem

- Christchurch, New Zealand, was affected by a series of earthquakes between 2010 and 2011.
- Much of the city centre, and suburban area, to the east of the city, was destroyed.
- There is debate about how Christchurch should be rebuilt. Who has the best ideas about Christchurch's future?

Section 1 – The earthquakes in Christchurch

Source 1a: Factfile on Christchurch, New Zealand

- Christchurch is located on South Island, New Zealand.
- It is New Zealand's third largest city, and the largest city on South Island.
- Christchurch is surrounded by a region called Canterbury, which is mainly farmland and small towns.
- About 360 000 people live in Christchurch itself, and another 100 000 in the Canterbury region.

Source 1b: The earthquakes of 2010-2011

- Several minor earthquakes occur every day in New Zealand.
- There were three big earthquakes between September 2010 and June 2011, and many others that were smaller (see map below).
- Earthquakes that occur in large numbers like this are known as 'earthquake swarms'.

Time sequence of the three main earthquakes:

- First earthquake - September 2010 in Canterbury.
- Second earthquake - February 2011 in Christchurch – smaller than the first, but affected the city of Christchurch itself.
- Third earthquake - June 2011 in Christchurch - many people and businesses had been evacuated by this time.

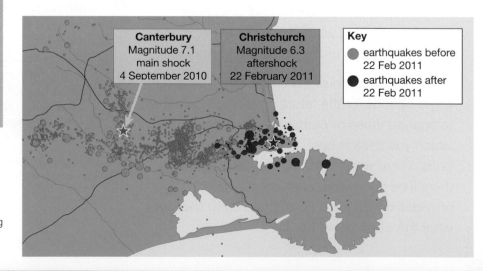

Canterbury
Magnitude 7.1
main shock
4 September 2010

Christchurch
Magnitude 6.3
aftershock
22 February 2011

Key
- earthquakes before 22 Feb 2011
- earthquakes after 22 Feb 2011

▶ The area around Christchurch showing the swarm of earthquakes that occurred between September 2010 and June 2011

Section 2 – What caused the earthquakes?

New Zealand lies across a plate margin. The plates are active and cause thousands of earthquakes every year.

Source 2a

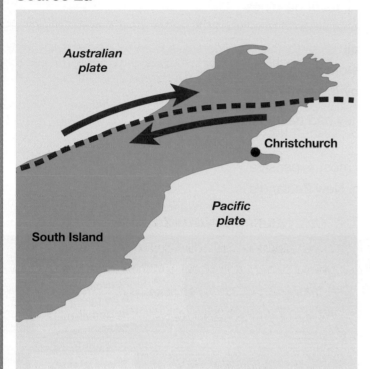

◀ The plate margin on South Island, New Zealand, which produced the earthquakes.

Source 2b

Key
▬ ▬ plate boundary

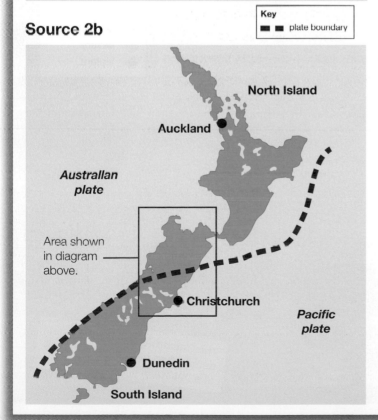

Source 2c: Earthquake prediction and protection

- A lot is known about where earthquakes are likely to happen.
- There is no known way of predicting when they will happen.
- People can prepare for earthquakes in different ways.
- Buildings can be designed to withstand earthquakes of different magnitudes.

◀ Map of New Zealand showing the plate margins

Section 3 – Impacts of the earthquakes in Christchurch

Source 3a

The city centre is oldest part of Christchurch. Two kinds of buildings suffered most in the earthquakes:

- the oldest buildings, (e.g. Christchurch Cathedral)
- the tallest buildings - of 220 buildings over five storeys high, half have been demolished due to damage caused by the earthquakes.

Further east, many houses were also destroyed. They had been built on softer sands beside the Avon River.

The three earthquakes together are New Zealand's most expensive natural disaster – costing as much as NZ$20 billion. New Zealand's total GDP was NZ$200 billion in 2010.

▼ Areas of Christchurch worst affected by the second earthquake on 21 February 2011.

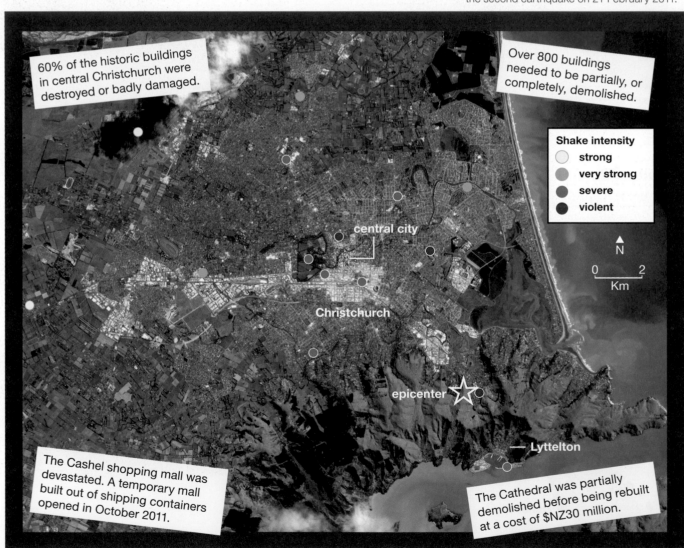

60% of the historic buildings in central Christchurch were destroyed or badly damaged.

Over 800 buildings needed to be partially, or completely, demolished.

Shake intensity
- ○ strong
- ○ very strong
- ● severe
- ● violent

N

0   2
Km

central city

Christchurch

epicenter ☆

Lyttelton

The Cashel shopping mall was devastated. A temporary mall built out of shipping containers opened in October 2011.

The Cathedral was partially demolished before being rebuilt at a cost of $NZ30 million.

Source 3b

▶ Christchurch cathedral, one of the city's oldest buildings, which was badly damaged in the second earthquake on 21 February 2011.

Source 3c

▼ Comparing the effects of the three largest earthquakes 2010-2011

	Earthquake 1: 3 September 2010	Earthquake 2: 21 February 2011	Earthquake 3: 13 June 2011
Location	Canterbury	Christchurch urban area	Christchurch to the east
Richter Scale	7.1	6.3	6.3
Deaths	0	181 115 of these were In the Canterbury TV building	0
Injured	100	6000-7000	46
Cost of damage	NZ$ 3 billion	NZ$ 15 billion	NZ$ 60 million
Buildings	Many buildings were weakened, but only a few were destroyed in the city centre.	Caused major damage. 1000 major buildings in the city centre and to the east were destroyed or had to be demolished later	Many buildings in the city centre were already damaged or had been evacuated and demolished.
Other points	Affected Canterbury with some damage in Christchurch	Affected the city centre badly	

Section 4 – Should Christchurch be abandoned?

Source 4a: Future earthquakes

- Christchurch lies in an area of active movement, with 10 000 aftershocks since September 2010. It will be affected by more earthquakes in future.
- 70 000 people – 20% of Christchurch's population – have left the city while re-building takes place.
- Scientists predict there is a 72% chance that Christchurch would be struck by an earthquake with a magnitude between 5 and 5.4 by the end of 2013.

Source 4b: Does Christchurch have a future?

Parker dismissed abandoning city

Christchurch Mayor Bob Parker has hit out at suggestions that Christchurch should be abandoned. In the Otago Daily Times, a Dunedin councillor said it was insane to rebuild Christchurch, and the money should be spent in developing Dunedin instead.

"Rebuilding Christchurch and hoping for no more earthquakes will doom Christchurch to loss of investment," he said. "It would be foolish to invest in a city that could be hit by a big earthquake."

Source 4c: Should Christchurch be re-built?

The government has divided Christchurch up into four zones: blue, red, orange and yellow (see map below). These show where re-building will, or could, take place.

▼ Map of central and eastern Christchurch to show whether areas are worth re-building.

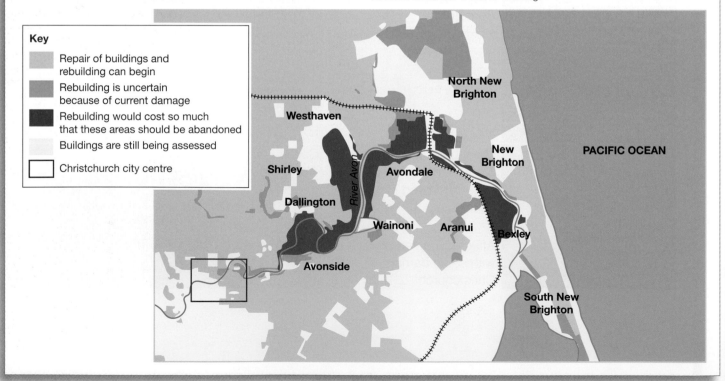

Key

- Repair of buildings and rebuilding can begin
- Rebuilding is uncertain because of current damage
- Rebuilding would cost so much that these areas should be abandoned
- Buildings are still being assessed
- ☐ Christchurch city centre

North New Brighton

Westhaven

PACIFIC OCEAN

New Brighton

Shirley

River Avon

Avondale

Dallington

Wainoni

Aranui

Bexley

Avonside

South New Brighton

Section 5 – What kind of city should Christchurch become?

Source 5a: The top-down plan

The New Zealand government and Christchurch City Council have formed the Canterbury Earthquake Recovery Authority (CERA). It believes that this is a chance to re-plan Christchurch, but with conditions:

- Re-building the city centre must be the priority.
- Many heritage buildings are dangerous, not earthquake-proof, and must be demolished.
- A new knowledge economy would attract high salary earners.
- All buildings should be built to resist earthquakes in future.
- No building in the city should be more than 28 metres high.
- Many residential areas in the east will have to be demolished and people moved out.

▼ The government plan for rebuilding Christchurch city centre.

Source 5b: The bottom-up plan

Our City, Our Say.... we want our city back!

In December 2012 there was a protest march held by people against the government plans for Christchurch. They wanted the following:

- **Housing areas should be re-built before the city centre.** Although the city centre is being re-planned, there is no plan for residential areas.
- **The red zones (Source 4c) are unfair.** People living in red zones are being forced to leave, but few will have new homes built in time.
- **Historic buildings should be re-built.** The finest heritage buildings have been demolished. "After months of demolition we are losing our heritage and character," said one resident.
- **The city should be sustainable.** Residents wanted a city centre based on sustainable principles that would 'green the city', with parks and good public transport, plus walkways for pedestrians and cycle paths.

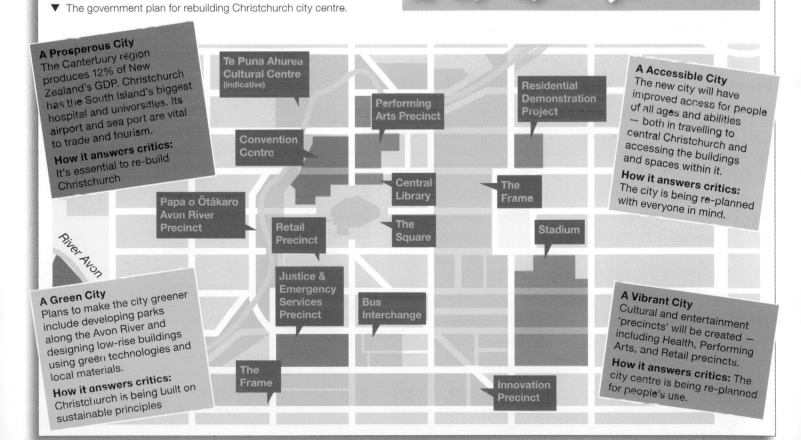

A Prosperous City
The Canterbury region produces 12% of New Zealand's GDP. Christchurch has the South Island's biggest hospital and universities. Its airport and sea port are vital to trade and tourism.
How it answers critics: It's essential to re-build Christchurch

A Green City
Plans to make the city greener include developing parks along the Avon River and designing low-rise buildings using green technologies and local materials.
How it answers critics: Christchurch is being built on sustainable principles

A Accessible City
The new city will have improved access for people of all ages and abilities — both in travelling to central Christchurch and accessing the buildings and spaces within it.
How it answers critics: The city is being re-planned with everyone in mind.

A Vibrant City
Cultural and entertainment 'precincts' will be created — including Health, Performing Arts, and Retail precincts.
How it answers critics: The city centre is being re-planned for people's use.

Map labels:
- Te Puna Ahurea Cultural Centre (indicative)
- Performing Arts Precinct
- Residential Demonstration Project
- Convention Centre
- Papa o Ōtākaro Avon River Precinct
- Central Library
- The Frame
- Retail Precinct
- The Square
- Stadium
- Justice & Emergency Services Precinct
- Bus Interchange
- The Frame
- Innovation Precinct
- River Avon

➕ These questions are based on the Resource Booklet (chapter 17.2) about the Christchurch earthquakes. You would be allowed 1 hour 30 minutes to do the exam. It's marked out of a total of 50, plus 3 marks for spelling, punctuation and grammar (SPaG). Two sets of questions are provided; one for Foundation and one for Higher.

Foundation Tier questions

Section 1 (page 290) (8 marks)

1 a Christchurch is in which **one** of the following countries? (1 mark)
   Australia, New Zealand, France, Nigeria

 b Christchurch is in which **one** of the following locations? (1 mark)
   The Northern hemisphere, On the Equator, The Southern hemisphere, In Antarctica

 c Describe **two** features of the 'swarm' of earthquakes shown in Source 1b. (2 marks)

 d Outline **two** reasons why the earthquakes in February 2011 had bigger impacts than those of September 2010 and June 2011. (4 marks)

Section 2 (page291) (10 marks)

2 a The arrows on the diagram of the plate margin (Source 2a) show that this is what type of plate margin? (1 mark)
   Conservative, Constructive, Destructive, Collision

 b Describe how this kind of plate margin can cause earthquakes. (3 marks)

 c Look at Source 2b. Give one reason why earthquakes are common across all of New Zealand. (2 marks)

 d Describe **two** ways in which people can prepare for earthquakes. (4 marks)

Section 3 (page 292-293) (10 marks)

3 a Look at Source 3a. Describe **two** ways in which people's daily lives were affected by the second earthquake in Christchurch on 21 February. (4 marks)

 b Look at Source 3b. Give **one** reason why the oldest buildings in Christchurch were more likely to collapse. (2 marks)

 c Look at Source 3c. Describe **(i) one** economic impact, and **(ii)** one social impact, of the earthquakes in Christchurch. (4 marks)

Section 4 (page 294) (5 marks)

4 a Look at Source 4a. Give **one** reason why earthquakes are very likely to happen again in Christchurch. (1 mark)

 b Look at Source 4c. Describe **two** features of the location of red zones (areas which should be abandoned). (4 marks)

Section 5 (page 295) (8 marks)

5 a Source 5a is about 'top-down' development. Which of the following is the best definition of top-down development? (1 mark)
   (i) development which is decided by the people, (ii) development which is wanted by the people, (iii) development which is decided by large companies, (iv) development which is decided by government and other large organisations.

 b Outline **one** feature of 'bottom-up' development. (1 mark)

 c Look at Sources 5a and 5b. Describe **three** differences between the 'top-down' plan for Christchurch (Source 5a) and the 'bottom-up' plan (Source 5b). (6 marks)

Section 6 (9 marks)

Different people have different ideas about how – or if – Christchurch should be re-developed. Three options have been put forward.

Option 1: Accept the top-down plan
Accept the City Council recovery plan for rebuilding Christchurch (Source 5a).

Option 2: Accept the bottom-up plan
Accept the ideas of the protestors (Source 5b)

Option 3: Abandon Christchurch
Christchurch should be left alone, and abandoned gradually.

6 Choose **one** of these options.
   a Explain your reasons for choosing this option
   b Outline **two** disadvantages of any one of the other options. (9 marks) plus 3 marks for SPaG.

Higher Tier questions

Section 1 (page 290) (8 marks)

1 a Describe the location of Christchurch. (2 marks)

b Describe the location of the 'swarm' of earthquakes shown in Source 1b. (2 marks)

c Compare the earthquakes of September 2010 with those of February 2011. (4 marks)

Section 2 (page 291) (7 marks)

2 a Identify the type of plate margin shown in Source 2a. (1 mark)

b Explain why earthquakes are common throughout New Zealand. (2 marks)

c Explain how people can prepare for earthquakes. (4 marks)

Section 3 (pages 292-293) (10 marks)

3 a Study Sources 3a and 3b. Describe the location of the greatest damage in Christchurch. (4 marks)

b Describe the economic and social impacts of the earthquakes in Christchurch between September 2010 and June 2011. (6 marks)

Section 4 (page 294) (6 marks)

4 a Explain why earthquakes are likely to occur again in Christchurch. (2 marks)

b Explain the decision to abandon areas shown as 'red zones' in Source 4c. (4 marks)

Section 5 (page 295) (7 marks)

5 a Give **one** feature of the plans in Source 5a which are typical of 'top-down' development. (1 mark)

b Explain the differences in viewpoint between those supporting the 'top-down' and 'bottom-up' proposals. (6 marks)

Section 6 Options for the future (12 marks)

Different people have different ideas about how – or if – Christchurch should be re-developed. Three options have been put forward.

Option 1 Accept the top-down plan
Accept the City Council recovery plan for rebuilding Christchurch (Source 5a).

Option 2 Accept the bottom-up plan
Accept the ideas of the protestors (Source 5b)

Option 3 Abandon Christchurch
Christchurch should be left alone, and abandoned gradually.

6 Choose **one** of these options. Justify your choice. (12 marks) plus 3 marks for SPaG.

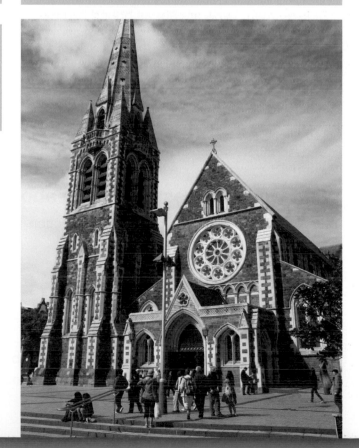

► Christchurch Cathedral before the earthquakes. Should heritage buildings like this be rebuilt or not?

✚ In this section you will learn what you have to do for Unit 4, the controlled assessment unit.

What's different about Unit 4?

The other units in this book help you to learn about 'Dynamic Planet' (Unit 1), 'People and the Planet' (Unit 2), and the 'Decision-Making Exercise' (Unit 3). These are each assessed by an exam, which requires preparation and learning, and are each worth 25% of your mark.

Like these, Unit 4 *Investigating geography* is worth 25% of your GCSE mark, but there is no exam.

- The work is done in school time and you have more control than in an exam. You can plan what to say and do, and go back and correct things. If exams stress you out, you'll probably do better in this unit.
- Although there are controls over how long you can spend (which is why it's called 'Controlled Assessment') you are not as bound by time as in an exam.

The unit is based on fieldwork.

- Your teacher organises a fieldwork investigation for your group.
- You spend class time writing a report of the investigation, which is marked.

What will the investigation be about?

Your teacher will choose a task from a selection based on these themes:

- coastal environments
- river environments
- rural environments
- town/city environments.

Whichever one your teacher selects, you will not be doing the same investigation as the year above you at school, or the year below.

▲ Your teacher will be able to choose between controlled assessment themes on rivers…

▲ …coasts…

▲ …the countryside…

▲ …or towns.

The hypothesis

The task will always be a broad statement like this:
The regeneration of the inner city has been successful.

Your challenge is to see whether the statement can be proven right or not. It is more accurate to speak of this statement as a **hypothesis**.

From this broad hypothesis, your teacher will devise a focus for your work, usually in a particular place, for example:

The regeneration of London's Docklands has been successful.

Your fieldwork will then be about proving, or disproving, the hypothesis.

> **+** A **hypothesis** is a statement intended as a prediction for an investigation. After the investigation, it can either be accepted (proved) or rejected (disproved).

The regeneration of London's Docklands has been very successful.
Hypothesis proved.
The regeneration of London's Docklands has not been at all successful.
Hypothesis disproved.
The regeneration of London's Docklands has been successful in some ways but not others.
Hypothesis only partly proved.

After the fieldwork, you'll spend several weeks putting the results together, presenting them using various methods e.g. graphs, maps and photos. You'll analyse these, and then draw conclusions from them.

You then carry out a final evaluation of your investigation.

This chapter will take you through an investigation so that you
- know what to do at each stage
- get ideas about how to present data
- know what to do when it comes to writing your analysis, conclusions, and evaluation.

How will I know what to do?

Don't worry – you're not on your own! Your teacher will make sure that you know what to do at all stages. You will have all the fieldwork organised for you and you will be helped closely in the early stages. Until you start doing the presentation of your data, you'll be allowed to work in a group, and to ask your teacher for advice.

Later, especially during the write-up, you will have to work alone, but even then your teacher can talk generally about what makes good data presentation or analysis. What you have to do is to make the decision about which styles you'll actually use.

+ In this section you will learn how to develop an introduction to your study, and how to write your aims.

Step 1

There are 7 steps to your investigation. Each is covered in the next few pages. Each stage requires a different amount of supervision by your teacher, or different level of 'control'. Take a look at this table to see what's involved in step 1.

Step 1	What do I do?	Level of supervision
Research and planning	Listen to the focus of your task, and make sure you understand what you are investigating, and where.	Low. All you need to do is listen!
	Find out about the area where you'll be doing your fieldwork. Research around the issue you are looking at.	Low. You'll probably do some of this in class, led by your teacher, together with some homework research which you do on your own. Note: if you do research at home you can't bring any material into school with you, and you can't take your work home to work on it there either.
	Start writing up your aims, with guidance from your teacher. You'll find that your aims look very similar to everyone else's.	Low. Your teacher is allowed to give you guidance on which websites would be useful for your research to make sure that you look in the right places, and to make sure you write up your aims accurately.

The introduction

Your teacher may have pictures or video material of the area you are going to study, and you will probably have been taught a bit about the theme already e.g. about inner cities or about coastal processes.

You need to write up a short introduction (perhaps 300 words) which could answer some of these questions.

- Where is the area located? Locally? Within which part of the UK? Outside of the UK?
- What is the area like? (Try using descriptions and photos, and look at digital maps from sources like Google Earth.)
- What issues are affecting the area, and why?
- What processes are affecting the area, and with what effects?

The aim and hypotheses

Remember that your teacher can help you in this section. You can also write this part before you do the fieldwork. Writing the aim is very simple. Look at the example hypothesis:

The regeneration of London's Docklands has been successful.

Your aim can be as simple as:

The aim of this investigation is to prove the hypothesis that 'The regeneration of London's Docklands has been successful.'

However, a broad hypothesis like this is usually better when split up. The success of the regeneration being investigated could be measured in different ways.

- Economically – is it creating jobs? Are they well-paid jobs? Does it offer economic opportunity to areas with high unemployment? Is there prosperity e.g. high numbers of car owners, home owners?
- Socially – does everyone gain? Do the poor gain as much as the wealthy? Does regeneration provide better housing, education (improved schools) or leisure facilities (e.g. parks)? Is public transport provided to help people who do not own cars?
- Environmental – is there an improvement in the environment? Are buildings well maintained, or streets free of litter? Is it safe and well lit at night?

You could now write:

I have split my main hypothesis into three smaller ones.

a) The regeneration of London's Docklands has been successful economically. By this I mean...

b) The regeneration of London's Docklands has been successful socially. By this I mean...

c) The regeneration of London's Docklands has been successful environmentally. By this I mean...

▲ Large office blocks overlook the old docks in London.

▲ New homes in east London close to Docklands. These are being built on the sites of old council housing built for low income groups. Many of these will be for private sale. Successful regeneration or not?

Add some colour

You could add photographs with captions designed to make the reader think. If you use other people's photos – e.g. from the internet – make sure you **always** give the source. Never try to pass something off as your own if it isn't.

✚ In this section you will learn some basic Ordnance Survey map skills, and ideas for using GIS in your Controlled Assessment.

Using OS Maps

A good way of demonstrating geographical skill in your Controlled Assessment is to make use of an Ordnance Survey (OS) map to locate your area of study, to show a route, or to identify key features. Combining this with the use of GIS (geographical information systems) could be a strong feature of your planning and research.

Finding places

Every OS map is divided into a grid, and each gridline has a number. You can use these numbers, the grid reference, to give the location of a particular place. To locate a grid square, use a 4-figure grid reference.

- Find the eastings gridlines first (the numbers along the top and bottom) e.g. Tower Bridge is between eastings 33 and 34. Take the lower value (33).
- Next, find the northings (the numbers running top to bottom) – Tower Bridge is between northings 80 and 81. Again take the lower value (80).
- So Tower Bridge lies in grid square 3380.

To pinpoint more specific locations, use 6-figure grid references.

- Again take the eastings first. Look at the square containing Tower Bridge (33 to 34). Imagine the square is divided into tenths. The bridge (not just its name!) lies about seven tenths across. So write down 337.
- Next, look at the northings and imagine the square to be in tenths between 80 and 81. The bridge is two tenths across. So write down 802.
- So the full 6-figure grid reference for Tower Bridge is **337802**.

Measuring Distance

The scale of this map is 1:50 000 so 2 cm equals 1 km. Measure the distance between two places on the map with your ruler. Divide this number by 2, and that is the distance in km. Use one decimal place for part-kilometres. The distance on the map between Tower Bridge and Liverpool Street station is about 3 cm. Divide this by 2, and we know the real distance is approximately 1.5 km, in a straight line.

Using Directions

On OS maps, north is always at the top of the map. To find the direction of Liverpool Street station from Tower Bridge, work from Tower Bridge. Move an imaginary line (or your ruler) like the hand of a clock around Tower Bridge, until it meets the station. Then look at the angle of this line compared to the compass. Liverpool Street station is NNW of Tower Bridge.

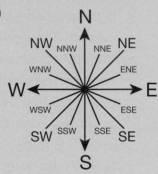

Geographical Information Systems (GIS)

Several GIS are available on the web. You have almost certainly used Google Maps or Google Earth. You should try to use GIS in your Controlled Assessment.

- Match an OS extract like this. Photos bring places alive.
- Update recent developments. GIS can be ahead of printed maps e.g. in showing new roads or housing.
- Provide screenshots of areas of your study to give more detail. An OS map can't show age or density of housing – but a digital image can.

Survey map of East London. This is 1: 50 000 scale, so 2 cm represents 1 km. The grid squares each measure 1 km².

your questions

1 Give 4-figure grid references for Canary Wharf, and Liverpool Street station.

2 Give 6-figure grid references for Canary Wharf, and Liverpool Street station.

3 Measure the straight line distance **a** from Canary Wharf to Liverpool Street station **b** from Canary Wharf to Tower Bridge.

4 Give the direction **a** from Canary Wharf to Liverpool Street station **b** from Tower Bridge to Canary Wharf.

5 Use Google Maps to find the area shown in this OS map. Take a screenshot, and paste it into a Word document.

6 Identify five things on the photo that can't be identified on the OS map.

7 Take screen shots of close-ups of **a** Canary Wharf **b** Shadwell **c** Victoria Park.

8 Describe what you can see about land use from these three areas.

✚ **In this section you will find out about the fieldwork activity, what you have to do, and how to get the best out of it.**

Step 2

This is the second step in your investigation. Take a look at this table to understand what level of control is required by your teacher.

Step 2	What do I do?	Level of supervision
Fieldwork activity	Take a full part in the field trip! Go out and collect data. Follow carefully all the instructions about what to do.	Low. You do this in groups. Your teacher does not have to be watching you all the time.
	Remember you are essential to successful data collection as everyone else is depending on you, and you on them.	Low. You can share your results with the class to get a better set of results.
	Your teacher will decide most of the fieldwork data collection methods e.g. land use maps or questionnaires.	Low. Your teacher can help you to make sure you have collected data in the right ways.
	You can show initiative by collecting some of your own data. For example, take photos and collect information that may be freely available e.g. from tourist offices, or estate agents.	Low. Your teacher will encourage you to do this anyway. But make sure any data you collect is relevant – don't just collect anything – and ask your teacher if you are unsure.

Fieldwork tips

Fieldwork methods depend upon the focus. You'll have very different tasks to do if you're studying a river compared to an inner city area. Whatever your focus, you need to make sure that you

- collect a range of data
- make sure you are clear about how every piece of data collected fits into what you are trying to investigate
- get experience in collecting all data
- remember that you're part of a team – the data that you collect could be important to everyone in your year group.

▲ Students collecting data in groups. Like these students, you're allowed to use each other's data and work together.

Collecting data

Look again at the earlier hypothesis: The regeneration of London's Docklands has been successful. The table below shows some possible data collection methods. You could use these ideas to suggest methods for your own investigation.

Method	Purpose
Land use map for your study area	Show what kinds of land use exist now, and how they may have changed – can you find a historic map from before the regeneration?
Age-of-buildings map for your study area	Identify the most recent buildings which have resulted from the regeneration.
Environmental Quality Survey (EQS) at different places	Compare environmental quality (building quality, noise, open space) in areas that have been regenerated with areas that haven't.
Questionnaire to find out people's ideas about the success of regeneration	Find out whether people think they have benefited from regeneration or not, and to see whether their ideas vary with age or gender.
Field sketches	Give your own impressions of the areas you visit – the annotations are more important than artistic quality!
Photographs	Take photographs of different areas to be used alongside the EQS, to give a visual reference.

▼ An Environmental Quality Survey (EQS).

Qualities being assessed	Good +2	Fairly good +1	Av. 0	Fairly poor −1	Poor −2	
Building quality 1. Well designed / pleasing to the eye						Poorly designed / ugly
2. In good condition						In poor condition
3. Evidence of maintenance / improvement						Poorly maintained / no improvement
4. Outside – land, gardens or open space are in good condition						Outside – no gardens, or land /open space in poor condition
5. No vandalism evident						Extensive vandalism

Collecting secondary data

Your own data may not be the only source that will help you. You can use secondary data – that is, data which has been collected by other people or organisations.

- Census data (www.ons.gov.uk) – although it may be out of date by a few years, it is still a valid, complete sample of everyone living in the area that you are studying.
- Data on house prices – either from estate agent publicity, or from upmystreet.co.uk, which will give you an idea of how much house prices have changed.

In this section you will find out about some data collection methods you could use and how to write about them.

Step 3

This is the third step in your investigation. Take a look at this table to understand what you have to do. You'll see that the level of control at this stage is still low.

Step 3	What do I do?	Level of supervision
Methodology	Describe, explain, and justify the methods used in your fieldwork to collect your data. Give some indication of how you sampled different places or people to survey.	Low. You can find out about different methods from your teacher to check that you understand them.

Writing the methodology

In your methodology, you need to do the following.

- Show on a map the places that you actually visited – or where you sampled people or features (e.g. the points along a river that you took readings from).
- Describe how and why you selected the places in which you recorded your data. Use the sampling table below to help you.

Method	What this means	When you might use it
Random	Places or people chosen at random – just like pulling numbers out of a hat.	• For a river study, when deciding how far apart to sample sediment.
Stratified	You base a survey on what you know to be there already e.g. if you know that 20% of the population is over 65, then 20% of the people sampled for your questionnaire should be over 65.	• Questionnaires. • EQS surveys of different land uses, making sure each land use is surveyed. • A river course with straight, winding and meandering sections, where you ensure that you include an example of each.
Systematic	You decide on a fixed interval between points, even if it means that you miss out certain interesting features. It might be places 50 metres apart, or every 4th person to walk past you.	• A river study where you sample places every 200 metres or every quarter of a mile, etc. • An inner city where you carry out an EQS survey every 100 metres.
Opportunistic	You take whatever comes your way, either because the sample size might be so small, or because something looks particularly interesting.	• Rural communities, where the number of people is so small that you interview everyone you see!

▲ Different sampling methods.

Writing up your method

You will need to describe and justify the methods you have chosen to use. You might find the following table useful as a format – make sure you write something in every box. The example here is for a study of inner city regeneration, but you could easily use the same idea for a river, coastal or rural study.

Data collected	Reasons why these data were needed	How you collected data	Problems you came across and how you tried to overcome them
Land use map			
Age of buildings map			
Environmental Quality Survey			
Questionnaire about people's views of the regeneration			
Field sketches			
Photos			
Property values			

▲ Using a table to describe and explain methodology.

Sharing your results

The important thing is to make sure that you have all the data from your group or class. Your teacher might be able to place them on the school website or VLE, which means you can access them from any point in or out of school.

➕ In this section you will find out about deciding on the best way to present your data.

Step 4

This is the fourth step in your investigation. Read this table so you understand what kind of help is allowed by your teacher.

Step 4	What do I do?	Level of supervision
Data presentation	Present your results in ways that make them easy for you and the reader to interpret. This will take place over several lessons. Data presentation is very time-consuming! Use varied methods of presenting data – the greater the variety, the better. Use maps, graphs, charts, and annotated photographs.	Low. Although you are expected to work on your own, you can be assisted by a teacher and can talk to friends to discuss ideas. You complete your presentation of results over a few lessons and can give work in for checking. Your teacher will tell you how long you have to complete this. Note: your teacher can't tell you which type of map or graph to use.

Follow these hints for productive lessons:

- Discuss with your teacher (and look through this book!) to see what the possibilities are for graphs, maps etc. Then you simply make the choice.
- If you don't know how to draw a particular kind of graph, you can ask people for help.
- You don't have to hand draw any graphs, though you can if you prefer – everything can be done on the computer.
- Go to each lesson with a plan or list of what you want to complete, so you make the best use of your time.

Presenting data

Choose varied ways of presenting your data.
- If you are dealing with statistics, produce graphs.
- If you are dealing with statistics used to compare places, then place graphs around a map, with lines to link your graphs to the relevant places.
- Find ways of linking different data together. For example, you could link together your EQS results on the same page as photographs that illustrate environmental quality (see page 297).
- Think creatively. Use a folded piece of A3 paper, if A4 paper is too small.

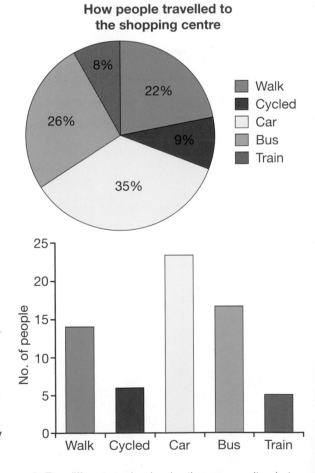

How people travelled to the shopping centre

- Walk
- Cycled
- Car
- Bus
- Train

▲ Two different graphs showing the same results – but which is the best way? You decide!

Graphs

Choose the right kind of graph for the data you want to display. You can use a computer to create them, or draw them by hand. These are just some examples, but there are many more you could use.

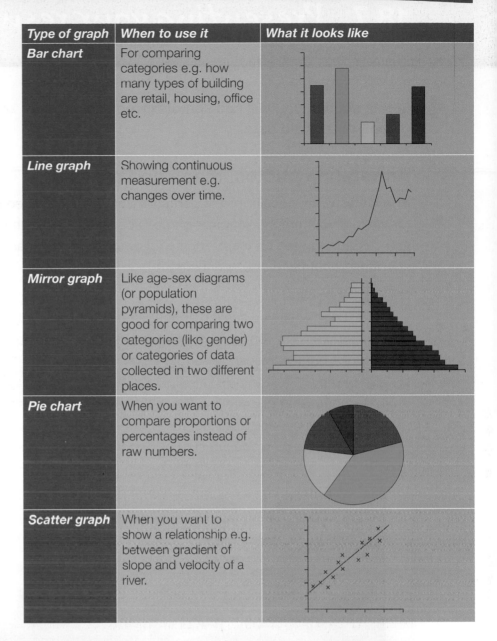

Type of graph	When to use it	What it looks like
Bar chart	For comparing categories e.g. how many types of building are retail, housing, office etc.	
Line graph	Showing continuous measurement e.g. changes over time.	
Mirror graph	Like age-sex diagrams (or population pyramids), these are good for comparing two categories (like gender) or categories of data collected in two different places.	
Pie chart	When you want to compare proportions or percentages instead of raw numbers.	
Scatter graph	When you want to show a relationship e.g. between gradient of slope and velocity of a river.	

Maps

Choose the right kind of map for the information you want to display. You can use GIS software to produce these, or other graphics software, but you can also draw them by hand.

Type of map	When to use it	What it looks like
Choropleth map	• Where there are categories e.g. building heights grouped by number of storeys.	Shaded areas with a key. Uses categories which range from light shades of a colour for low values to dark shades of the same colour for high values.
Dot map	• To show distribution e.g. particular kinds of shop in a city or town.	Dots displayed on a map to indicate an occurrence. The dots might be coloured e.g. a red dot represents a café, a blue one shows an antiques shop.
Isoline	• Rainfall map with areas between isolines coloured in. • Building heights within a city centre. • Pedestrian counts.	A map showing lines which join values of the same amount e.g. rainfall, or temperature. Often, the areas in between each line are shaded in like choropleth maps.

✚ In this section you will see some examples of well-presented data.

Annotating photos, maps and graphs

Annotations are specific to the particular item they are linked to. Therefore, for each of your graphs, maps or photos you should try to annotate what you can actually see on them. Try the following as a guide, and look at the annotated photo and chart on the right.

a) On a set of data
  • describe what you see e.g. Place X has a high percentage of offices…
  • compare what you see e.g. Place Y has the highest percentage of shops but the lowest percentage of offices…

b) On a photo
  • describe what you see e.g. Place X has a high percentage of offices…
  • combine evidence from the photo with information you have collected elsewhere e.g. in an EQS survey. See the photo below.

> ✚ An **annotation** is a piece of information that you add to some text, or an illustration, to describe or explain a particular aspect of it.

The buildings are quite well designed and are in good condition. There is evidence of good maintenance.

There is no litter. The precinct is well maintained and close to public transport, shops and amenities.

Although there is a lot of traffic noise from the nearby main road, this precinct is traffic-free and safe for people.

Although there is no green space or parkland, this is used by young people for biking and leisure. Very few trees and shrubs are visible.

▲ A well-annotated photo from Shadwell in east London.

Combining your results

Sometimes presenting results in combination with each other can work really well. In this excellent example, a student has annotated the photos by linking them to the EQS results from the same place (Padstow, in Cornwall). Text in green type shows positive comments, and red shows negative.

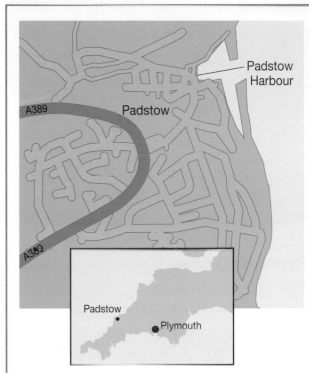

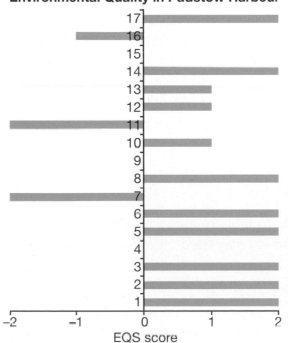

▲ Padstow Harbour.

The Harbour, Padstow

- Got a total score of 14.
- Received high scores on most criteria.

Good points include:

- The old buildings are attractive, and well designed.
- There is little traffic congestion in this area.

Negative points include:

- There is little open space in or near this area.
- Few houses have any gardens.

Environmental Quality in Padstow Harbour

(EQS score bar chart, values 1 to 17, EQS score axis from −2 to 2)

EQS key

1 – Design
2 – Condition
3 – Maintenance
4 – Open space
5 – Vandalism
6 – Traffic congestion
7 – Parking
8 – Traffic noise
9 – Safe for people
10 – Pollution smell
11 – Large gardens
12 – Trees and shrubs
13 – Public parks
14 – Litter
15 – Road maintenance
16 – Public transport
17 – Amenities and services

In this section you will learn about the analysis – what it is, and how to do it well.

Step 5

This is a vital step in your investigation. It's at this stage that you have to work on your own and cannot ask for help. The analysis is worth a high percentage of your marks and so should be given plenty of thought.

What is analysis?

Analysis is about using the data you've collected to see whether your hypotheses are correct. In your analysis you need to cover a number of things.

- **Describe** results that you have not already covered in your annotations e.g. place X has high environmental quality.
- **Compare** places for the highest and lowest e.g. place C has the highest buildings of all our sample points in Chichester city centre.
- **Explain likely** reasons for what you see e.g. place H has the highest pedestrian count because this is where all the large stores are such as Debenhams and M & S.
- **Explain possible** reasons for what you see e.g. place Q probably has a higher number of people working part-time because ….
- **Link** different graphs or photos e.g. place Z has the lowest environmental quality score (Graph 2c) which is due to the amount of derelict land (Photo 1A).

Step 5	What do I do?	Level of supervision
Analysis	Look at your results and interpret what they mean, describe what they show, and explain the reasons for what you observe.	High. This section must be done on your own, supervised by a teacher. This will be done over a few lessons, and you must hand your work in at the end of each one. Your teacher will tell you how long you have to do each section but they cannot help you or give you feedback. You can't talk to your friends.

Where do I start?

1 Start with general points.
Generally, the inner city areas I studied prove my hypothesis correct, because environmental quality is poorer in places which have not been regenerated than in those which have been.

2 Then go for more specific points to show differences between places. Always illustrate with data.
Canary Wharf has the highest percentage of people in professional and managerial jobs (78%) whereas Shadwell has the lowest (16%). However, Shadwell has the greatest percentage of people in work (only 5% unemployed) so the pattern is not definite.

3 Then compare data within places.
Shadwell may have the lowest unemployment, but it also has the highest proportion of those in part-time work (21%).

Good analysis

Here is an extract of some analysis that a student wrote.

Disposable income and home ownership are also key indicators. We see in Fig. 9 that the majority of people in all three areas own a car, over 85% in both Tintagel and Padstow, and that most families in the settlements own more than one car, with every family in St. Teath owning two. We can see from Figure 4.6, as we saw in the social analysis, that the majority of people in all three areas own their own homes, with very few people (less than 30% in each area) who rent their homes, suggesting available housing, and wealth in the areas. This is emphasised by Figures 4.7 and 4.8, where we see that Cornwall's housing prices are some of the lowest in the UK, especially compared to London and the UK average. We can also see that people in the three areas have a large amount of disposable income available to them, with most people (around 30% for St. Teath and Padstow, and over 40% for Tintagel) being able to afford 7 out of the 10 possessions. This would suggest economic welfare is not dependent on settlement.

This is well structured analysis because the student has
- referred to data and figure numbers in the presentation (yellow)
- used a range of geographical terminology (green)
- quoted data from their results to support their ideas (blue)
- used words to show links, strengths, weaknesses, suggestions and assumptions (pink).

Know the mark scheme!

Level 3 quality is the highest on the mark scheme. Marks are awarded at this level if you
- illustrate your work with data
- compare more than one set of data e.g. compare environmental quality scores with home ownership
- refer to other sources such as your own interviews or research e.g. People I interviewed in Canary Wharf showed that... .

✚ **In this section you will find out how to write your conclusion and evaluation.**

Step 6

Your conclusion is worth doing well, because, together with analysis, it counts for a high percentage of the marks. Follow these pointers to help you write a strong conclusion.

Writing the conclusion

In your analysis, you look at your separate hypotheses (the examples here were on the economic, social, and environmental effects of regeneration). In your conclusion you need to return to your main hypothesis – our example is: The regeneration of London's Docklands has been successful.

Your conclusion must take this statement and try to decide whether it is true or not. Consider the possible outcomes.

- The statement in your aim might be correct, partly correct or incorrect. You have to decide which it is.
- Whichever you decide must be on the basis of evidence – not just from general impressions you have in your mind.

To help you do this, try summarising everything in a table (see right) before the lesson so that your mind is clear when you go into class to write it up.

If you are not sure about whether the statement is correct or not, it is alright to say so. But you must give plenty of examples to support each side of the argument, to show why you cannot decide.

At the end, list the main findings of your work. What firm conclusions can you come to?

Step 6	What do I do?	Level of supervision
Conclusion	Here you draw out the answer to your main aim e.g. How successful has regeneration in the inner city been?	High. Both sections must be done on your own in lesson time, supervised by a teacher. You can't take the work home. Your teacher will tell you how long you have to do each section. They can't help or give you feedback. You can't talk to your friends.
Evaluation	Here you look back at the whole task, and comment on the strengths and weaknesses of your investigation.	

Evidence that regeneration has been successful	Evidence that regeneration has not been successful
1. The large number of office blocks, shops and restaurants shows that employment ought to have increased.	1. Most of the people we talked to and questioned were low paid, and told us many of the offices were empty.
2.	2.

Step 7 Writing the evaluation

In your evaluation, you assess the **validity** of what you have done. Is what you have found out reliable? Can your data be trusted? Did the study go well? You need to be able to assess the success of the whole investigation. Again, before your lessons when you will be writing this up, think about the following questions.

In paragraph 1 consider your methods. Don't use the excuse: if we had more time, we could have done more. Think about these points instead.

- Did your data collection methods work well? Did some work better than others?
- Would any results be different at another time of the day? Of the week? Of the year? For instance, would environmental quality data be different in summer compared to winter?
- Were the results accurate? If so, why? If not, why not? If you went back at different times, would you get similar conclusions?
- Do you have any **anomalies** (strange results) that cannot really be explained?

▲ Cornwall. How would this place look in a different season or at a different time of day?

+ An **anomaly** is an individual result that does not seem to fit into the general trend being displayed.

In paragraph 2 consider the reliability of your conclusions.

- Even though you took a small sample of places or people to be studied, are your conclusions valid?
- If you went to other similar places, would you find out similar things, or is your study area unique?
- Would you go about the investigation in the same way again?

In paragraph 3, think about the relevance of the investigation.

- How is your investigation useful to others?
- Who might be interested in your investigation, and why? What might they do with the information?

Finally, think about whether the investigation could be extended.

- Could your investigation be taken any further?
- Which parts might be worth developing, and how?

In this section you will find out how to present your report. Remember to leave plenty of time – if you complete all the points in the checklist, you'll need a few days! Don't leave it until the last hour before handing it in.

Final Checklist

Follow this checklist of things to do before you hand in your work.

1. Section headings

Are the following clearly identifiable by name?

a Introduction and aims

b Methodology

c Presentation of results

d Analysis

You can combine these if you wish but results **must** be clearly shown.

e Conclusions

f Evaluation

2. Presentation of results

- Give all your maps, diagrams, photos and graphs figure numbers with captions e.g. Figure 1 Map showing the location of the River Skerne.
- Any photo should be given a caption. Try and connect it to some other results.
- Refer to figure numbers in your writing.
- Don't take figure numbering to extremes! If you have, for example, an EQS graph with a photo next to it, the whole thing can be Figure 3.6.

3. Numbering and index

- Number your pages. All pages need to be numbered in sequence. Do this last!
- Create a contents page.
- Include an index of all your figures.

4. General presentation

- All material must be handed in as A4 size. Anything that is A3 size must be trimmed down, hole-punched, and be accessible without having to undo clips or ring binders.
- Stick down any loose material (e.g. acetate overlays) so they don't fall out and get lost.
- All secondary sources (including photos taken by other student or data found on a website) must be given in a bibliography at the end, quoting the source and what you used the source for e.g. www.bbc.co.uk/news for inner city projects in Manchester.
- Your coursework must be handed in as a loose-leaved project in a card or plastic wallet file. Every page must be in the correct order. You are not allowed to hand in ring binders or individual plastic sleeves.

Who am I writing this for?

Your report can be written either in the first person (e.g. I carried out a questionnaire in...) or third person (e.g. a questionnaire was carried out in...). Write for an intelligent adult, but one who does not know anything about the area you are investigating. Therefore, make all your explanations really clear.

Does it have to be a written report?

Your report does not have to be continuous writing. You can choose to create short movies, PowerPoint presentations, web pages, or perhaps a GIS map.

But if you want to do this **you must talk to your teacher** so that you can gain maximum marks from your work. The important thing is that a movie has to be just as well scripted as a piece of written analysis. In the same way, a PowerPoint must have notes which make crystal clear what it is you are saying about each slide. Remember – the information must be clear to someone who doesn't know you and who does not know your study area.

And one other thing

Check through the mark scheme. Make sure you understand it properly, so that you know what is needed to hit the highest levels.

Remember

+ Make a plan for yourself ahead of each lesson so that you make the most of the time available.

Remember

+ All secondary sources must be properly acknowledged.

Remember

+ Everything you include must be clear to somone who has no previous knowledge of your study.

Remember

+ Always keep a photocopy or electronic copy of your whole project.

If you want to be successful in the exams then you need to know how you will be examined, what kinds of questions you will come up against, how to use what you know, and what you will get marks for. That's where this chapter can help.

What is Edexcel Geography specification B like?

Edexcel's specification B includes some up-to-date ideas and material to help you understand the world – such as our changing climate – and offers a new take on some familiar subjects – such as coastal change and conflict. It is so broad that you're bound to find several interesting topics to give you a good understanding of the world.

Your course consists of four units.
- The taught part of the GCSE specification consists of two units – Dynamic Planet (chapters 1 to 8) and People and the Planet (chapters 9 to 16). Each unit consists of 8 topics, but you only need to study 6 of them. Make sure that you know which options you have to study.
- Unit 3 is a decision-making exercise. It draws on what you have been taught in units 1 and 2. Chapter 17 tells you all about this unit.
- Unit 4 is called Controlled assessment – what teachers and students used to call coursework. Look at chapter 18 for lots of information on this.

How will you be assessed?

Each unit counts for 25% of your final grade. Units 1, 2 and 3 are assessed by an exam. Unit 4 is done over a period of about one term. You create a report on your fieldwork, which is then marked.

The tables on the next two pages show what the exams for units 1, 2 and 3 will be like. Look through them carefully so that you know what to expect.

Units 1 and 2

Foundation tier	Higher tier

For each tier in each paper:

- the exam is one hour 15 minutes long
- it is worth 78 marks in total – 72 for questions on topics you have learnt and another 6 for spelling, punctuation and grammar (SPaG).
- it counts 25% towards your final grade
- it has three sections – A, B and C.

Any resources you need are included in with the questions – there won't be a separate booklet.

You must answer all parts of questions 1-4 in section A (each question is worth 12 marks, making a total of 48 marks) plus

- **one** question from questions 5 and 6 in section B (worth 12 marks, plus 3 for SPaG)
- **one** question from questions 7 and 8 in section C (worth 12 marks plus 3 for SPaG)

Section A

Questions 1 to 4 are broken up into short questions, varying between 1 and 4 marks.

- There are resource materials (data, photos, cartoons etc.) on which you'll be asked questions.
- You'll be expected to know what these resources are getting at from what you've learnt in class.
- Detailed case study knowledge is NOT needed in this section – though you will get marks for using an example or two.
- Section A is point marked.

Sections B and C

Questions 5-6 in section B and 7-8 in section C allow you to write more about what you've learnt. You will need to have learnt examples and case studies to complete the longer answers.

- The longest answer you will have to write is for 6 marks.
- 6 mark answers are marked by levels – level 1 (lowest) to level 3 (highest).

Section A

This has several short questions, worth from 2 to 6 marks.

- There are resource materials (data, photos, cartoons etc.) on which you'll be asked questions.
- You'll be expected to know what these resources are getting at from what you've learnt in class.
- Detailed examples are needed in this section for questions carrying 4 or 6 marks, based on what you've learnt in class.
- Questions carrying 5 marks or fewer are point marked. Those carrying 6 marks are marked by Levels – Level 1 (lowest) to Level 3 (highest)..

Sections B and C

Questions 5-6 in section B and 7-8 in section C allow you to write more about what you've learnt. You will need to have learnt examples and case studies to complete the longer answers.

- The longest answer you will have to write is for 8 marks.
- 8 mark answers are marked by Levels.

Unit 3 Decision-making Exercise (DME)

Foundation tier	Higher tier

For each tier in each paper:

- the exam is one hour 30 minutes long

- it is worth 50 marks in total, plus 3 marks for SPaG

- it counts 25% towards your final grade

- there is no choice of questions – you have to answer **all** the questions in the paper.

- the exam paper has six sections – numbered 1-6. These will follow the same sequence as the sections in the Resource Booklet – ie Section 1 will be about the resources in Section 1, and so on.

- the options for your final decision in Section 6 will normally be printed in the exam paper. It will ask you to make a decision based upon the information provided in the Resource Booklet.

- each section will vary in the number of marks – though Section 6 will always be worth the greatest number of marks.

The Resource Booklet a separate booklet – given to you in the exam. You won't have seen this booklet before, and it will be on a topic that you may know nothing about. Don't worry – it's your ability to understand the resources, which is being assessed. You'll be encouraged just to make notes on the exam paper for the first 30 minutes. You don't have to follow this advice – but it will help you to do so.

Section A	**Section A**

Section A

Section A normally looks at the background to the issue in the Resource Booklet.

Questions are generally short – between 1 and 4 marks.

- Some questions will ask you to interpret the resource materials (data, photos, cartoons etc.) so you must look carefully to find the answers.

- You'll be expected to know what these resources are getting at from what you've learnt in class before the exam.

- Section A is point marked.

Section B

Section B normally looks at the problems and issues in more depth. Like section A, questions are based on the resources, and whatever you have learnt in class before the exam.

- Questions vary between 1 and 4 marks

- Section B is point marked.

Section C

Section C involves you in making a decision about which is the best way forward. This is usually the longest piece of writing you'll have to do in exam conditions.

Usually, you will be presented with options to choose between. These will normally be a choice of schemes that will be set out clearly in the Resource Booklet.

- Questions vary between 3 and 8 marks.

- Questions of 6 marks or more are marked by levels – Level 1 (lowest) to Level 3 (highest).

Section A

Section A normally looks at the background to the issue in the resource booklet.

Questions are generally short – between 3 and 6 marks.

- Some questions will ask you to interpret the resource materials (data, photos, cartoons etc.) so you must look carefully to find the answers.

- You'll be expected to know what these resources are getting at from what you've learnt in class before the exam.

- Questions of 5 marks or under are point marked. Those of 6 marks or more are marked by levels – Level 1 (lowest) to Level 3 (highest).

Section B

Section B normally looks at the problems and issues in more depth. Like section A, questions are based on the resources, and whatever you have learnt in class before the exam.

- Questions vary between 3 and 6 marks

- Questions of 5 marks or under are point marked. Those of 6 marks or more are marked by levels – Level 1 (lowest) to Level 3 (highest).

Section C

Section C involves you in making a decision about which is the best way forward. This is usually the longest piece of writing you'll have to do in exam conditions.

Usually, you will be presented with options to choose between. These will normally be a choice of schemes that will be set out clearly in the Resource Booklet.

- Questions vary between 6 and 10 marks.

- Questions of 6 marks or more are marked by levels – Level 1 (lowest) to Level 3 (highest).

How are exam papers marked?

Examiners have clear guidance about how to mark. They must mark fairly, so that the first candidate's exam paper in a pile is marked in exactly the same way as the last. You will be rewarded for what you know and can do; you won't lose marks for what you leave out. If your answer matches the best qualities in the examiner's mark scheme then you will get full marks.

The tables on the previous two pages tell you which parts of the exam are point marked and which parts are level marked. Are you clear about what that means?

Point marking

Look at this question.

> Forests are ecosystems. Suggest two ways in which humans could protect ecosystems. (2)

There are two marks for this question and you have to suggest two ways of protecting ecosystems. So, you get one mark (or point) for each way that you give. The mark scheme tells examiners which methods they can accept as correct.

Level marking

Look at this question.

> Using examples, describe the key physical characteristics of either polar or hot arid areas. (6)

There are six marks for this question and examiners award these using the following mark scheme.

Level	Mark	Descriptor
Level 1	1-2	Describes a few physical characteristics with no detail; does not move beyond 'very cold' or similar. No link to chosen location. Lacks focus and organisation. Basic use of geographical terminology, spelling, punctuation and grammar.
Level 2	3-4	Some structure. Describes a range of characteristics with some detail for some areas, and with some use of appropriate examples. Some detailed description of chosen examples although this is variable. Examples are appropriate, with a range generally, or one or two areas more specifically. Clearly written, but with limited use of geographical terminology, spelling, punctuation and grammar.
Level 3	5-6	Structured answer. Describes a good range of characteristics in some depth with appropriate details/data. Uses examples to illustrate descriptions. Accurate and clear link to chosen extreme environment. Well written with good use of geographical terminology, spelling, punctuation and grammar.

So to gain the most marks your answer needs to meet the criteria for Level 3.

Choosing and answering the questions

Command words

When you look at an exam question, check out the command word – that is, the word that tells you what the examiner wants you to do. This table gives you some of the most commonly used command words.

Command word	What it means	Example
Account for	Explain the reasons for – you get marks for explanation rather than description.	Account for the earthquakes along destructive plate margins.
Compare	Identify similarities and differences between two or more things.	Compare flood management methods in Sheffield with those in Darlington.
Define	Give a clear meaning.	Define 'erosion.'
Describe	Say what something is like; identify trends.	Describe the trends shown in the graph.
Explain	Give reasons why something happens.	Explain, using examples, why the floods in Sheffield affected some parts of the city more than others.
How far?	You need to put both sides of an argument.	How far were the Sheffield floods due to human causes?
Justify	Give evidence to support your statements.	How should Christchurch be redeveloped following the devastating earthquakes?
List	Just state the factors; nothing else is needed.	List the different ways of protecting coasts affected by erosion.
Outline	You need to describe and explain, but more description than explanation.	Using the graph, outline the trend in population.
To what extent?	The same as 'How far?'	'Coastal protection – throwing good money after bad.' To what extent is this true?

Interpreting the questions

You have probably already been told by your teacher to read the exam questions carefully, and answer the question set, not the one you think it might be. That means you need to interpret the question to work out exactly what it is asking. So, look at the key words. These include:

Command words – these have distinct meanings which are listed above.

Theme or topic – this is what the question is about. The examiner who wrote the question will have tried to narrow the theme down, and you need to spot how so that you do not write everything you know about the theme.

Focus – this shows how the theme has been narrowed down.

Case studies – look to see if you are asked for specific examples.

Here is an example of a question that has been interpreted using the key words opposite.

Using named examples, explain how rapid erosion on coastlines can be managed using traditional and more modern strategies. (6)

> **Command words**
> 'explain how' You must give reasons.
> 'Using named examples' You are being asked to include specific examples you have studied.

> **Theme and focus**
> This question is for Coastal Change and Conflict, looking specifically at how coastlines can be managed.

> **Case studies**
> You can name examples from one or more places you've studied – the examples (plural) are for strategies, not for places you've studied. So three different strategies in one place is fine.

Choose the right question for yourself

> Do not be seduced by an attractive cartoon or photo – how difficult is the whole question?

> If you choose the wrong question, have the guts to change in the first 5-6 minutes. If you find yourself running out of time, bullet points are better than no points at all.

> *Choose the right question*

> Always choose questions from the part of the specification you have studied – the other questions only look easy because you haven't studied them.

> Read all the questions carefully and then choose. At this stage, you might want to sketch out a few points in pencil, to make sure that you include examples you think you might forget.

Using case studies

Look again at the question on the previous page. There is one case study in this book that you could use to answer this question – Christchurch Bay. You might have studied another example – perhaps seen a video of Holderness. You may also have studied coastlines for your controlled assessment – perhaps somewhere like Walton-on-the-Naze or Norfolk.

Throughout your course you will look at lots of different case studies, and will need a good way to learn and remember them. Try this method – draw a spider diagram with all the key information on it. An example is shown below, but it's only a start – you would need to fill in more details.

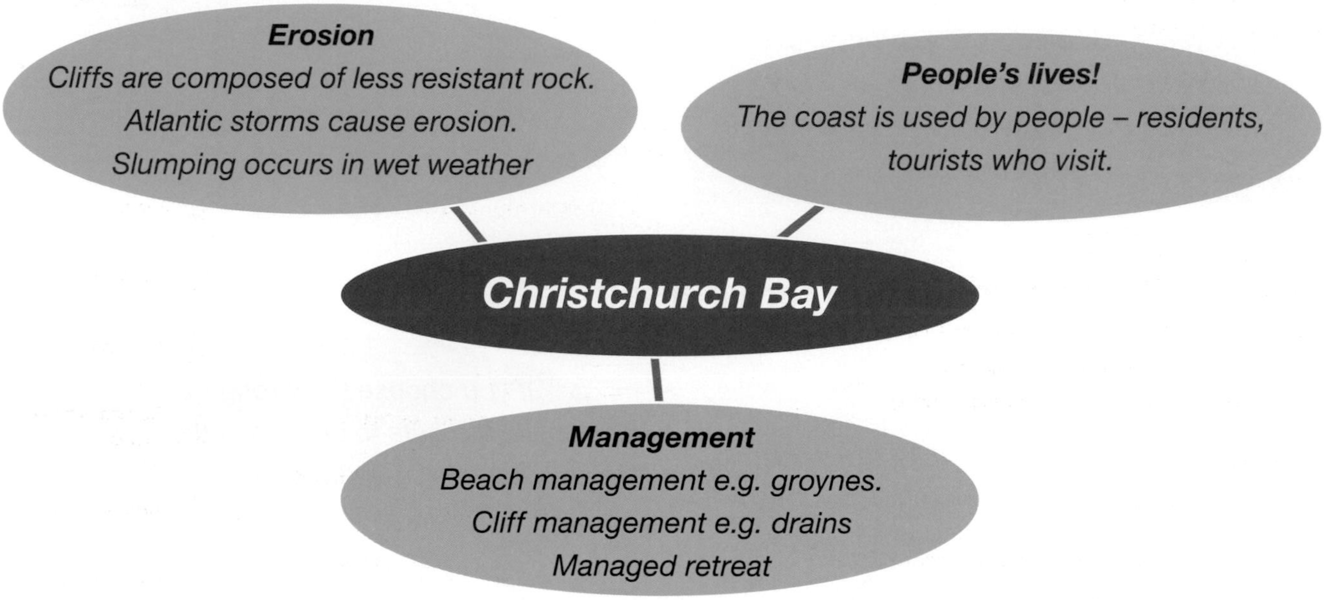

Erosion
Cliffs are composed of less resistant rock.
Atlantic storms cause erosion.
Slumping occurs in wet weather

People's lives!
The coast is used by people – residents, tourists who visit.

Christchurch Bay

Management
Beach management e.g. groynes.
Cliff management e.g. drains
Managed retreat

Once you have revised and learned your case study you can use it to answer exam questions. Look again at the example question. It asks you how rapid erosion on coastlines can be managed. It asks you to explain how the coastline can be managed using traditional and more modern strategies.

The ways are:
- hard engineering
- soft engineering
- holistic management.

Ordnance Survey symbols

ROADS AND PATHS

M1 or A6(M)	Motorway
A 35	Dual carriageway
A 31(T) or A 35	Trunk or main road
B 3074	Secondary road
	Narrow road with passing places
	Road under construction
	Road generally more than 4 m wide
	Road generally less than 4 m wide
	Other road, drive or track, fenced and unfenced
	Gradient: steeper than 1 in 5; 1 in 7 to 1 in 5
Ferry	Ferry; Ferry P – passenger only
	Path

PUBLIC RIGHTS OF WAY

(Not applicable to Scotland)

1:25 000	1:50 000	
		Footpath
		Road used as a public footpath
+++++++		Bridleway
	-+-+-+-+-	Byway open to all traffic

RAILWAYS

	Multiple track
	Single track
	Narrow gauge/Light rapid transit system
	Road over; road under; level crossing
	Cutting; tunnel; embankment
	Station, open to passengers; siding

BOUNDARIES

—+—·—+—	National
—+—+—+—	District
	County, Unitary Authority, Metropolitan District or London Borough
	National Park

HEIGHTS/ROCK FEATURES

50	Contour lines
·144	Spot height to the nearest metre above sea level

outcrop   cliff   850   600   scree

ABBREVIATIONS

P	Post office	PC	Public convenience (rural areas)
PH	Public house	TH	Town Hall, Guildhall or equivalent
MS	Milestone	Sch	School
MP	Milepost	Coll	College
CH	Clubhouse	Mus	Museum
CG	Coastguard	Cemy	Cemetery
Fm	Farm		

ANTIQUITIES

VILLA	Roman	⚔	*Battlefield* (with date)
Castle	Non-Roman	*	*Tumulus/Tumuli* (mound over burial place)

LAND FEATURES

ruin	Buildings
	Public building
	Bus or coach station
	Place of Worship { with tower / with spire, minaret or dome / without such additions }
○	Chimney or tower
	Glass structure
Ⓗ	Heliport
△	Triangulation pillar
	Mast
	Wind pump / wind generator
	Windmill
+	Graticule intersection
	Cutting, embankment
	Quarry
	Spoil heap, refuse tip or dump
	Coniferous wood
	Non-coniferous wood
	Mixed wood
	Orchard
	Park or ornamental ground
	Forestry Commission access land
	National Trust – always open
	National Trust, limited access, observe local signs
	National Trust for Scotland

TOURIST INFORMATION

P	Parking
P&R	Park & Ride
V	Visitor centre
i	Information centre
☎	Telephone
⚓	Camp site/ Caravan site
⚑	Golf course or links
	Viewpoint
PC	Public convenience
✕	Picnic site
	Pub/s
	Museum
	Castle/fort
	Building of historic interest
	Steam railway
	English Heritage
	Garden
	Nature reserve
	Water activities
	Fishing
☆	Other tourist feature

WATER FEATURES

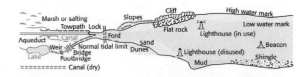

Marsh or salting   Slopes   Cliff   High water mark   Low water mark   Towpath   Lock   Flat rock   Lighthouse (in use)   Aqueduct   Canal   Ford   Normal tidal limit   Sand   Lighthouse (disused)   Beacon   Weir   Dunes   Lake   Footbridge   Bridge   Mud   Shingle

========== Canal (dry)

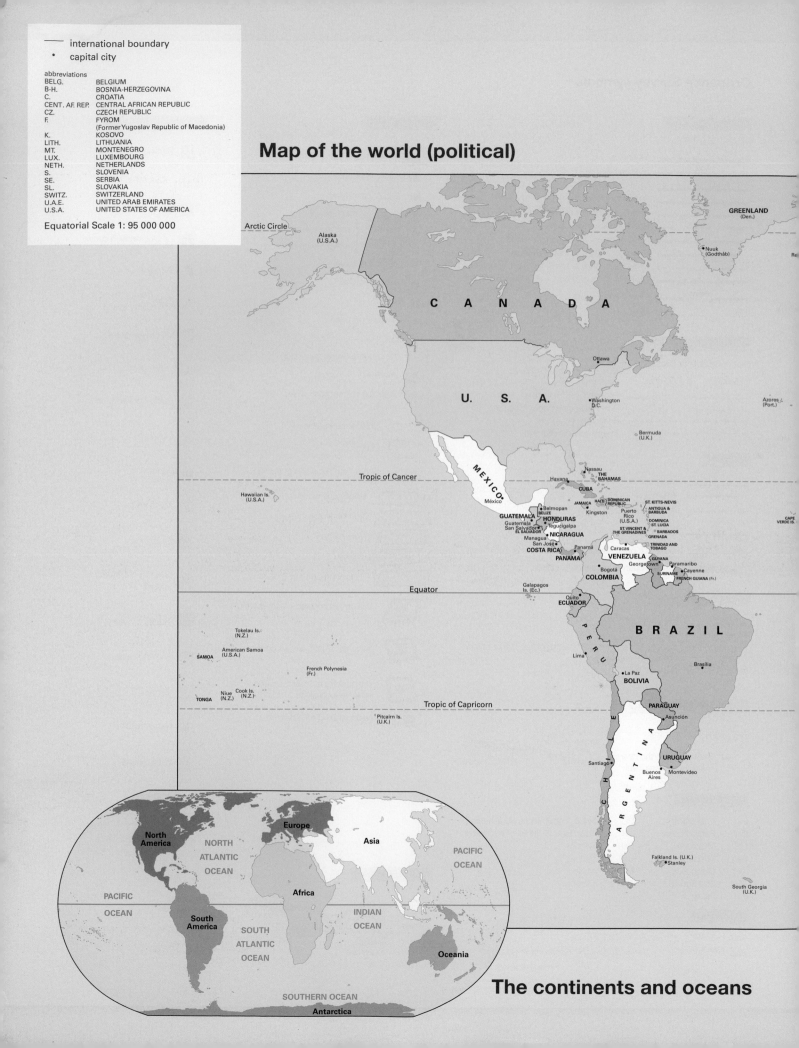

Map of the world (political)

The continents and oceans

international boundary
• capital city

abbreviations
BELG.   BELGIUM
B.-H.   BOSNIA-HERZEGOVINA
C.   CROATIA
CENT. AF. REP.   CENTRAL AFRICAN REPUBLIC
CZ.   CZECH REPUBLIC
F.   FYROM
    (Former Yugoslav Republic of Macedonia)
K.   KOSOVO
LITH.   LITHUANIA
MT.   MONTENEGRO
LUX.   LUXEMBOURG
NETH.   NETHERLANDS
S.   SLOVENIA
SE.   SERBIA
SL.   SLOVAKIA
SWITZ.   SWITZERLAND
U.A.E.   UNITED ARAB EMIRATES
U.S.A.   UNITED STATES OF AMERICA

Equatorial Scale 1: 95 000 000

GREENLAND
(Den.)

Arctic Circle

Alaska
(U.S.A.)

C A N A D A

• Nuuk
(Godtháb)

• Ottawa

U.   S.   A.

• Washington
D.C.

Azores
(Port.)

Bermuda
(U.K.)

Tropic of Cancer

Hawaiian Is.
(U.S.A.)

MÉXICO

• México

Nassau •
THE
BAHAMAS

Havana •

CUBA

CAPE
VERDE IS.

• Belmopan
BELIZE

ST. KITTS-NEVIS

HAITI

DOMINICAN
REPUBLIC

ANTIGUA &
BARBUDA

GUATEMALA
HONDURAS

JAMAICA

Kingston •

Puerto
Rico
(U.S.A.)

DOMINICA
ST. LUCIA

Guatemala •
San Salvador •
EL SALVADOR

• Tegucigalpa

NICARAGUA

ST. VINCENT &
THE GRENADINES

BARBADOS

GRENADA

Managua •
San José •

COSTA RICA

• Panamá

PANAMA

Caracas •

TRINIDAD AND
TOBAGO

VENEZUELA

Georgetown •

GUYANA

Paramaribo •

SURINAME

Cayenne •

FRENCH GUIANA (Fr.)

• Bogotá

COLOMBIA

Galapagos
Is. (Ec.)

Equator

Quito •
ECUADOR

B R A Z I L

Tokelau Is.
(N.Z.)

P
E
R
U

SAMOA

American Samoa
(U.S.A.)

Lima •

• Brasília

French Polynesia
(Fr.)

• La Paz

BOLIVIA

TONGA

Niue
(N.Z.)

Cook Is.
(N.Z.)

Tropic of Capricorn

PARAGUAY

Pitcairn Is.
(U.K.)

• Asunción

C
H
I
L
E

A
R
G
E
N
T
I
N
A

URUGUAY

Santiago •

Buenos
Aires •

• Montevideo

Falkland Is. (U.K.)
• Stanley

South Georgia
(U.K.)

North
America

NORTH
ATLANTIC
OCEAN

Europe

Asia

PACIFIC
OCEAN

PACIFIC
OCEAN

Africa

INDIAN
OCEAN

South
America

SOUTH
ATLANTIC
OCEAN

Oceania

SOUTHERN OCEAN

Antarctica

Glossary

*cross reference

A

abrasion the scratching and scraping of a river bed and banks by the stones and sand in the river

air mass a huge body of air with uniform temperature and humidity

alluvium all deposits laid down by rivers, especially in times of flood

antecedent rainfall the amount of moisture already in the ground before a rainstorm

artesian water water that rises out of the ground by natural pressure

atmosphere the layer of gases above the Earth's surface

asthenosphere part of the Earth's *mantle. It is a hot, semi-molten layer that lies beneath the *tectonic plates. It is between 20 and 70 km thick

attrition the wearing away of particles of debris by the action of other particles, such as river or beach pebbles

B

bankful the discharge or contents of the river which is just contained within its banks. This is when the speed, or velocity, of the river is at its greatest

bar an accumulation of sediment below the water which may be exposed at low tide. It is caused by heavy wave action along a gently sloping coastline. The crest of a bar normally runs parallel to the coast and can extend across a bay or estuary

basalt a dark-coloured volcanic rock. Molten basalt spreads rapidly and is widespread. About 70% of the Earth's surface is covered in basalt *lava flows

base flow the usual reliable level of a river

biodiversity the varied range of plants and animals (flora and fauna) found in an area

biofuels any kind of fuel made from living things, or from the waste they produce

biogas a gas produced by the breakdown of organic matter, such as manure or sewage, in the absence of oxygen. It can be used as a biofuel

biome a very large ecosystem. The rainforests are one biome. Hot deserts are another

bioproductive area the area of land, sea and air required to provide the goods and services we consume. Our *eco-footprint shows how much of this area we are using

biosphere the zone where life is found. It extends 3m below ground to about 30m above ground and up to 200m deep in the oceans

bleaching the whitening or fading of the bright coral colours, due to the loss of algae from the coral system. Bleaching is a sign that the coral reef is undergoing stress

blue water water that is stored in dams and then used for domestic purposes

borehole a deep hole drilled in the ground, especially to find water or oil

brownfield site an area of land that has been built on before and can be built on again

C

carbon footprint a measurement of all greenhouse gases we individually produce

carrying capacity the maximum number of people (or plants or animals) who can be supported in a given area

climatologist a scientist who is an expert in climate and climate change

closed system where there are clear boundaries with no movement of energy across them. The *hydrological system is a closed system. The water goes round and round

collision zone where two tectonic plates collide – forming mountains

community forests areas in England where there is a programme of environmental improvement which involves planting and protecting trees and other natural habitats. The idea is to improve the quality of life of people living there

compact community a community where the best use of space is made

conflict matrix a technique for identifying possible areas of disagreement by putting different interest groups into a table or matrix. It is also known as a 'clash table'

congestion charge a fee for motorists travelling within a city. The main aims are to reduce traffic congestion and to raise funds for investment in the city's transport system. London's congestion charge was one of the first to be introduced

conservation farming farming methods which aim to conserve soil and water while at the same time providing a sustainable livelihood for the farmer

conservative boundary where two tectonic plates slide past each other

constructive plate boundaries where two tectonic plates are moving apart and new crust is constructed. Also known as 'plate margins'

continental climate a dry climate, with hot summers and cold winters – found inland, far from the coast

continental crust the part of the Earth's crust that makes the continents; it's between 25 km and 100 km thick

convection currents transfer heat from one part of a liquid or gas to another. In the Earth's mantle, the currents which rise from the earth's core are strong enough to move the tectonic plates on the earth's surface. Convection currents also occur in the atmosphere

convergence there are two meanings. a) the coming together of *tectonic plates and b) when air streams flow to meet each other

cost-benefit analysis looking at all the costs of a project, social and environmental as well as financial, and deciding whether it is worth going ahead

counter-urbanisation when people leave towns and cities to live in the countryside

cultural dilution when traditional / cultural beliefs, customs and values are lost, due to increasing contact with people from other cultures (such as tourists)

cycle of poverty a vicious spiral of poverty and deprivation passing from one generation to the next

D

dependency ratio the ratio between the number of people in a population of working age (15-64 years) and the number of people of non-working age (0-14 years and 65 years and over — often call the 'dependent population')

deprived areas an area, usually in a developed nation, where there is high unemployment and crime, and poor health and education services and housing

desertification the gradual change of land into desert

development the use of resources to improve the standard of living of a nation

diguette a line of stones which are laid along the contours of gently sloping farmland to catch rain water and reduce soil erosion

discordant coast a coast which alternates between bands of hard rocks and soft rocks. Discordant coastlines will have alternating headlands and bays

diversification when a business (e.g. a farm) decides to sell other products or services in order to survive or grow

E

ecological footprint (eco-footprint) the area of land and sea that supplies all of the 'stuff' that you need to live – land for your home, your food, the energy you use, plus all the materials you buy.

ecosystem a unit made up of living things and their non-living environment. For example a pond, a forest, a desert

El Niño occurs in the Pacific Ocean every three to seven years. Unusually warm ocean conditions off the western coasts of Ecuador and Peru cause climatic disturbances. El Niño can affect climates around the world for more than a year

energy a source of power

environmental refugee a displaced person caused by environmental disasters as a result of climate change

environmental sustainability what we need to consume in order to protect the Earth's environment. We need to reduce our demands on the planet to a level where future generations will not suffer

epicentre the point on the ground directly above the focus (centre) of an earthquake

eutrophication the process by which ecosystems, usually lakes, become more fertile as fertilizers and sewage flow in. The resultant loss of oxygen in the water kills off all species that need oxygen to survive, such as fish

evacuate when people move from a place of danger to a safer place

evaporation the changing of a liquid into a vapour or gas. Some rainfall is evaporated into water vapour by the heat of the sun

F

fault a fractured surface in the Earth's crust along which rocks have travelled relative to each other

fauna the animals or animal life of any particular region or time period

fell running the sport of running and racing, off road, over upland country with steep climbs and descents. It is called after the fells of northern Britain especially those in the Lake District

finite resource a resource that is limited — and will one day run out

flood plain flat land around a river that gets flooded when the river overflows

flora the plants or plant life of any particular region or time period

focus the point of origin of an earthquake

food chain a chain of names linked by arrows, showing what species feed on. It always starts with a plant

friction the force which resists the movement of one surface over another

G

genetic modification any alteration of genetic material (DNA or RNA) of an organism by means that could not occur naturally. It is also called 'genetic engineering'

glacial a long period of time during which the Earth's glaciers expanded widely

global warming the way temperatures around the world are rising. Scientists think we have made this happen by burning too much fossil fuel

goods products that are of value to us

greenfield site an area of land which has not previously been built upon

greenhouse effect the way that gases in the atmosphere trap heat from the sun. Like the glass in a greenhouse – they let heat in, but prevent most of it from escaping

greenhouse gases gases like carbon dioxide and methane that trap heat around the Earth, leading to global warming

gross domestic product (GDP) the total value of goods and services produced by a country in one year

H

HDI a standard means of measuring human development

helical flow a continuous corkscrew motion of water as it flows along a river channel

holistic management in coastal management this means looking at the requirements of a long stretch of coast when planning, rather than just a single beach or short stretch of coastline. This is also known as 'integrated coastal zone management' (ICZM)

hot spot there are two meanings a) where volcanoes occur away from plate margins; probably due to strong upward currents in the mantle b) areas with rich animal and plant life

hunting and gathering a form of society with no settled agriculture or domestication of animals, which has little impact on the environment

hydraulic action the force of the water within a stream or river

hydrological cycle the movement of water between its different forms; gas (water vapour) , liquid and solid (ice) forms. It is also known as the water cycle

hydrosphere all the water on, or close to, the surface of the Earth. 97% is in the seas and oceans

I

impermeable doesn't let water through

industrial societies where the main industries are in the manufacturing sector

industrialization where a mainly agricultural society changes and begins to depend on manufacturing industries instead

infiltration the soaking of rainwater into the ground

infrastructure the basic services needed for an industrial country to operate e.g. roads, railways, power and water supplies, waste disposal, schools, hospitals, telephones and communication services

integrated transport policy a government policy aimed at improving and integrating public transport systems, and of making cars and lorries more environmentally acceptable and more efficient

interception the capture of rainwater by leaves. Some *evaporates again and the rest trickles to the ground

interglacials a long period of warmer conditions between *glacials

interlocking spurs hills that stick out on alternate sides of a V-shaped valley, like the teeth of a zip

intermediate technology (sometimes called appropriate technology) technology devised for the developing world which is simpler, cheaper, and more environmentally friendly than that of the developed world, and so is better suited to local resources and knowledge

internal migration the movement of people within a country, in search of seasonal or permanent work or for social reasons

L

lagoon a bay totally or partially enclosed by a *spit or reef running across its entrance

land degradation a decline in the quality of the land – especially the soil – due to overuse or mistreatment by humans

landfill an area of ground where large amounts of waste material are buried under the earth

landslide a rapid *mass movement of rock fragments and soil under the influence of gravity

lava melted rock that erupts from a volcano

lava flows *lava flows at different speeds, depending on what it is made of. Lava flows are normally very slow and not hazardous but, when mixed with water, lava can flow very fast and be dangerous

longshore drift the movement of sand and shingle along the coast

low-income the World Bank's main criterion for classifying economies is gross national income (GNI) per capita. Based on its GNI per capita which is adjusted every year, every economy is classified as low-income, middle-income (subdivided into lower-middle and upper-middle), or high-income

M

magma melted rock below the Earth's surface. When it reaches the surface it is called lava

magnitude of an earthquake, an expression of the total energy released

mantle the middle layer of the Earth. It lies between the *crust and the *core and is about 2900 km thick. Its outer layer is the *asthenosphere. Below the asthenosphere it consists mainly of solid rock

marine hotspots areas rich in marine species. These are small areas and highly vulnerable to extinction

maritime climate a wet climate, with little extremes in temperature – found close to the sea

marram grass grasses that are found on coastal sand dunes. They have extensive systems of creeping underground stems which allow them to survive under conditions of shifting sands and high winds

mass movement the movement downslope of rock fragments and soil under the influence of gravity. A *landslide is a rapid mass movement

meander a bend in a river

mega-city a many centred, multi-city urban area of more than 10 million people. A mega-city is sometimes formed from several cities merging together

middle course the journey of a river from its source in hills or mountains to mouth is sometimes called the course of the river. The course of a river can be divided into three main sections a) *upper course b) middle course and c) lower course

Milankovitch cycles the three long-term cycles in the Earth's orbit around the sun. Milankovitch's theory is that *glacials happen when the three cycles match up in a certain way

multi-cropping the practice of producing two or more crops at the same time on the same parcel of land during a 12-month period

N

national park an area that is protected from human exploitation and occupation. Most developed countries have national parks

natural increase the birth rate minus the death rate for a place. It is normally given as a % of the total population

newly industrialised countries (NICs) countries which were recently less developed but where *industrialization has happened quickly

NGO – non-governmental organisation NGOs work to make life better, especially for the poor. Oxfam, the Red Cross and Greenpeace are all NGOs

nomads people who move with their animals from place to place in search of pasture

nutrient cycle the transfer and storage of nutrients by living organisms from their physical surroundings and back again in a continuous cycle

O

oceanic crust the part of the Earth's crust which is under the oceans; it's made of basalt and is between 5 km and 10 km thick

outsourcing where part or all of a project, such as the design or the manufacturing, is handed over to a third-party company. This is normally done to save time or costs or both

overhang the part of a cliff above the reach of waves which is unaffected by wave action

ox-bow lake a lake formed when a loop in a river is cut off by floods

P

Pangea a supercontinent consisting of the whole land area of the globe before it was split up by continental drift

peak oil the point at which oil production reaches its maximum level and then declines

percolate to move gradually through a surface that has very small holes or spaces in it

permafrost permanently frozen ground

permeable letting water pass through

plate boundaries where *tectonic plates meet. There are three kinds of boundary a) *constructive – when two plates move apart b) destructive – when two plates collide c) conservative – when two plates slide past one another

plumes upwellings of molten rock through the *asthenosphere to the lithosphere, forming a *hot spot

point bar the accumulation of river sediment on the inside of a meander

population balance where births and deaths are almost equal

population decline where the number of live births is less than the number of deaths

population increase where the number of live births exceeds deaths

post-industrial societies where service and high-technology industries are dominant, and where heavy manufacturing industries are less important

poverty line the minimum level of income required to meet a person's basic needs ($1.25)

predict saying that something will happen in the future. A scientific prediction is based on statistical evidence

pre-industrial the situation before the Industrial revolution. It can be used to describe poor countries which are mainly agricultural

primary effects the direct impacts of event, usually occurring instantly

primary industry where people extract raw materials from the land or sea. e.g. farming, fishing and mining

Q

quality of life a measure of how 'wealthy' people are, but measured using criteria such as housing, employment and environmental factors, rather than income

quaternary industry where people are employed in industries providing information and expert help. For example IT consultants and researchers

Quaternary the last 2.6 million years, during which there have been many *glacials

R

Ramsar sites wetlands of international importance. They are named after The Convention on Wetlands, signed in Ramsar, Iran, in 1971

recurved hooked

remittance payments money sent home to their families by people working overseas. Remittance payments are an important source of income for some less developed countries

re-urbanisation when people who used to live in the city and then moved out to the country or to a suburb, move back to live in the city

river basin system a river basin system is the area of land drained by a river and its tributaries. It is also known as a drainage basin

rock outcrop a large mass of rock that stands above the surface of the ground

rural idyll when people move to the countryside because they think it will offer them a better quality of life and lower crime, particularly when they have children

rural-urban migration the movement of people from the countryside to the cities, normally to escape from poverty and to search for work

S

salt marsh salt-tolerant vegetation growing on mud flats in bays or estuaries. These plants trap sediments which gradually raise the height of the marsh. Eventually it becomes part of the coast land

sand dune a hill or ridge of sand accumulated and sorted by wind action

saturated soil is saturated when the water table has come to the surface. The water then flows overland

scree shattered rock fragments which gather below free rock faces and summits. It is also used to refer to the slope below a rock face which is made up of these fragments

secondary effects the indirect impacts of an event, usually occurring in the hours, weeks, months or years after the event

secondary industry where people make, or manufacture, things. For example turning iron ore into steel, making cars and building houses

seismometer a machine for recording and measuring an earthquake

services things that satisfy our needs

Shoreline Management Plan (SMP) this is an approach which builds on knowledge of the coastal environment and takes account of the wide range of public interest to avoid piecemeal attempts to protect one area at the expense of another

siltation to become filled with silt

sink a container, a pool or a pit into which waste goes

socio-economic a term used to classify people depending on what they do for a living

soil creep the slow (sometimes very slow) gradual movement downslope of soil, *scree or glacier ice

source a place, person or thing that you get something from

spatial relating to space and the position, size and shape of things in it

sphere of influence area around a settlement (or shop, or other service) where its effect is felt. London has a very large sphere of influence

spit a ridge of sand running away from the coast, usually with a curved seaward end

storm hydrograph a graph which shows the change in both, rainfall and discharge, from a river following a storm

stratosphere the layer of air 10-50km above the Earth's surface

stratovolcano a cone-shaped volcano formed from layers of different kinds of lava

sub-aerial processes occurring on land, at the Earth's surface, as opposed to underwater or underground

subduction the transformation into *magma of a denser *tectonic plate as it dives under another less-dense plate

subsistence farming where farmers grow food to feed their families, rather than to sell

suburbanisation when people leave cities and towns to live in suburbs

surface run-off rainwater that runs across the surface of the ground and drains into the river

sustainable energy - energy sources which are unlikely to run out (also called renewable energy) — e.g. solar, wind, wave, hydroelectric, geothermal and tidal power

sustainable management meeting the needs of people now and in the future, and limiting harm to the environment

T

tectonic plate the Earth's surface is broken into large pieces, like a cracked eggshell. The pieces are called tectonic plates, or just plates

tertiary industry where people are employed in providing a service. For example the health service (doctors, nurses, dentists) and education (teachers)

thalweg the line of the fastest flow along the course of a river

thermal expansion expansion as a result of heating. When sea water warms up, it expands

through-flow the flow of rainwater sideways through the soil, towards the river

tombolo a *spit, resulting from *longshore drift which joins an offshore island to the mainland

transnational companies companies which operate across more than one country. They are similar to multinational companies but may operate in only two countries

transpire when plants lose water vapour, mainly through pores in their leaves

U

unsustainable use ways of catching wild fish that threaten the fish stock itself by overfishing, or threaten the environment the fish need to thrive

upper course the journey of a river from its source in hills or mountains to mouth is sometimes called the course of the river. The course of a river can be divided into three main sections a) upper course b)*middle course c) lower course. In the upper course the water flows fast and straight

urban heat island where temperatures in cities are much higher (up to 4°C) than the surrounding countryside due to heat released from buildings including factories and offices and from air pollution

W

water table the level below which the ground is *saturated

water wars conflicts between countries for access to and control over water. Global warming and population increase increases the chances of this happening

wave-cut notch an indentation cut into a sea cliff at water level by wave action

wave-cut platform the flat rocky area left behind when waves erode a cliff away

whole ecosystem approach an approach that aims at protecting the whole marine ecosystem and that acts first before the damage is done

windpump a type of windmill used for pumping water from a well, or for draining land

Index